London Topographical Record

London Topographical Record

VOL. XXVI

EDITED BY
ANN LOREILLE SAUNDERS, Ph.D., F.S.A.

Publication No. 141
London Topographical Society
1990

©
LONDON TOPOGRAPHICAL SOCIETY
3 Meadway Gate
London NW11 7LA
(081-455-2171)
1990

ISBN 0 902087 28 2

Printed in Great Britain by the University Press, Cambridge

CONTENTS

I The Topography of St Paul's Precinct, 1200–1500
   by RODERICK MACLEOD   page 1
II Four Plans of Southwark in the time of Stow
   by MARTHA CARLIN   15
III Whitehall Palace and King Street, Westminster
   by GERVASE ROSSER and SIMON THURLEY   57
IV Moorfields, Finsbury and the City of London in
   the Sixteenth Century
   by ELEANOR LEVY   78
V The Broadway Chapel, Westminster: a forgotten
   exemplar
   by PETER GUILLERY   97
VI Some Notes on Hollar's *Prospect of London and
   Westminster taken from Lambeth*
   by PETER JACKSON   134
VII Guide Books to London before 1800: a survey
   by DAVID WEBB   138
VIII Sources for London History at the India Office
   Library and Records
   by MARGARET MAKEPEACE   153
IX London Illustrations in the *Gentleman's
   Magazine*, 1746–1863
   by PETER JACKSON   177
X Richard Horwood's Plan of London: a guide to
   editions and variants, 1792–1819
   by PAUL LAXTON   214
XI The Museum in Docklands
   by CHRIS ELMERS   264
XII Centre for Metropolitan History
   by HEATHER CREATON   268
XIII William Francis Grimes: An Obituary
   by RALPH MERRIFIELD
   with a Bibliography compiled by
   ORTRUN PEYN   271
XIV Philip David Whitting: An Obituary
   by CHRISTOPHER THORNE
   with a Bibliography compiled by
   R. H. THOMPSON   294

XV Priscilla Metcalf: An Obituary
    by CAROLINE BARRON
    with a Bibliography compiled by
    DAVID WEBB   *page* 308

Prospectus   312

Council and Officers, 1990   313

Rules   314

Secretary's Report, 1985–89   315

List of Publications   317

List of Members   327

Index to Volume XXVI   357

London Topographical Society

I. THE TOPOGRAPHY OF ST PAUL'S PRECINCT, 1200–1500

By RODERICK MACLEOD

THe area around St Paul's Cathedral was generally known in the Middle Ages as the Churchyard or the Cemetery, and contained not only burial ground but a great number of buildings serving the religious community. The years from 1200 (Fig. 1) to the early sixteenth century (Fig. 2) saw a steady change to the topography of this area: first the construction of residences for church officials, then the leasing out of parts of these buildings to laymen, and finally the destruction of much that remained expressly religious as part of the general dissolution of the old Church. The following is a survey of tenements, gardens, chapels and palaces which existed during this period; or at least of the very few about which contemporary deeds say anything at all.[1] It does not attempt to trace the historical development of diocesan property in London or the social change to the area implied by the Reformation. Any such study, however, would be the richer for taking Topography into consideration.

The precinct was a neat rectangle, in keeping with the relatively regular layout of London streets. It was framed by four of these: Paternoster Row, Ave Maria Lane, Carter Lane, and Old Change. Whether these streets existed before the precinct or developed in relation to it is a matter of debate. The area appears to have been surrounded by walls with large gates which could be closed at night, although these defences were probably not built at once. Much of the circumference of the precinct, especially the southern side, may have been closed off merely by rows of buildings. There were two major thoroughfares running through the Churchyard: one from Paternoster Row in the north which

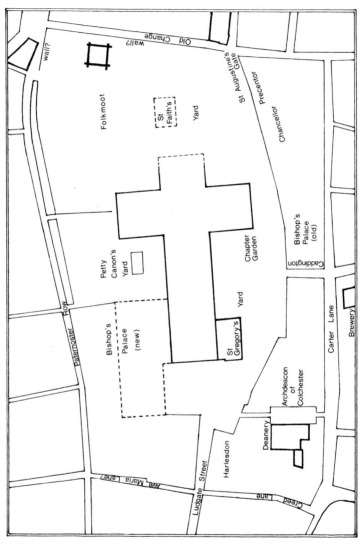

Fig. 1. St Paul's Precinct c. 1250.

TOPOGRAPHY, 1200–1500

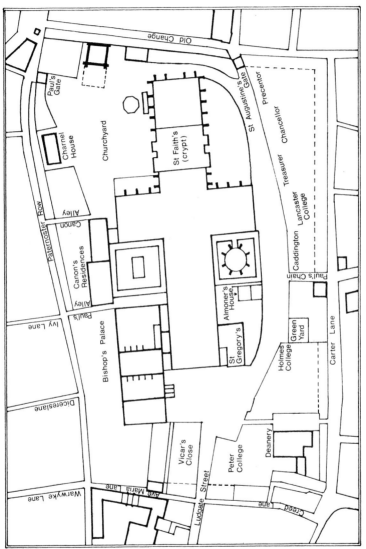

Fig. 2. St Paul's Precinct c. 1500.

passed through St Paul's Gate and ran south behind the east end of the cathedral, and the other from Watling Street which passed through St Augustine's Gate in the south-east corner of the precinct and ran west to the main door of St Paul's and out to Ludgate.

St Augustine's Gate in the south-east corner is a good place to start a tour of the precinct. The Precentor's Inn is the first building of significance inside the gate. It is not known whether specific residences went with specific offices during this period, but it is safe to assume this tended to be the case. In the 1240s, the current precentor, Roger de Orseth, inhabited certain houses in the precinct 'next to the east gate'.[2] Orseth left his houses to prebendaries, and there is no indication that any precentor occupied the Inn afterwards. In 1285 a new house was built for the precentor, which could mean that a different residence now went with the office, or that the old one was rebuilt.[3] A house belonging to John Drury, precentor in 1441, sounds as though it lay in roughly the same corner of the precinct: 'next to the gate of St Paul's'.[4] The gate in question could have been the one in the north-east corner of the precinct, but was more likely St Augustine's, given references to the public road lying to the north of Drury's Inn. To the east of the Inn there lay at least one tenement, so the building could not have stood directly against the gate.

Drury's Inn was of fair size, what is often called a Courtyard House; it appears to have had some buildings facing the street, others to the rear, and gardens sandwiched between. Two tenements lay on either side of a ten foot wide gate; the east one 28 feet wide and 22 feet deep, the west one 28 foot square, giving the property a total street frontage (facing north into the Churchyard) of some 66 feet. In 1441 Drury leased out the two front tenements but reserved for himself all of the gardens save for a 5 foot wide strip behind the east tenement and access to a latrine located in the west tenement. He also retained a right of way through the gate, which suggests that there were other buildings further back, to the south of the property, in which the precentor continued to live.

The Chancellor's Inn, lying to the west of the Precentor's Inn,[5] appears to have been inhabited consistently by chancellors since 1230. In that year the current chancellor, Henry de Cornhill, bequeathed a house 'to the south of the church' for the use of his successors.[6] What sort of house it was is harder to say. There is a reference in 1309 to the Chancellor's Inn containing a 'room with a chimney on the wall towards the South Street' which was severely damaged.[7] The wall in question may have been the precinct wall running along Carter Lane, and if so the structure of the Inn must have been that of a courtyard house. The chancellor at the time, Robert de Clothale, had paid for the construction of a hall with two solars at the west end, a chapel and garderobe adjoining the other end to the east, and a study on the south. Because of these expenses Clothale was exempt from having to contribute towards the repair of the damaged room, but his successors were not. One wonders whether they were still paying their annual two marks in 1422 when a reference to the Chancellor's Inn confirms its site to the west of St Augustine's Gate and 'directly to the east of the Inn of the Treasurer'.[8]

This, the third residence associated with a major church office in the south-east corner of the precinct, is not referred to in deeds until 1422 when parts of it were leased off. The deed[9] mentions a chapel, some 'new houses' with a great stable underneath, a latrine, and some vacant land or gardens, all of which suggest a simpler shape than that of the Chancellor's and Precentor's Inns. Exactly how new the new houses were is impossible to say, although it may be that the Treasurer's Inn was once part of the lands and buildings belonging to the 'Old Paleys'. In 1403 the bulk of the bishop's former palace, by that time perhaps more vacant land than habitable building, was given over to provide a chantry to John of Gaunt.[10] This area measured $36\frac{1}{2}$ feet along the Churchyard but expanded 20 feet to the rear to a width of 44 feet, which might indicate a slight curve in the churchyard frontage. To the east of this land lay a tenement held by a canon, John Wyke.[11] Wyke's tenement either lay between the Lancaster Chantry land and the Treasurer's Inn or was in some way part of the Inn itself. This would depend upon the stages at which the Old Palace was divided up and parcelled out.

The Bishops of London had long since moved from this area across from the south door of the cathedral to the more extensive palace to the north-west. The move may have been as early as the late thirteenth century when a great number of changes took place in the precinct. One can only assume that the Old Palace was relatively large; there are no statistics apart from those concerning the Lancaster Chantry, but in 1596 the lands which comprised the Old Palace were sold off and there are references to a fair number of tenements and shops occupying the original area.[12] To the west of the Lancaster Chantry lands lay the residence for the Prebendary of Caddington, the only prebendary whose property within the precinct can be identified. As early as the late twelfth century there lay a house made of stone donated to the bishop by the current prebendary, a Master Alard who later became Dean of St Paul's.[13] In 1311 it was let temporarily to some of the bishop's officials,[14] but was still referred to as the 'mansions for the prebends called Caddyngton' in 1403.[15] This last reference places it beside land from the Old Palace.

Paul's Chain, the great south gate to the precinct which Stow presents as a rather daunting barrier,[16] must have had some early precedent. Unfortunately there are no clear references in the early deeds to any gate or wall along the south side of the precinct. One can only speculate as to where the Old Palace and the Inns fitted into the puzzle in relation to such a gate.

Stow is not much help either in locating the site of the Dean's House. It was built by Dean Ralph de Langdon in 1145[17] and established by the bishop as the dean's official residence by 1185.[18] The house built for the dean after the Great Fire of 1666, which in part still stands, lies to the south-west of the cathedral, and so it is likely that the medieval residence lay on that site also. Stow describes it as a 'fayre olde house directly against the pallace'.[19] If by palace he meant the bishop's new residence, this makes sense only if 'directly against' means on the other side of Ludgate, and even then there ought to have been a number of buildings in between.

A series of references to buildings around the Dean's House help form a picture of the south-west corner of the precinct. In 1274 the Archdeacon of Colchester, Fulk Lovel, had a house with a garden next to the Deanery which he granted to the dean in exchange for some other property.[20] In the 1190s the Prebendary of Harlesden, Richard de Stortford, granted a house in the Churchyard for the use of his successors. Another reference to this house places it next to the Dean's House;[21] yet another says it lay 'in atrio australi'.[22] If 'atrio' meant the whole precinct – and sometimes it seems to – the Dean's House must have lain to the south; if 'atrio' meant specifically the space in front of the cathedral, then the Harlesden mansion lay just south of Ludgate and the Dean's House just south of it. There were two residences for chantry priests in this area. One was Peter College, which after the Reformation became Stationers' Hall, which lay beside the Harlesden mansion.[23] The other was Holmes College, established in 1386 south of the churchyard.[24] Beside it lay a carpenter's yard, sometimes called the Tymbar Yard or Greene Yard. By 1465 the yard seems to have been surrounded by shops, and several chambers built on top of Holmes College.[25] These buildings about the Deanery were most likely what Stow had in mind when he spoke of the 'divers large houses...which yet remayne, and of olde time were the lodgings of Prebendaries and Residenciaries, which kepte great Householdes'.[26]

Next to the cathedral along its south side lay the parish church of St Gregory and its burial ground, and a small piece of land which, at least by the fourteenth century, belonged to the rector of the church of St Mary Staininglane. In 1315 this piece of land was leased to the Almoner of St Paul's, William de Coleshunt, as his residence.[27] In 1332 work began on that glory of early perpendicular architecture, the St Paul's Cloister and Chapter House; the site chosen was part of the chapter's garden between the cathedral's south transept and a wall marking the extent of the almoner's land. There is reason to believe that later almoners lived in this residence; again the evidence is that by the fifteenth century they were leasing out parts of the building. In 1441 the front tenement, which comprised shops and chambers and ran 42 feet along the road, was leased out to a Robert Aliard.[28] The almoner retained a portion to the rear with a courtyard measuring nine foot square; it was reached by a three foot wide passage next to the cloister wall. The passage also led to a latrine, to which the dean and chapter appeared to have access; its dimensions are given very precisely as 10 feet 10 inches by 8 feet 5 inches. Another reference to these tenements in almoners rentals of 1526 confirms their location between the cloisters and St Gregory's cemetery.[29]

The area to the north of Ludgate, to the west of the cathedral, contained more residences for chantry priests and a row of houses for the vicars of St Paul's. For this reason the area was known as the Vicarage or Vicar's Close; the arrangement may have been similar to the existing close in Wells. The residence for the Waltham chantry priest lay some 40 feet inside the gate to this close.[30] The row was probably longer than this, however, judging by the number of other names, such as Mondene and Donyon, mentioned as lying near the Waltham chantry and the vicar's mansions.[31] After the dissolution of all such offices and the vacancy of the tenements they occupied, the area seems to have been considered as a unit. In 1599 the 'mansion for the vicar', located north of the Churchyard and south of the Bishop's Palace, was described as a stately manor known as 'Purlewe'; it sported 'a house, a buttery, a kitchen, a shop, chambers, garden yard and a well'.[32] There is a reference to a long alley going into the vicarage and a small house to the south near the garden wall which was sometimes known as 'the Vicar's stillioury house'. No doubt the alley was what once had been the close; how long the brewery had been there is unfortunately unclear. It may well have been this building, and not the Deanery, to which Stow referred when he spoke of the 'fayre olde house directly against the pallace'.

Further to the north lay the New Palace.[33] From the thirteenth century this was the bishop's main residence, although the land may have belonged to the bishop from a much earlier date. No doubt the move from the Old Palace was a gradual one, as space was needed. Many deeds to properties on Ivy Lane and Dicers Lane to the north of the precinct refer to a wall along Paternoster Row. Earlier deeds[34] tend to say it is the wall of the bishop's 'curia', which may mean the bishop's court was held in a building in the north-west part of the precinct. Deeds from the fourteenth century[35] tend to refer to the wall along Paternoster Row as that of the bishop's house. The Bishops of London continued to live in this palace until the Reformation. It was not until 1661 that an Act of Parliament formally sold off the land, although by that date the area had long since been divided up into a great many shops and tenements.[36]

To the east of the palace and north of the cathedral lay the Pardon Churchawe or Pardon Churchyard, a burial ground for the canons. Somewhere inside it lay a chapel founded by Gilbert Becket, St Thomas' father.[37] There do not seem to have been any other noteworthy buildings in the area until 1394 when King Richard II founded a College of Petty Canons, which was completed by 1408.[38] Most likely the College lay on the north side of the yard, forming a rough square with the cathedral nave and transept. In 1424 Dean Thomas More built a cloister around the Pardon Churchyard, although it is not clear whether this was an actual architectural structure, like the southern cloister around the Chapter House, or merely the enclosing of a space by additional buildings. More was also responsible for the 'Prestehous', another residence for canons which lay 'to the north and east of the cloister'.[39] Most likely the College of Petty Canons and the Prestehous formed the north side of this 'square'. On the east side there was a library, built by a canon, Walter Sherington, and on many of the walls was depicted a Dance of Death which seems to have impressed Stow.[40]

The Prestehous may have extended quite far towards the north walls of the precinct, over lands which were either burial grounds or gardens.[41] A parcel of land granted in 1431 running 40 feet along the walls and yet only 8 feet from them also touched the Prestehous on the east.[42] Either that building was unusually long or else the name 'prestehous' was applied to many such buildings. The area does seem to have accommodated a number of residences for chantry priests by the end of the fifteenth century. These were sold off at the time of the Reformation and made into shops and tenements. All the buildings around the Pardon Churchyard were torn down.

The north-east quarter of the precinct was distinct in several ways. Until at least the fourteenth century it was the site of the periodic gathering known as the Folkmoot, and may have been the largest open public space in the City. A bell to summon citizens to such meetings hung in a tower next to the eastern precinct wall since at least 1220.[43] These gatherings continued to take place until Stow's day; in the sixteenth century St Paul's Cross, next to the north wall of the apse, appears to have been a sort of Speaker's Corner. If our present understanding of medieval parish boundaries is correct, this area was the only part of the precinct that lay outside the parish of St Gregory. The parish of St Faith included lands outside the precinct, although its church lay inside the walls. (Until the middle of the thirteenth century it lay directly behind the cathedral; it was then torn down to make way for the extension to St Paul's apse and relocated in the cathedral crypt.) It is possible that the parish boundary represented, at least in the early centuries, some sort of division between an area under the cathedral's jurisdiction and land that was in some way public.

This area was frequently the focus of disputes between clergy and citizenry. In 1285 Edward I completed or strengthened the precinct walls, ostensibly for the protection of its residents; the precinct could be closed off at night to prevent the 'divers robberies, homicides and fornications' which Dugdale lamented.[44] Then, during the Iter of 1321, the citizens of London complained to the king that their passage through the Churchyard was at times restricted.[45] The dean and chapter replied by waving a charter they appeared to have received from Henry I in 1111 which granted them rights to the precinct lands. But what of the Folkmoot? Until that time it may not have been considered part of the precinct. It is possible to interpret the completion or strengthening of the walls, as well as references from the fourteenth century on to the 'St Paul's Churchyard' instead of 'Folkmoot', as an attempt on the part of the dean and chapter to enclose public land. No doubt this issue was never properly resolved.

The earliest references to property holding in this area are from the late thirteenth century. Most of these are to lands outside the walls or against the bell tower. In 1285 Mayor Henry le Waleys attempted to build along Paternoster Row against the north walls of the precinct.[46] This plan was opposed by the bishop, possibly because it was seen as an encroachment on cathedral property. In the end the mayor built his tenements, and in return paid for the construction of a chapel on the cathedral side of the wall. These two structures must have stood almost back to back, as it was stipulated that rainwater from the outside tenements should not trickle onto the roof of the chapel. This chapel and the burial vaults below it became known as the Charnel House. Nearby lay some houses which by 1312 appeared to have been old and dilapidated. In that year they were assigned to John of Langton, keeper of the New Work, who was expected to undertake the repairs.[47] Langton's houses are mentioned again in a 1336 St Paul's rental which confirms their location in the north-east corner of the precinct.[48] These may be the tenements referred to in 1637 as lying to the east of the Charnel House, although the reference is confusing as the Charnel House had been dismantled at the time of the Reformation.[49]

The area around the bell tower itself seems to have been the site of some building from the fourteenth century. As early as 1269 a parcel of vacant land at the base of the tower was leased to a Henry de Keles.[50] In 1322 a shop was leased to a John Flege and his son Henry.[51] This shop may well have been built on the land given to Keles; both are described as lying at the base of the tower, outside the walls. If this is a reference to the precinct walls, then neither land nor shop were ever part of the precinct; and yet, both transactions involved the dean and chapter of St Paul's. The walls in question might be those of the tower itself, the term 'outside' used to distinguish from space within the tower. In 1353 John de Grafton, a parchment maker, and his wife Mary were granted a room inside the tower measuring 25 by 5 feet; not a convenient sounding space, but a sign that ecclesiastical buildings were being used by laymen by this time. The Graftons were also granted two shops lying beside the tower, measuring $25\frac{1}{2}$ by $22\frac{1}{2}$ feet.[52] This time it is clear these lay within the precinct: to the west of the tower, to the east of the cemetery, to the south of the precinct gate and to the north of another shop owned by John Gretehened. It may have been these same shops and this same room that were granted to John Page, a carpenter, and William Cotgave, a tailor, in 1375; the description, in the cemetery, outside the belfry wall, recalls that of the Grafton's properties.[53]

The dimensions given for these shops seem enormous; surely buildings that wide would have been a great inconvenience to traffic passing from Paternoster Row to the south Churchyard. It is also worth noting the presence of artisans inside the precinct boundaries. Indeed, as the fifteenth century progressed, the precinct became much more cluttered; tenements sprung up at the expense of religious institutions, and even, it would seem, at the expense of the natural flow of traffic. In 1441 the front buildings of the Precentor's Inn were leased to a haberdasher, and another haberdasher lived next door. In 1422 the treasurer leased part of his Inn to another layman. Long before the Reformation there were laymen living and working in what had always been used exclusively by the cathedral community. By the later sixteenth century there were even tenements stuck between the buttresses of the famous east end of St Paul's itself.[54]

The secularization of the St Paul's precinct comes as no surprise in the light of the history of later medieval London. All the same, this process, like the dissolution and selling off of tenements at the time of the Reformation, is a pity. Many of the features of the medieval precinct of St Paul's, especially the Cloisters and Chapter House, and the vicars and chantry priests' residences, recall the spiritual charm that remains in many rural spots across Europe. The culprit in London's case, of course, is not the change in religion, but the sheer numbers of people. By the later sixteenth century the city was a capital, with all that implies, good and bad, and the demand for space became chronic. In many ways it became a very different city, and the precinct of St Pauls's just another part of London.

1 Most of the cited material is deeds originally held in the collection of St Paul's Cathedral and now reorganized at the Guildhall Library, London. These documents have been abbreviated GL MS.
2 GL MS 25121/466.
3 GL MS 25121/3061.
4 GL MS 25121/417.
5 There is a reference to it in Drury's deed, GL MS 25121/417.
6 GL MS 25121/1948.
7 GL MS 25121/1368.
8 GL MS 25121/519.
9 GL MS 25121/519.
10 GL MS 25121/1941.
11 Wyke had recently been the Precentor; that he moved might suggest that the residence did tend to go with the office.
12 Public Record Office MS E 211/686.
13 MS reproduced in Marion Gibbs, *Early Charters of St Paul's, London*, Camden Third Series, Vol. LVIII, 1939, No. 105.
14 GL MS 25121/1800.
15 GL MS 25121/1941.
16 John Stow, *A Survey of London*. Reprinted from the volume of 1603. Introduction by C. L. Kingsford, 1908, Vol. I, p. 328.
17 GL MS 25121/3017.
18 GL MS 25121/3015.
19 Stow, ibid., Vol. II, p. 20.
20 GL MS 25121/676.
21 GL MS 25121/242. Stortford's deed was amended by a slightly later hand with the additional information.
22 GL MS 25121/243.
23 Stow, Vol. II, p. 20.
24 GL MS 25121/1737.
25 GL MS 25121/1095.
26 Stow, Vol. II, p. 20.
27 GL MSS 25121/1078, 1986.
28 GL MS 25121/520.
29 GL MS 25173: List of rents from 1526.
30 GL MS 25121/1938.

31 GL MSS 25121/1917, 1918, 1922, 1923.
32 GL MS 25121/524.
33 For a detailed description of the topography of the palace based on archaeological excavations, see W. Sparrow Simpson, 'The Palaces and Town Houses of the Bishops of London', *London and Middlesex Archaeological Society Transactions*, New series, Vol. I, 1905.
34 For example, GL MSS 25121/341, 511.
35 For example, GL MSS 25121/425, 515.
36 GL MS 25765/14.
37 Stow, Vol. I, p. 328.
38 Stow, Vol. I, p. 327.
39 GL MS 25121/1960. This deed makes references to More's building activities in the Pardon Churchawe.
40 Stow, Vol. I, p. 327.
41 GL MS 25121/1858.
42 GL MS 25121/1950.
43 GL MS 25121/494.
44 Sir William Dugdale, *A History of St Paul's Cathedral*. Republication by Henry Ellis, 1818, p. 5.
45 GL MS 25121/1756.
46 GL MS 25121/1802.
47 GL MS 25121/1937.
48 GL MS 25127: rentals to the chapter of St Paul's from 1336.
49 GL MS 25121/1831.
50 GL MS 25121/1495.
51 GL MS 25121/1079.
52 GL MS 25121/1090.
53 GL MS 25121/525.
54 GL MS 25121/1807.

II. FOUR PLANS OF SOUTHWARK IN THE TIME OF STOW

By MARTHA CARLIN

IN January 1583 the mayor of London, Sir Thomas Blancke, wrote to Lord Burghley, the Lord High Treasurer, with a novel proposal: that a one-page street guide to London be posted in Westminster for the convenience of those visiting the city. This idea apparently arose in the course of a recent mayoral task, the revision of the catalogue of infected victualling houses. Blancke had been aided in the revision by a 'Mr Norton' – perhaps John Norden, the topographer – who, wrote Blancke, 'informeth me that he hathe herein had speciall regard to two thing*es*, the one to give suche plaine discription and note of the stretes and places as may s*e*rve for easye notyce to suche as repaire to this Citie, the other that it be in suche shortenes as maye be brought into Lesse than one face of a shete of papre, to be fixed in places Convenient'. Blancke enclosed a list of proposed posting sites for the new guide, which he had compiled 'havinge respecte to Westm*inster* and the waye thither, and the entrance alwaies into this Citie', and asked for Burghley's opinion.[1]

Nothing seems to have come of this scheme – at least, no such guides are mentioned in contemporary visitors' accounts, nor in the Stationers' Register – although if 'Norton' was indeed Norden, this may have been the genesis of the small plan of London that Norden first published a decade later.[2] The incident is relevant to this paper only as a reminder of the growing interest in the topographical description and representation of London that developed in the sixteenth and early seventeenth centuries, and as a possible example of one such work that may not have survived.

Students of the period are fortunate in the abundance of extant written records. The numerous surviving municipal, parochial, and guild records; historical accounts, most notably John Stow's monumental *Survey of London*; estate records such as deeds, rentals, wills, and inventories; diaries and travellers' accounts, such as those of Henry Machyn, John Manningham, Thomas Platter, and Paul Hentzner; and literary texts, such as *Eastward Ho!* and *Bartholomew Fair*, form only part of the great wealth of surviving documentary material.

Cartographical and pictorial sources, while few in number compared with the written sources, also contain crucial information on the city, its streets, buildings, and inhabitants. Of particular value to the topographer are the earliest extant maps of London: the copperplate map (only two plates of which survive) and its derivatives (the woodcut map formerly attributed to Ralph Agas, and the engraved map-view published by Braun and Hogenberg in 1572), as well as Norden's own plan of 1593. These have been studied by Adrian Prockter and Robert Taylor,[3] Martin Holmes,[4] and Stephen Marks.[5] Valuable collections of graphical sources have been reproduced in books on London such as those by Felix Barker and Peter Jackson,[6] Christopher Hibbert,[7] John Schofield,[8] and Philippa Glanville;[9] in the recent volume on medieval English maps and plans by P. D. A. Harvey;[10] and in the publications of the Museum of London and the London Topographical Society. Still others are scattered in more specialist works, such as W. G. Groos's edition of the English travel-diary of Baron Waldstein,[11] or histories of the London stage.

For Southwark, London's ancient southern suburb, there is a considerable volume of documentary material surviving from Stow's day but only a handful of contemporary graphical sources. Most of these have been reproduced in modern publications: the 'Agas' and Braun and Hogenberg maps;[12] Anthonis van den Wyngaerde's mid-century panorama of London, Westminster and Southwark;[13] the bird's-eye view of London, Westminster and Southwark in William Smith's manuscript *Particular Description of England*,

1588;[14] a sketch of the marketplace in the Borough High Street in Hugh Alley's 'Caveatt for the Citty of London' of 1598;[15] Ralph Treswell's plans of two blocks of tenements, surveyed in 1611;[16] Norden's plan of London and Southwark;[17] a simplified plan of the western part of Southwark made in 1618 and an estate map of the manor of Paris Garden made in 1627;[18] the 'Fête at Horselydown' painting at Hatfield House; and Johannes de Witt's sketch of the Swan playhouse.[19] Glimpses of Southwark also can be seen in such views as the Cowdray wall painting of Edward VI's coronation procession[20] and the view of London Bridge by Richard Garth.[21]

Especially useful for the topographer are two detailed sketch-plans of Southwark. The earlier plan, unsigned and dating from about 1542, shows the major buildings and other landmarks in the central part of Southwark, from Park Street on the west to Bermondsey Street on the east. This plan, now in the Public Record Office, was published in reduced facsimile by William Rendle in his volume on *Old Southwark and Its People* (1878). The north-west portion of it was reproduced photographically in the Bankside volume of the *Survey of London* (vol. 22) and in an article by I. A. Shapiro on the Bankside theatres;[22] another small segment was reproduced in volume 25 of the *Survey of London*. The later plan, unsigned and surviving now only in three nineteenth-century copies (each with varying details), shows Southwark's eastern riverside district, known as Horselydown. One of these copies was published by G. R. Corner in the first volume of the *Surrey Archaeological Collections* (1858); the other two have not heretofore been published.

There are, in addition, two sketch-plans of Southwark that have not been published at all. One, unsigned and dating from the early 1550s, is a rough property-map of the western part of Blackman Street; the other, signed 'Ralphe Treswell', is a simple plot-plan of Horselydown. Together, these plans form an important source for the topographical history of Southwark in the late sixteenth and early seventeenth centuries.[23]

Plate 1. Sketch-plan of *Horsey Downe*, signed *Ralphe Treswell*. N.d. [*post*-1612; perhaps *post*-1616]. Society of Antiquaries, Harley Collection, vol. v, fol. 28. Pen and ink and water colour on paper. Dimensions: H 30 cm × W 21 cm.

This small sketch-plan was not included in John Schofield's list of Ralph Treswell the elder's known surveys.[24] Two details suggest the possibility that it may have been drawn by Treswell's second son, also named Ralph. The plan itself must date from 1612 or later, because the almshouses shown here were ordered to be built in 1612.[25] The published London plans of the elder Treswell show that by 1612 he commonly signed his plans *Rad'us Treswell senior*,[26] although in one instance he omitted the word *senior*.[27] In no case does he use the English spelling *Ralph(e)*. Also, in those London plans that concerned almshouses, Treswell senior used the plural *Almeshouses* or *Almes howses*; not *Almeshousen*, as here.[28]

These details are, however, more suggestive than conclusive, especially without further information on the life of Ralph Treswell the younger. It is known that, following the elder Treswell's death (which occurred between July 1616 and March 1617), Ralph junior was appointed the administrator of his father's estate.[29] If this plan is, in fact, by the younger Treswell, it presumably dates from 1616 or later, since the signature lacks the epithet *junior*.

The plan represents the open pasture ground known as Horselydown, which lay on the south bank of the Thames opposite the Tower. We see the irregular outline of the down, with its bordering ditches, the pathways that traversed it, the unnamed roadway leading north to the riverside, and the *waie to Deptforde* (now Jamaica Road) leading east. An annotation gives the total acreage as 12 acres, 3 roods, and 29 perches. In the north-west corner is the detached *churchyarde* (established 1583) of St Olave, Southwark, and the row of *Almeshousen* (commissioned 1612) beside it. Another row of houses along the western side of the ground is the only other encroachment shown. These few details cor-

PLANS OF SOUTHWARK

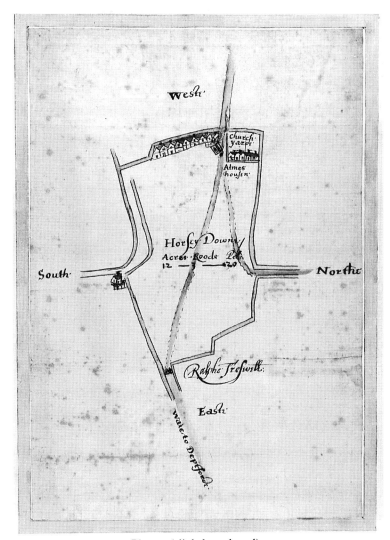

Plate 1 (slightly reduced).

respond well with those shown in Plate 2a. This plan, like that of Plate 2a, probably was drawn for the churchwardens of St Olave's, Southwark, or for the governors of St Olave's Grammar School.[30]

Plate 2a. Sketch-plan of *Horseye Downe*, unsigned. Facsimile, described as a 'Plan of HORSEYE DOWNE. Reduced from the Original Dated 1544. In the possession of the Governors of St. Olave's Grammar School.'[31] Published by George R. Corner in *Surrey Archaeological Collections*, vol. 1 (1858), facing p. 171. The published facsimile measures 24·2 cm in height by 17·2 cm in width.

Two additional nineteenth-century copies of similar plans are reproduced here as Plates 2b and 2c. The three copies agree in general but differ in a number of major and minor details. Several of the people named in the plan probably can be identified with persons mentioned in the St Olave's churchwardens' accounts of the 1540s and 1550s. However, all three copies contain features, such as the churchyard (established 1583) and almshouses (commissioned 1612), that considerably post-date the year 1544. In addition, the crowned hats shown on two of the boatmen in 2b, and the script used in the annotations (in so far as one can judge it from the copies), seem more suggestive of a date in the early seventeenth century than in the mid sixteenth century.

Corner stated that the original of 2a was in the possession of the governors of the parish's free school (founded 1560–1[32]), while a note on 2b states that its original was owned by a Mr Jones. 2c is said to be a tracing by Corner from 'the old parish plan'. Both 2b and 2c contain features that appear to have been sketched in at a later date; 2a contains no suggestion of this, but may have suppressed such distinctions. It seems likely that the three surviving copies derive from exemplars that were used as working plans and updated to reflect new property developments. For the variant details, and for a fuller discussion of the presumed exemplars, see under Plates 2b and 2c.

The documented history of Horselydown dates from 1206, when the Knights of St John of Jerusalem purchased $7\frac{1}{2}$ acres of land in *Horsemead* or *Horscimead*, and one acre in *Horscidune*. By 1327 they owned three watermills, three acres of land, one acre of meadow and twenty acres of

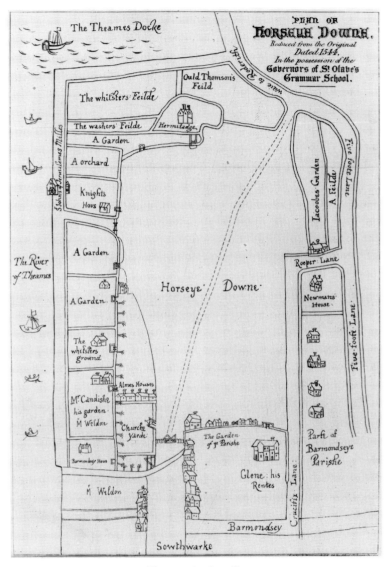

Plate 2a (reduced).
(East at top of page.)

pasture in *Horsedoune*, but by 1511 they had alienated much of this property – approximately the western half – to the archbishop of Canterbury. The knights did, however, retain a mill called *Saynt Johns Mylle* and about seven acres of grazing common in the eastern half of the down, and these passed to the Crown at the suppression of the English branch of the order in 1540. In 1538 Thomas Cranmer sold the Canterbury estate in Southwark and Horselydown to Henry VIII. Soon after, Sir Roger Copley and his wife Elizabeth seem to have acquired all or most of the Horselydown estate, for in 1545 they sold fifteen acres of land there to three Southwark residents: Adam Byston, Henry Goodyere, and Hugh Eglesfeld. Eglesfeld acquired sole ownership of the Horselydown estate by 1553, when he leased the fifteen acres to the churchwardens of St Olave's, Southwark; his son Christopher conveyed the property to the governors of St Olave's Grammar School in 1581. In 1583 the vestry of St Olave's ordered that a pit be dug at Horselydown for use by the scavenger as a laystall (dung heap). The parishioners were to be allowed to use the down 'for their recreacion to walke in' and for archery practice by the 'youthe' of the parish. Parishioners also were to be allowed to pasture cattle on the down for a weekly fee of 2*d*. in the summer and 1*d*. in the winter, and horses at twice that rate. Hogs and sheep were not allowed, nor were non-parishioners allowed to graze their animals there. Two years later, in 1585, the vestry increased the fees for those wishing to pasture more than one cow and one horse, and leased the down (apparently on a year-to-year basis) for £6 a year to John Alderton, who was to be allowed to dig for sand and gravel there at the discretion of the churchwardens.[33] In the seventeenth century the southwestern part of the down evidently was used as a mustering-ground; it is called the 'Marshall Yarde' in 2*b* and 2*c*, and 'Artillery Yard' in Ogilby and Morgan's map.

This important plan shows not only the down itself, but also the ditch-lined roadways and properties that surrounded it. While the general outlines of the plan reflect the actual shapes and relationships of the streets and other features shown, it is inaccurate in scale and sometimes also in delineation. For example, Bermondsey Street is shown with a straight north–south alignment rather than its true NW–SE alignment.

The textual annotations read:

[at the east end of the river bank]
[2.1]   *The Theames Docke*
This is St Saviour's Dock, so called from the nearby abbey of St Saviour, Bermondsey (see below, Plate 4, no. [4.65]), which owned the eastern half of the dock until the suppression of the house in 1538.[34]

[in the centre of the down]
[2.2]   *Horseye Downe*

[on the north side, in the river]
[2.3]   *The River of Theames*

[at the west end of the river bank]
[2.4]   *Mr Weldon*
A survey of Southwark made in 1555 records sixteen tenements and certain grounds here, owned by the heirs of Edward Weldon. A second survey, dating from *c.* 1564–5, notes that at that time the property, containing sixteen tenements, brewhouses, wharves, gardens, and Weldon's mansion house, extended eastward as far as the mill (see [2.11], below).

[2.5]   *Barmondsey Hous*

[2.6]   *Mr Candishe his garden*
In 1553–4 Christopher Candishe was fined 40*s.* by the parish for a trespass done in Horselydown.[35]

[2.7]   *Mr Weldon*
Presumably the same Mr Weldon as above.

[2.8]   *The whitsters ground*
Note the use of Horselydown for the drying and bleaching of laundry (see also the entries for the 'washers Feilde', the 'whitsters Feilde', and the churchyard, below).

[2.9]   *A Garden*

[2.10]   *A Garden*

[2.11]   *St Johns of Jerusalemes Milles*
In 1531 the Knights of St John had let this mill for a term of sixty years at £8 a year to Christopher Craven. Craven remained the

tenant until at least 1540, but in 1544, when the king granted the mill to his servant John Eyre, it was in the tenure of Hugh Eglesfeld. By 1555 the property had come into the possession of Cuthbert Beeston and was occupied by James Dixon. A mill survived on the site as late as 1803.[36]

[2.12] *Knights Hous*
Probably so called from the Knights of St John (see above), or possibly from one William Knight of Southwark, brewer, who in 1617 was sued by the governors of St Olave's Grammar School for a trespass upon the down.[37]

[2.13]   *A orchard*

[2.14]   *A Garden*

[2.15]   *The washers Feilde*

[2.16]   *The whitsters Feilde*

[at the eastern end]
[2.17]   *Ould Thomsons Feild*

[2.18]   *Hermitadge*

[2.19]   *waie to Roderith*
The 'way to Rotherhithe'; now Jamaica Road.

[along the south side]
[2.20]   *Five foote Lane* (named twice)
Now Tanner Street.

[2.21]   *A Feilde*

[2.22]   *Iacoobes Garden*

[2.23]   *Rooper Lane*
Recorded in 1390 as *Ropereslane*; now disappeared.[38]

[2.24]   *Newmans House*
Possibly to be identified with Thomas Newman, who in the 1540s and 1550s together with his wife Margaret worked for the church of St Olave, doing odd jobs such as cleaning the church, scouring the candlesticks and carrying sand.[39]

[2.25]   *Parte of Barmondseye Parishe*
Crucifix Lane/Rooper Lane marked the boundary between the parishes of St Olave, Southwark, and St Mary Magdalen, Bermondsey.

[on the west side]

[2.26]   *Crucifix Lane*
The Bermondsey Street end of this lane still survives under the same name.

[2.27]   *Barmondsey*
Bermondsey Street.

[2.28]   *Glene his Rentes*
The row of houses on the northern side of this plot originally had been built by Sir John Fastolf in the 1440s–1450s. To an earlier house at the corner of Bermondsey Street (known as *la Cruch Huse* from the cross that stood in the street nearby) he added five additional tenements and paved the road in front of them, which was known as Horselydown Lane. In the 1450s these five tenements consisted of three dyeshops and two fullers' establishments. The portion to the east of the five tenements was described in the survey of *c.* 1564–5 as consisting of four other tenements with 'certain grounds', lying in the 'Isle of Duk*is*'. The Isle of Ducks is marked on Morgan's map of 1681–2; it may have been so called from a nearby millpond. Fastolf's extensive estate in Southwark passed to his executor William Waynflete, who used much of it to endow Magdalen College, Oxford.[40]

[2.29]   *The Garden of ye Parishe*

[2.30]   *Churche Yarde*

[2.31]   *Almes Houses*
In 1583 the churchwardens of St Olave's established a new churchyard [2.30] 'vpon the lefte hand of horsey downe as you doo goo into the same'. It was to have on the south and east sides a brick wall seven feet high (the north and west sides probably were enclosed by ditches), and a keeper was hired in 1585 at a salary of 13*s*. 4*d*. per annum. In 1586 those parishioners who had subscribed to the purchase were permitted to dry clothing there (four pieces at a time) on payment of a penny, so long as they or their servants did not misbehave themselves by 'scoldinge fytinge or breakinge of any rosmary or other thinge that is set in the sayd churche yarde'. In 1596 the vestry of St Olave's decided that some young elm trees should be planted at Horselydown 'in suche convenient places as they Churchewardens shall thinke conveniente' – possibly it is these trees that are depicted on the north and west sides of the plot containing the churchyard and almshouses. The almshouses themselves were commissioned in 1612.[41]

[2.32] *Sowthwarke*
This marks the eastern end of Tooley Street.

> Plate 2*b*. Nineteenth-century copy of plan of Horselydown similar to that described above under Plate 2*a*. A note at the bottom reads: 'Fac-Simile of an ancient Plan of Horselydown traced from the original in the possession of [–] Jones, Esqu. The original is without Title and the only date appended to it consists of the Figures 1544 marked on the back of the Vellum on which it is drawn. RJC[?]. September 19: 1852.' Southwark Local Studies Library, Map 100. Pencil, pen and ink, and water colour on paper. Dimensions: H 54·2 cm × W 40·3 cm.

This plan clearly is closely related to 2*a*. Both in general appearance and in most individual features the resemblance is exceptionally strong. For example, the plans agree virtually entirely in spelling, capitalization, form of script (note especially the letters 'g' and 'e' and the ligatures in 'whitsters'), and in the placement and orientations of the annotations, as well as in the style of the drawing. One major difference between the two copies is that 2*b* includes an outer border containing the words *NORTH SOVTH EAST WEST* and a scale of feet; the scale, however, is very inaccurate and has the appearance of a later addition. 2*b* also differs from 2*a* in a number of other respects: the annotation for 'Barmondsey Hous' is missing and the single letter 'R' appears in its place; Mr Weldin's name is spelled with an 'i' instead of an 'o', and his western property is not fully enclosed; there is an extra rowing boat at the north-east corner; the boats and their passengers are shown in greater detail; there is a set of river stairs near the mill and there are additional bridges sketched in over the boundary ditches; the 'Hermitadge' has an additional ditch; 'Ould Thomsons Feild(e)' contains a house; the annotation for 'Rooper Lane' is missing; houses to the east of the parish garden are surrounded by a gated enclosure; there is a ditch or path dividing the parish garden from Glene's rents; the word

PLANS OF SOUTHWARK

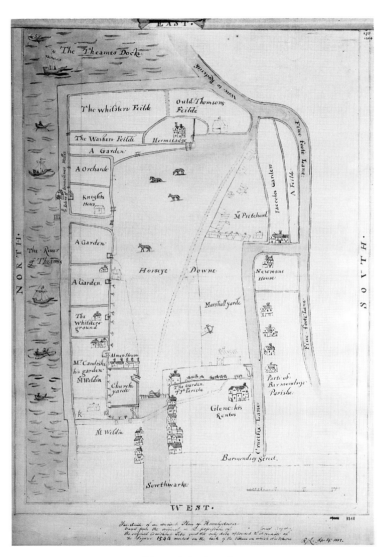

Plate 2b (reduced).

'Street' is added to 'Barmondsey' in the south-west corner; and the down itself, which is shown virtually featureless in 2*a*, contains four horses, a number of houses within enclosures labelled *Mr Pritchard* and *Marshall yarde*, and a round-topped structure that perhaps represents an ornamental gateway leading to the big house in Marshall Yard. A number of these features seem to have been sketched in at a later date or dates; they include the properties in the southern part of the down, the house in Thomson's field, the two houses included in 'Knights' house, the house on the northern side of the down, the house labelled 'R', and the extra ditch at the Hermitage.

> Plate 2*c*. Nineteenth-century copy of plan of Horselydown similar to those described above under Plates 2*a* and 2*b*. A note on the back, evidently a later annotation, reads: 'HORSELY DOWN. SKETCHED FROM THE OLD PARISH PLAN BY M[r] GEORGE CORNER VESTRY CLERK ST OLAVES SOUTHWARK'. Southwark Local Studies Library, Map 101. Pencil, pen and ink, and water colour on paper. Dimensions: H 53 cm × W 38·7 cm.

2*c* closely matches 2*b* in size and delineation; when the two are superimposed, the lines of their features nearly (but not exactly) match. 2*c* resembles 2*a* in lacking the scale of feet and the outer border containing the names of the compass points; in omitting the river stairs, the extra rowboat, the extra ditch at the Hermitage, the house in Thomson's field, and the ditch or path between the parish garden and Glene's rents; and in including the annotation for 'Rooper Lane'. 2*c* does contain most of the additional houses and other features shown on the down in 2*b*, but they appear with less detail; 'Mr Pritchard', for example, appears as 'Mr Put ... d' and the horses are much more crudely drawn. 2*c* also lacks a number of features common to both 2*a* and 2*b*: the features west of Crucifix Lane; Weldon's two properties and the house between them; the trees and enclosure to the west of the churchyard; the trees, enclosures and house on the north

PLANS OF SOUTHWARK

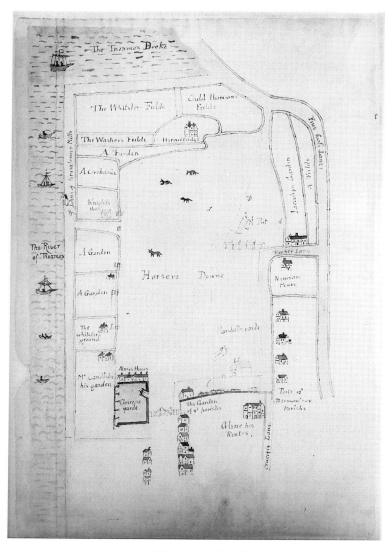

Plate 2c (reduced).
(East at top of page.)

side of the down; the entrances to the Hermitage and the garden to the north-west of it; the annotation for the way to Rotherhithe; and the western annotation for Five Foot Lane. There are also a few minor spelling variants: 2*c* adds a final 'e' to 'Marshall', omits the final 'e' from 'Barmondseye', and twice uses 'Fielde' instead of 'Feilde'.

What is one to conclude about the exemplars of these three copies? To begin with, both 2*a* and 2*b* contain unique details. In the case of 2*a*, the unique detail is the annotation 'Barmondsey Hous'. 2*b* uniquely contains a scale of feet (probably a later addition), an outer border with compass names, and a number of lesser details (the extra boat, extra bridges, the river stairs, the extra ditch at the Hermitage, the annotation 'R' where 2*a* has 'Barmondsey Hous', the word 'Street' in 'Barmondsey Street').

2*c* contains no unique details (other than the spelling variants mentioned above), and therefore could be interpreted as a conflation of 2*a* and 2*b*. Two points, however, argue against this. Firstly, why is it so incomplete? So many details are missing, particularly from the western end but also elsewhere (such as the trees, house, and enclosures on the north side of the down), as to suggest that the exemplar of 2*c* was itself incomplete or damaged. Secondly, it seems unlikely that 2*c* could have derived from 2*b* or from 2*b*'s exemplar, because if 'RJC' (the copyist of 2*b*) could read the name 'Pritchard', then so, presumably, could Corner (the publisher of 2*a* and alleged copyist of 2*c*). But if 2*b* did not derive from 2*b* or 2*b*'s exemplar, then what is the source of the horses and other features sketched in on the down in 2*c*? Evidently not 2*a*: Corner makes no mention of the existence of such features in 2*a*. This suggests that 2*c* derives from yet a third exemplar.

The origins, as well as the fates, of the exemplars are unknown. From the evidence of some of the names (such as Weldon and Candishe) it seems most likely that this plan originally was drawn in the 1540s or early 1550s (perhaps in 1544, which the exemplars of 2*a* and 2*b* both mentioned). Sometime in the early seventeenth century (after 1612) the original plan was copied, and new features such as the churchyard and almshouses were incorporated.

Probably three copies – the exemplars of 2a, 2b, and 2c – were made at this time, all annotated by the same hand. The exemplar of 2a was made for the governors of St Olave's Grammar School, who kept it until at least the 1850s, when Corner saw it and copied it. This copy seems to have received one later annotation: Bermondsey House. The other two copies – the exemplars of 2b and 2c – probably were made for the churchwardens and vestry of St Olave's . Both of these copies were updated with the added features on the down, and at least one copy (the exemplar of 2b) also had a house and the letter 'R' sketched in on the site of 2a's Bermondsey House. The exemplar of 2b either was updated more carefully or for a longer period than that of 2c (see 2b's additional bridges, ditches, scale, etc.). Perhaps the exemplar of 2b was meant to be the 'deluxe' copy, and that is why it also was given the outer border and (perhaps) the most detailed horses and boats. The exemplar of 2b subsequently strayed from the parish muniments and by the 1850s had come into the possession of Mr Jones. The exemplar of 2c evidently survived in a damaged condition amongst the parish muniments. There it was found and copied by Corner (who was the vestry clerk), presumably after his publication of 2a (1858), since he does not refer to it in his article.

> Plate 3. Sketch-plan of part of the western side of Blackman Street, Southwark, and of St George's Fields; unsigned. N.d. [c. 1550 × 1555]. Corporation of London Records Office, MS 39 C. (Southwark court book, 1539–64), fols. 147^v–148^r. Pen and ink on paper. Dimensions: H 30·5 cm × W 38·4 cm.

This rough sketch plan, drawn on a single sheet bound as two facing leaves, forms part of the court book of the Southwark manors that were under the jurisdiction of the City. The plan follows a written survey made in 1555 of the three Southwark manors – the Guildable manor, the

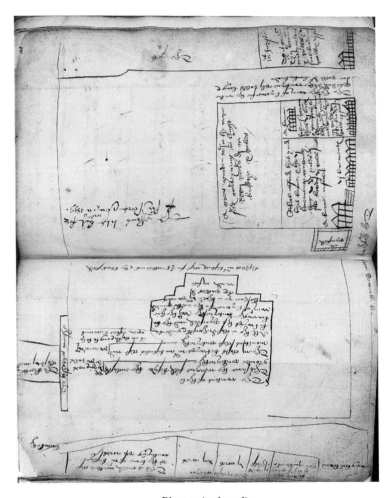

Plate 3 (reduced).
(North at top of page.)

Great Liberty manor, and the King's (in 1555 known as the Queen's) manor – that the City acquired from the Crown in 1550.[42] The last page of the survey (fol. 146^r) is written on the recto of the plan, but the information in the plan seems slightly to antedate the information in the survey (see, e.g., entry on Mrs Juxson, below). Perhaps this is a rough copy of an earlier plan (possibly made at the time of the acquisition of the three manors by the City in 1550), copied as an *aide-mémoire* for the local jurymen who compiled the survey. One later annotation is dated 1599. The textual annotations to the plan read:

[east side]
[3.1] *The highe way frome Sowarke to Newigton towen*
This road from Southwark to Newington also was known by the mid thirteenth century as *Blakemannestrete* (modern Blackman Street).[43]

[enclosed property at north-east corner of plan]
[3.2] *Mrs Juxson*
[3.3] *This garde[n] [and] ye ij teneme[n]t[es] conteyne long x polles [and] vj foot, brode iiij pooles [and] vj fote*
This property, formerly part of St George's Fields, was let to William Jackson by Bermondsey Abbey in 1522 for a term of sixty years at 10s. 4d. p.a. According to the survey of 1555, after the death of Jackson's wife it paid only 10s. p.a.[44] Since note [3.2] implies that Mrs 'Juxson' was still alive, the plan presumably pre-dates the survey.

[on west side of Mrs Juxson's property]
[3.4] *The p[ar]ke*
The park was attached to Suffolk Place, which is described below under Plate 4 no. [4.73].

[roadway on south of the park]
[3.5] *The highe way of lycenc' for the raker p[ro]vided dothe conteyne xlj polles long [and] one polle di[midiam] [broke struck through] brode*
This was a special right-of-way for the rakers of the Guildable manor to cross to the new dunghill in St. George's Fields that had been provided by the mayor of London *c.* 1550. Now Great Suffolk Street.[45]

[four enclosed properties to the south of the rakers' highway]
[3.6] *A barne*
[3.7] *John Sporwe thes vj ten[ementes] [and] gardens to theme [and] a barne behynd ye gardens conteyne long iiij Poles xiij fote brode viij poles xj foote*
In 1555 this property ([3.6] and [3.7]) was held by John Sparowe's executors, who paid to the City the annual rent of 15s. 8d.[46]

[3.8] *A barne*
[3.9] *vj tenement[es]*
[3.10] *Robert Alford this yard this barne [and] the vj tenement[es] conteneth long [v struck through] vj poles iij fote brode vj poles lacke a foote*
[3.11] *A great garden w[ith]in the mayne feld[es] conteyning in lenght xxxiij pooles di[midiam] in bredthe x polles*
In the survey of 1555 these properties ([3.8]–[3.11]) were described as *formerly* held by Alford (another indication that this plan antedates the survey), and currently held by one Eton, a carpenter, who paid the annual rent to the City of 3s. 4d.[47]

[3.12] *A pynfold*
An animal pound; the keeper was responsible for seeing that the rakers' highway was barred to other traffic.[48]

[to the west of the great garden, in different hand]
[3.13] *The p[ar]sons tithe but for [of struck through] six score [and] on acre a[nno] 1599*
The tithes of grain payable to the rector of St George's in 1545 comprised two quarters each of wheat, barley, and rye; four quarters of oats; and four cartloads of hay. St George's was the only Southwark rectory to receive agricultural tithes in the sixteenth century.[49]

[large open ground to south]
[3.14] *The content of thole*
[3.15] *This same the mayne feld besyde the enclosed ground conteynethe v^{xx} xix acres di[midiam] xv poles*

[3.16] *The iij closes betwyxt M^{rs} [m struck through] bostockes close and moultons close conteynethe xv acres wch is inclosed [and] oughte to ly open after lammas*
These three closes probably are to be identified with a property of fifteen acres described in the survey of 1555 as claimed by the heirs of Ralph Mydleton, in the tenure of Robert Gaynsboroughe (who paid the annual 5s. quitrent to the City), which fifteen acres were

enclosed and were supposed to lie open after Lammas.[50] Mrs Bostock presumably was the widow of the Bostock whose house is shown in Plate 4 (no. [4.60]). For Moultons Close, see [3.26], below.

[3.17] *All the in closed houses [and] gardens betwyxt the pynfold and the [bo struck through] boruwghe conteynethe w^{t}h the hie way of Lycence for the raker p[ro]vided So have ye in thole vij^{xx} xv acres ix polles di[midiam] and iij fote*

[3.18] *bysydes m^{rs} Bostock[es] close for y^t I recken' not [and] the breck feld*

The brickfield is mentioned below. The survey of 1555 records the main field as containing 4 furlongs and 155 acres, of which some 89 acres were held by 17 different tenants and claimants, and the remainder by the City as lord of the fee.[51] In 1539–40 39 acres of land here were held by various tenants as 2s. 8d. the acre.[52]

[to the west of the main field]
[3.19] *The bysshoppes marshe*
Lambeth marsh.

[3.20] *The bricke close vj acres*

[five enclosed properties at south end of plan]
[3.21] *Mrs Bostock[es] close is of it self paing her quitrente*
On Mrs Bostock, see above.

[3.22] *Thes iij closes contayne*
[3.23] *iij acres a half*
[3.24] *vj acres*
[3.25] *viij acres*

[3.26] *This is called moultons close in the tenure of Mr Basleye conteyni[n]g xv acres*
William Baseley held Moultons Close by virtue of royal letters patent dated 7 March 1546, paying 40s. p.a. In the 1530s and 1540s he was the tenant of the mansion house, grounds and pastures in the neighbouring manor of Paris Garden.[53]

[at south-west corner of plan]
[3.27]] *Lambethe*

Plate 4*a*. Sketch-plan of Southwark, unsigned. N.d. [*c.* 1542]. Public Record Office, MPC 64. Dark brown ink over pencil, with touches of blue colour, on paper. Dimensions: 83·06 cm × 58·42 cm. A reduced photo-

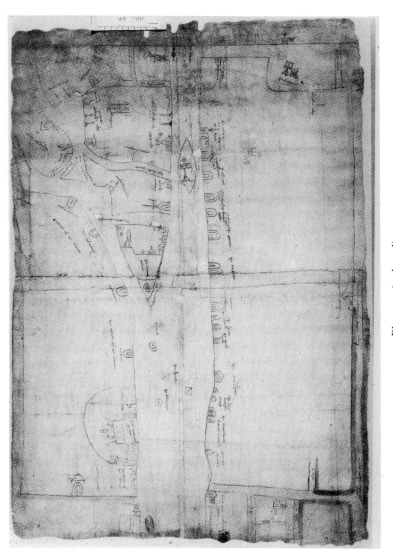

Plate 4a (reduced).
(North at top of page.)

graph of the north-west portion of this plan was published in the *Survey of London*, XXII, *Bankside*, plate 8, and by I. A. Shapiro in 1948.[54] Another small section, showing St Margaret's Hill, was reproduced in a reduced photograph in the *Survey of London*, XXV, *St. George's Fields*, plate 1A. A reduced facsimile of the entire plan, with added annotations, and with the placements of the original annotations often altered, was published by William Rendle as the frontispiece to *Old Southwark and its People* (London, privately printed, 1878). Rendle's added annotations number the features shown, a useful aid to the present study, and so his facsimile is reproduced here as Plate 4*b*.

This important plan represents the central area of Southwark, extending from the foot of London Bridge on the north to Suffolk Place and Long Lane on the south, and from Bermondsey Street on the east to Winchester Palace on the west. Some eighty features are shown or named, including inns, bridges, churches, gates, wells, houses, alleys, the pillory, the cage, and the courthouse. Some of the named houses (such as the Ram, the Swan, and the Bull's Head) are shown with appropriate signs. Particular note is made of manor boundaries: the plan centres on the Guildable manor, which lay at the foot of London Bridge, and which until 1550 was the only one of Southwark's five manors to be under the jurisdiction of the City. The boundaries of the Guildable manor are designated both by textual annotations and by representations of London's symbol, the sword used on the City seal. London Bridge, shown with its two southern boundary posts or *stulpes*, is not named, nor are the major Southwark streets.[55] The spellings used in the annotations suggest that the writer was a Lowlander, a common nationality for Southwark residents at this time. The origins and purpose of this plan are unknown, but it was included among the contents of a box of Duchy of Lancaster maps and plans listed in 1869 as 'chiefly descriptive of boundaries, extents and quantities of premises viewed under Commissions of Surveys, and for that purpose directed; and

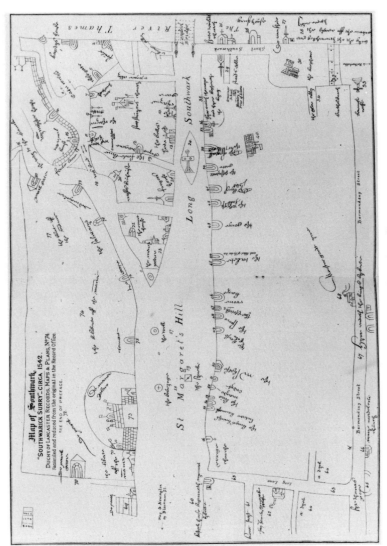

Plate 4b (reduced).
(North at top of page.)

principally elucidating the claims of parties in disputes pending in the Duchy Court, or otherwise contested'.[56]

The features (following Rendle's numeration with the addition of the Plate number, '4') are as follows:

[4.1] *baptys house*
This was a dyehouse, formerly the laundry of the priory of St Mary Overy (see [4.8], below). The priory had let it in 1537 to Robert Colyns for a term of forty years at 5s. p.a., and Colyns sublet it to John Baptist for 33s. 4d. p.a.[57]

[4.2] *fouler*
A house and brewhouse, also formerly part of the priory precinct, leased by the priory to Christopher Fowler by indenture.[58]

[4.3] *beare alley*
So called from the adjoining Bear Tavern, which stood at the foot of London Bridge from at least 1356 (the name is first documented in 1429) until it was pulled down in 1761, when London Bridge was widened.[59]

[4.4] *Gate to close*
This stone-faced gateway survived until the early nineteenth century. Together with [4.5] it controlled access to the riverside portion of the former priory precinct, where the conventual buildings had been. See also [4.38], below.[60]

[4.5] *close gate*
This gate, like [4.4], controlled access to the domestic portion of the former priory precinct. It probably is to be identified with a timber-framed gatehouse with a jettied solar for which the building contract, dated 1529, survives.[61]

[4.6] *peper ally*
The unnamed extension to the riverside also was known at this time as Pepper Alley.[62]

[4.7] *the church door*
The western doorway of the former priory church (see below).

[4.8] *sint savyors church*
The church of the former Augustinian priory of St Mary Overy (founded *c.* 1106, dissolved 1539). Perhaps the site of a pre-Conquest minster mentioned in Domesday Book. In 1205–38 a chapel dedicated to St Mary Magdalen was built against the south aisle to serve as a parish church for the lay residents of the precinct

and for those who dwelt on the east side of the high street opposite the priory. Wills of 1486–1520 mention parish fraternities of St Katharine and St Anthony. In 1540 the parishioners of St Mary Magdalen and St Margaret united to purchase the former priory church from the Crown for 400 marks to serve as the church of their new joint parish of St Saviour. Much of the medieval building (twelfth century and later) survives in the present Southwark Cathedral.[63]

[4.9] *the chayne gatte*
This controlled access to the churchyard, which lay along the south side of the church.

[4.10] *the chayne gate*
This controlled access to the 'close of the churchyard'. There were some twelve or thirteen tenements within this chain gate in 1555.[64]

[4.11] *wyt wonte[?] house*
Possibly to be identified with Thomas West; see under [4.18], below.

[4.12] *gryne dragone*
In the late fourteenth and early fifteenth centuries this property was the London townhouse of the Cobham family; in 1431 it was given as part of the endowment of Lingfield College, which retained it until the college was dissolved in 1544. In the 1540s the Green Dragon was a public inn kept by John Eves and his wife Edith, and later by Edith and her second husband, Matthew Smithe.[65]

[4.13] *the bolles hede*
A tavern, owned in 1555 by the heirs of Thomas Mackerell.[66]

[4.14] *wynchister house*
The London townhouse of the bishop of Winchester, in whose diocese Southwark lay. The site was purchased by Bishop Henry of Blois in the 1140s. The house, which was inhabitable by 1170, was largely rebuilt in the thirteenth century, with only limited construction thereafter. The western gable wall of the hall and the foundations of the hall and of other buildings survive.[67]

[4.15] *wavoral[?] house*
In 1299 the bishop of Winchester granted this site to the prior of St Swithun's, Winchester, and this became the site of the prior's London townhouse until the Reformation. It then was acquired by Lord Russell, the Lord Admiral, who exchanged it with the bishop of Rochester for Chiswick Place in 1542. The name given here may derive from a confusion with Waverley Place, the former

townhouse of the abbots of Waverley, which lay opposite the prior of St Swithun's house on the south side of the street.[68]

[4.16] *the way to the banck*
Also known in 1457-8 and in 1503-4 as *Cherchewey* or *le Church wey*;[69] now Park Street.

[4.17] *foule lane*
Rocque's 'Dirty Lane'; now disappeared.

[4.18] *wis lane*
Also known as West's Rents, after Thomas West, a carpenter, who in the 1520s-1530s built some 30 tenements here.[70]

[4.19] *the foule lane*
Rocque's 'Foul Lane'; now disappeared.

[4.20] *crosses bruhouse*
This was the Red Lion, an inn and brewhouse owned in 1555 by John Crosse, who probably had been the owner or proprietor since the 1530s.[71]

[4.21] *frogets house*
Richard Froggett, a brewer, is mentioned in the Guildable manor court book from 1539 to 1549. His house probably was the smaller building shown here. The larger building, marked with a sword to signify that it stood at the boundary of the City's Guildable manor, probably represents a tavern owned by St Thomas's Hospital and known as the Cross Keys; in 1540 it was kept by the vintner William Gylberd.[72]

[4.22] *the court house*
Until 1540 this was the parish church of St Margaret (first recorded 1107-1129), the rectory of which belonged to the priory of St Mary Overy. Two parish guilds are recorded: the sisterhood of St Anne, mentioned in 1485, and the more important fraternity of the Assumption of the Virgin, incorporated in 1449 and possibly to be identified with an unnamed fraternity mentioned in 1361. When the parishioners of St Mary Magdalen and St Margaret purchased the former priory church and united their parishes in 1540, St Margaret's church was taken over for use as a courthouse or 'Sessyons Howse' and prison by the Justices of the Peace for Surrey, and probably continued to be so used until at least 1563.[73]

[4.23] *the mart place*
The market place, with a well with two buckets at its southern tip.

Formerly the site of St Margaret's churchyard, after the deconsecration of the church in 1540 it evidently was added to the market area of the Guildable manor. The main marketplace lay in the high street (see [4.24], below).[74]

[4.24] [*the pillory*]
The pillory stood in the marketplace, which had been established in the Guildable manor in the fifteenth century, rivalling an older market held within the precincts of St Thomas's Hospital (see [4.38], below). The high street marketplace extended southward from the foot of London Bridge on both sides of the street as far as the manor boundaries; these were the former St Margaret's churchyard (see above) on the west side of the street, and the Swan (see below, [4.36]) on the east side. The pillory and market were depicted by Hugh Alley in 1598.[75] The pillory remained in this location until 1620; the market until 1755.[76]

[4.25] *synte toulus church*
The church of St Olave, dedicated to St Olaf of Norway (d. 1030), is first recorded in 1096. It measured 69 feet square, with a western tower and turret 95 feet high, and it contained four aisles terminating in chapels dedicated respectively to the Virgin, St Clement, St Anne, and St Barbara, each maintained by its own fraternity or (in the case of St Anne) sorority. In 1533 these were augmented by the foundation of a Guild of the Name of Jesus.[77]

[4.26] *brusthouse*
This was the site of the Bridge House, the headquarters of the trustees who administered London Bridge and its estates. First recorded *c*. 1222–3, in the 1540s the Bridge House consisted of domestic buildings, a garden, a yard, storehouses and granaries, and eleven ovens built in 1522 for baking bread from the grain stored there for the relief of poor citizens. At the north-east corner of the property was a brewhouse (formerly a townhouse), given to the City in 1524 by the alderman George Monnox.[78]

[4.27] *de ranshede*
The Ram's Head, on the east side of which ran the boundary between the Guildable manor and the King's manor.[79] In the early 1550s the Ram probably was kept by Geoffrey Wolff; the churchwardens of St Olave's had a breakfast at Wolff's establishment in 1552–4.[80]

[4.28] *hyer endith the lyberte off the mayre and beghinnith the [liberty of] the king*

This marks the boundary of the Guildable manor with the Great Liberty manor.

[4.29] *smits alle*
Smiths Alley is first recorded in 1508, when it consisted of a capital messuage and 22 gardens. The representation of the City sword indicates that it marked the boundary of the Guildable manor.[81]

[4.30] *the borghene*
The 'Burgony' or 'Petty Burgen', a district so called by 1394, when a tenement there was said to lie *in loco vocato Burgoyne*. The origin of the name is unknown.[82]

[4.31] [*the cage and pillory*] (number missing from Rendle's facsimile; beside [4.30])
In 1539 and again in 1540 the jury of the Guildable manor ordered the churchwardens of St Olave to build a 'cage' (lock-up) next to the pillory. By 1549 the cage was being cited as an eyesore, a public nuisance (because of its noisy inmates), and an obstruction to traffic (because it was used as a hitching post for horses).[83]

[4.32] *berthol burch*
Battle Bridge, so called from the nearby mills and townhouse formerly owned by Battle Abbey. The mills, whose millstream the bridge crossed, were alienated by the abbey in the early thirteenth century, and the new owner was required to maintain the bridge. The townhouse, however, which lay between the mills on the east and the Bridge House on the west, remained in the possession of the abbey until its suppression in 1538, when the property passed to the Crown.[84]

[4.33] *barmese crosse*
Bermondsey cross was in existence by 1312, and perhaps by *c.* 1240–50; it is mentioned as late as 1572.[85]

[4.34] *the glene ally*
Glean Alley, which in 1555 contained some 38 tenements and a mansion house, evidently took its name from the latter, which by 1504 was known as 'le Glene'. The alley name is first recorded in 1506.[86]

[4.35] *hyer endith the mayre and hyer beghineth the kyng*
See [4.28], above

[4.36] [*unnamed*]
The representation of the City sword labels this building as lying

on the boundary of the Guildable manor. As such it can be identified with a tenement known in the mid sixteenth century as the Swan or the Swan with Two Necks.[87]

[4.37] [*unnamed*]
This probably is to be identified with a house known in the 1530s–1550s as the Angel on the Hoop.[88]

[4.38] *The ospitall ch[u]rch dore*
The hospital of St Thomas the Martyr (rededicated in 1552 to St Thomas the Apostle) originally lay within the precinct of the Southwark priory of St Mary Overy. About 1212, however, much of the priory was destroyed in a fire, and in 1213–15 the bishop of Winchester refounded the hospital on a new site across the road where there was more space, cleaner air and a better water supply. By *c.* 1240 the hospital probably had acquired most of its precinct. The church stood on the north side of a lane called Trenet Lane (now St Thomas's Street) and its nave served as the hospital ward. The conventual buildings of the Augustinian brothers and sisters formed a rectangle on the south side of Trenet Lane, behind the hospital's churchyard. An infirmary lay on the northern edge of the precinct in the later thirteenth century, and an eight-bed maternity ward for unwed mothers was built in the fifteenth century, reputedly by Richard Whittington. In the early sixteenth century a large new almshouse for men was built on the northern edge of the precinct, and [4.38] may actually refer to this building rather than to the church itself, which was set back from the street (see [4.40], below). The remainder of the precinct contained tenements and gardens. The hospital had maintained a market for 'grain and other things' by the late twelfth century, and this market transferred with the hospital to the new Trenet Lane site, where it was held near the hospital gate (see [4.39], below) on Wednesdays, Fridays and Saturdays. The hospital's market seems to have declined in the late fifteenth century (evidently from competition from the new Guildable manor market; see [4.24], above), but it later revived. The hospital, whose precinct had become an independent parish at the end of the fifteenth or the beginning of the sixteenth century, was surrendered in January 1540, and from 1541 to 1551 it was in the keeping of royal grantees. Reopened by the City in 1552, it was transferred to its present site in Lambeth in 1862.[89]

[4.39] *the hospytall gatte*
This gate was built in the early thirteenth century across Trenet Lane to control access to the hospital precinct.[90]

[4.40] [*unnamed*]
This sketch probably is meant to represent St Thomas's Hospital, the precinct of which was set back from the street.

[4.41] *The kyng[es] hede*
A public inn by 1379. Known in 1467 as the Cardinal's Hat, and in 1507–36 as the Pope's Head.[91]

[4.42] *the whyght hart*
By 1400 a public inn called *le Herteshed*; known as the *Herte* in 1450, when it was used by Jack Cade as his headquarters.[92]

[4.43] *the gorge*
In 1509 described as 'tenements called *le Syrcote*'; by 1544 apparently a public inn known as the St George. One wing of this inn, rebuilt in the seventeenth century, today is the sole London survivor of the galleried inns.[93]

[4.44] [*unnamed*]
This perhaps is meant to represent the former townhouse of the abbot of Hyde. The abbey acquired this property, which included the site of the later Tabard Inn (see below), in 1304–6, and the abbots used the townhouse from 1307 until shortly before the dissolution of the abbey in 1538. It lay to the north of the Tabard, behind some streetfront tenements. The townhouse remained in the possession of the Crown until 1544, when it was sold.[94]

[4.45] *the tabete*
The Tabard Inn, where Chaucer's pilgrims stayed. The site was owned by Hyde abbey from 1304–6 until the dissolution of the abbey in 1538; it was sold by the Crown in 1544 (see above). An inventory of 1538 mentions two parlours, a hall, a kitchen with a well in it, two cellars, seven chambers, and a 'drynkynge bower'. In 1546 the lease of the Tabard was worth £21 p.a.[95]

[4.46] *crone kaye*
In the fifteenth century this was the townhouse of the Ponyngs family and was known as Ponyngs Inn. In 1508 it was described as 'tenements called le Crown Key'. From 1517 to 1536 Henry VIII rented the Crown Key from Sir Edward Ponynges for use as an armoury; by 1555 it was in the possession of the heirs of one Boyer.[96]

[4.47] *the [chris]otofor*
The Christopher, a brewhouse by 1565, was among a group of Southwark tenements sold by the Fishmongers' Company in 1550.[97]

[4.48] *the spore*
The Spur; a public inn so called by the 1530s.[98]

[4.49] *the horse hede*
So called by 1391; a public inn by the 1430s.[99]

[4.50] *the m[ar]shalse*
The Marshalsea prison (the prison of the Marshal and Steward of the king's household), probably built in 1373. An inventory of 1483 lists a *grete Warde*, a *litill Warde*, a *Womans warde*, a *Chapell*, a *Parlour*, a [chamber called the?] *sprus*, two chambers, a *wacchehous*, a *Taphous*, and a *logge att the Gate*. The Marshalsea was broken open by rebels in 1381 and 1450 and by sanctuary men in 1470. The prison remained on this site until 1798.[100]

[4.51] [*unnamed*]
This may represent a messuage known in 1555 as the Half Moon, which probably is to be identified with a brewhouse called the *Mone* that was the subject of a title dispute in the 1440s. The Half Moon was bequeathed in 1514 to Jesus College, Cambridge by Roger Thorney in remainder upon the death of his wife Eleanor (d. 1530).[101]

[4.52] *the blue mayde endyt*
The survey of 1555 places the boundary between the Great Liberty manor and the King's manor on the south side of 'Blewemeade' Alley, presumably named after a house (or drinking-house or inn) called the Blue Maid.[102]

[4.53] [*unnamed*]
This may represent an inn called the Mermaid, owned in the first half of the sixteenth century by John and William Kyrton.[103]

[4.54] *lirtate barmese*
The 'liberty of Bermondsey', a reference to the fact that this block of properties lay in the manor formerly formerly owned by the abbot of Bermondsey (after 1550 known as the King's or Queen's manor).

[4.55] *the kinges benchs*
The King's Bench prison occupied this site from the 1360s (perhaps earlier) until the 1750s. Like the Marshalsea, it was sacked in 1381, 1450 and 1470. On the latter occasion some 99 prisoners escaped, of whom nine were women and three were priests. The prison was enlarged in the reign of Henry VIII. Excavations on the site revealed an early sixteenth-century pit full

of waste from a bone-working industry, but no traces of the medieval prison buildings were recovered.[104]

[4.56] *goroge churche*
The parish church of St George the Martyr, granted in 1122 to the monks of Bermondsey by Thomas de Ardern and his son Thomas. The church, which was rebuilt at the end of the fourteenth century, is shown in Anthonis van den Wyngaerde's panorama of London as a plain building with several tall south windows and a large western tower. It was heavily restored in 1629, after which it measured 69 feet in length 'to the altar rails' and 60 feet in width. The tower, with eight bells, was 98 feet high. There was one parish fraternity, dedicated to the Virgin and St George, and in existence by the 1460s.[105]

[4.57] *the well*
A second well, known as 'the well with two buckets', is drawn but not named to the south of the market place.[106]

[4.58] *the bolrynge*
This bullring is otherwise undocumented but may have been the predecessor of the Bankside bull- and bear-baiting rings shown in the Braun and Hogenberg and 'Agas' maps. It gave its name by the early sixteenth century to an alley known as 'Bulrynge Alley', which lay nearby on the west side of the street.[107]

[4.59] *the syncke*
A sink, or gulley-hole. Another survey of the Southwark manors, dating from *c.* 1564–5, records a 'syncke or common Swear' in Blackman Street to the south of the pound.[108]

[4.60] *bostock house beginneth kynges*[?] *lyberte*
This house marked the boundary between the King's manor and the Great Liberty manor. The 'Mrs Bostock' who held the close shown in Plate 3 (no. [3.16]) presumably was Bostock's widow.

[4.61] *kent stryt*
Kent Street, so called by the 1270s;[109] now Tabard Street and the Old Kent Road.

[4.62] *Jan Jonck* [insert: *Puggelis*(?)] *houshe*
Otherwise unknown.

[4.63] [*unnamed street*]
This is Long Lane, so called by *temp.* Hen. VII.[110]

[4.64] *a dych* (twice)
A ditch.

[4.65] *syr thomas pope*
The site of the Cluniac priory (from 1399, abbey) of St Saviour, Bermondsey, founded in 1082 as a cell of La Charité-sur-Loire. The conventual church has been identified with the 'new and handsome church' in Bermondsey mentioned in Domesday Book, but the first colony of monks did not arrive from La Charité until 1089. The church was completed in 1339. Many pilgrims visited St Saviour's to see an ancient crucifix found near the Thames in 1117 and placed in the conventual church. Queens Katharine de Valois and Elizabeth Woodville spent their last years in the abbey. Following the surrender of the house in 1538 the site was granted in July 1541 to Sir Robert Southwell, who sold it the following month to Sir Thomas Pope. (This provides an apparent *terminus post quem* for this plan.) Pope pulled down the church and other buildings and built himself a mansion from the materials, and then reconveyed it to Southwell in 1554–5. Excavations have revealed part of the plan of the church and of a portion of the precinct and its buildings; a fragment of the gatehouse survives.[111]

[4.66] *marye madelene church*
The parish church of St Mary Magdalen, Bermondsey. Probably established by the monks of Bermondsey (see above), who held the advowson until the surrender of the house in 1538. The church was rebuilt in the seventeenth century, but the lower stages of the tower and of part of the west wall of the present church date from the fifteenth century.[112]

[4.67] *Hyer endith the kings lyberte*
Two unnamed buildings adjoin this annotation. The survey of 1555 records the house at the manor boundary (Roper Lane) as the tenement of one Sampson, occupied by Ralph Maynley, and next to it the dwelling of William Chare.[113]

[4.68] *Maestur goudt yere*
The house of Henry Goodyer or Goodyere, alderman of London (d. 1556). This house, which lay opposite Roper Lane, had formed part of a gift by John Scraggys of Southwark to the Leathersellers' Company in 1531. A company terrier of 1535 valued the house at £5 p.a. and listed Goodyere as the tenant. The company sold the house c. 1543–4. Goodyere's house is shown here because it marked the boundary between the Great Liberty manor and the

manor of Bermondsey. It also lay on the boundary between the parishes of St Olave, Southwark, and St Mary Magdalen, Bermondsey.[114]

[4.69] *De ponth*
The pound; also depicted in Plate 3 (no. [3.12]) as *the pynfold*.

[4.70] *De parck gate*
This gave access to the park of the bishop of Winchester (see [4.72], below).

[4.71] *the liberte off the manor*
Part of the King's (or Queen's) manor, formerly the manor of the abbot of Bermondsey.

[4.72] *De parck*
The park (consisting of a meadow and three pastures) of the bishop of Winchester. The bishop leased it to Sir Thomas Brandon for a term of years in 1502, and the lease passed to the Crown with Suffolk Place in 1536 (see below). Queen Mary surrendered the unexpired portion of the lease to the bishop in 1555.[115] The full extent of the park is shown in the plan of 1618.[116]

[4.73] *De manor place*
Also known as Suffolk Place, the elaborate mansion built by Charles Brandon, duke of Suffolk, on the site of a smaller estate originally established by his grandfather Sir William Brandon. Suffolk enlarged the property in 1516 and the mansion was built *c.* 1518–20. In 1536 Henry VIII acquired the mansion and the adjoining park (see [4.72], above) by exchange with Brandon for Norwich Place near Charing Cross. The following year the property was granted to Queen Jane Seymour, but it reverted to the king on her death in October 1537. Henry VIII enlarged the park and stocked it with game, but used the house infrequently, if at all, until he established a mint there in 1545 (closed 1551). In 1556 Queen Mary granted the house to the archbishop of York to use as a townhouse, but the following year the archbishop purchased Norwich Place to use instead, and sold the Southwark house, and by 1562 the back part had been cut up into gardens and the front part into some 20 tenements. A view of the mansion appears in the foreground of van den Wyngaerde's panorama of London.[117]

[4.74] *the liberte off the manor*
See [4.71], above.

[4.75] *clement*
The Clement is mentioned as a brewhouse in the first half of the fifteenth century; it seems to have been part of a block of properties acquired by Merton priory between 1369 and 1394 and retained by the priory until its dissolution in 1538.[118]

[4.76] *the Gotte*
Described in the survey of 1555 as a messuage or inn called the Goat, formerly in the possession of Merton priory. At the priory's dissolution in 1538 the Goat was in the tenure of William Kellett, a brewer, at a rent of £8. 3s. 4d. p.a.[119]

[4.77] *the liberte off the manor*
See [4.71], above.

[4.78] *the salutacyn*
Described in the survey of 1555 as a messuage or inn called the Salutation, formerly in the possession of the priory of St Mary Overy.[120]

[4.79] *dedma[n]place*
Deadman's Place was a messuage and a cluster of gardens (so called by 1457-8) on the edge of the manor of the bishop of Winchester. The origin of its name is unknown.[121] On Ogilby and Morgan's map the name is used for the streets that bordered the precinct of Winchester House on the west and south (see [4.16] and [4.17], above).

[4.80] *parck Gate*
See [4.70], above.

Acknowledgements:

The above plans are published, respectively, by the kind permission of The Society of Antiquaries, the Southwark Local Studies Library, the Corporation of London Records Office, and the Public Record Office (with permission of the Chancellor and County of the Duchy of Lancaster).

Notes:
1 British Library, MS Lansdowne XXXVII, fol. 8ʳ. The list of posting sites is missing.
2 On Norden's plans of London, see Ida Darlington and James Howgego, *Printed Maps of London, circa 1553-1850* (London:

George Philip and Son, 1964), no. 5. Norden himself is otherwise first mentioned in 1593; see *Dictionary of National Biography*, s.n. 'Norden, John'.
3 *The A to Z of Elizabethan London* (LTS, no. 122, published concurrently with Harry Margary of Lympne Castle, Kent, in association with the Guildhall Library, 1979).
4 *Moorfields in 1559* (London: H.M.S.O., 1963).
5 *The Map of Mid Sixteenth Century London* (LTS, no. 100, 1964).
6 *London: 2000 Years of a City and Its People* (London: Macmillan, 1974).
7 *London: The Biography of a City* (Harlow: Longmans, 1969; New York: William Morrow and Company, 1970).
8 *The Building of London from the Conquest to the Great Fire* (London: Colonnade, 1984).
9 *London in Maps* (London: The Conoisseur, 1972).
10 *Local Maps and Plans from Medieval England* (Oxford: Oxford University Press, 1986).
11 *The Diary of Baron Waldstein: A Traveller in Elizabethan England* (London: Thames and Hudson, 1981).
12 In Prockter and Taylor, *The A to Z of Elizabethan London* (above, n. 3).
13 Published by the Topographical Society of London in 1881–82 and by Barker and Jackson, pp. 48–53.
14 British Library, Sloane MS 2596, fol. 52; published in Philippa Glanville's *London in Maps*.
15 Folger Library, MS V.a. 318, fol. 21; published in *Hugh Alley's Caveat*, ed. Caroline Barron, Vanessa Harding and Ian Archer (LTS, publication no. 137, 1988), pp. 76–7, 96–7.
16 Guildhall Library, MS 12805, Christ's Hospital Evidence Book, fols. 5–6; published in *The London Surveys of Ralph Treswell*, ed. John Schofield (LTS, publication no. 135, 1987), pp. 138–41.
17 The 1593 edition was published by the LTS in 1899; the edition of 1600 was published by the LTS in 1961 and in David Johnson's *Southwark and the City* (London: Oxford University Press, for the Corporation of the City of London, 1969), plate 5.
18 Both published in the L[ondon] C[ounty] C[ouncil], *Survey of London*, vol. 22, *Bankside* (1950), plates 1 and 65.
19 Both published in Barker and Jackson, pp. 98, 103.
20 In Barker and Jackson, pp. 66–7.
21 Reproduced in Christopher Hibbert's *London: The Biography of a City*, facing p. 43.
22 I. A. Shapiro, 'The Bankside Theatres: Early Engravings', *Shakespeare Survey*, 1, ed. Allardyce Nicoll (Cambridge University Press, 1948), 25–37 and plates.
23 For a complete topographical gazetteer of Southwark to circa 1550 see Martha Carlin, 'The Urban Development of Southwark to

c. 1550' (unpublished Ph.D. thesis, University of Toronto, 1983), chapters 4–11.
24 *The London Surveys of Ralph Treswell*, pp. 5–8.
25 Information on the building of the almshouses kindly supplied by Mr Stephen Humphrey of the Southwark Local Studies Library.
26 *London Surveys of Ralph Treswell*, plates 5, 6, 9, 10, 11; pp. 51, 65, 69, 71, 76, 81, 98, 101, 105, 125, 133.
27 *London Surveys of Ralph Treswell*, p. 117.
28 *London Surveys of Ralph Treswell*, pp. 89, 109, 130.
29 *London Surveys of Ralph Treswell*, p. 2; *Testamentary Records in the Archdeaconry Court of London*, ed. Marc Fitch, I (British Record Society, vol. 89, 1979), p. 378.
30 For the churchyard and almshouses, see discussion under Plate 2*a*, below.
31 This description is reiterated in Corner's discussion of the plan on p. 171.
32 Southwark Local Studies Library, St Olave's, Southwark, Vestry Book, 1551–1604, fols. 18^v, 21^r.
33 Carlin thesis, pp. 319–23; Southwark Local Studies Library, St Olave's, Southwark, Vestry Book, 1551–1604, fols. 66^v, 71^v.
34 Unless otherwise noted, the commentary on this plan is derived from Carlin thesis, pp. 317–23. The later descent of St Saviour's Dock is traced in *VCH, Surrey*, vol. 4, pp. 20–1.
35 Southwark Local Studies Library, St Olave's Churchwardens' Accounts, 1546–1610, p. 24.
36 *Letters and Papers, Foreign and Domestic, of the Reign of Henry VIII*, XIX, ii, no. 340 (22).
37 Paper on the foundation of St Olave's Grammar School published by G. R. Corner in the *Gentleman's Magazine*, 1836, Part I, pp. 137–44; rpt in *The Gentleman's Magazine Library, English Topography, Part XVII* (London, vol. III), ed. G. L. Gomme (London: Eliot Stock, 1905), pp. 124–34 (William Knight: p. 127).
38 Magdalen College, Oxford, Southwark deed 12C.
39 Southwark Local Studies Library, St Olave's Churchwardens' Accounts, 1546–1610, pp. 1, 2, 25, 28, 34, 40, 41, 53.
40 Carlin thesis, pp. 310–17.
41 Southwark Local Studies Library, St Olave's, Southwark, Vestry Book, 1551–1604, fos. 61^v, 71^v, 73^v, 97^v. Information on the building of the almshouses kindly supplied by Mr Stephen Humphrey of the Southwark Local Studies Library.
42 There were five Southwark manors in all: Paris Garden manor (known until the fifteenth century as 'Wideflete', 'Withyflete' or the 'Wiles'), which lay on the west, adjoining Lambeth; the bishop of Winchester's manor (also known from the late fifteenth century as the Clink manor), which abutted the eastern side of Paris Garden; the King's manor, so called after Henry VIII acquired it from

Bermondsey abbey in 1536, and lying on the west side of the high street, in St George's Fields, and around the parish church of St George; the Great Liberty manor, acquired by Henry VIII from the archbishop of Canterbury in 1538, and lying in the eastern part of Southwark; and the Guildable manor, lying at the foot of London Bridge. London had obtained the farm of the Guildable manor in 1327, and acquired fuller rights by royal charter over the following century and a half. In 1550 the City purchased outright from Edward VI the King's manor, the Great Liberty manor, and the Guildable manor, for a total cost of about £1000. See Carlin thesis, pp. 37n., 58, and chap. 13. A modern plan of the manors of Southwark is published as an endpaper in David Johnson's *Southwark and the City*.

43 BL, Stowe MS 942 (cartulary of St Thomas's Hospital, Southwark), fols. 103^v–104^r (*Blakemannestrete*, mid thirteenth century), 93^v, 94^r (*Blakmanstrete*, late fifteenth or early sixteenth century).
44 Carlin thesis, pp. 183–4; C[orporation of] L[ondon] R[ecords] O[ffice], MS 39 C., fol. 138^r.
45 Carlin thesis, pp. 183, 595. This way was the subject of litigation in 1618–20 between the high street innkeepers, who wished to stop its use as a public right-of-way to the Bankside, and the Bankside innkeepers, who wished to keep it open. See L[ondon] C[ounty] C[ouncil], *Survey of London*, vol. 22, *Bankside*, pp. 133–5 and Plate 1.
46 Carlin thesis, p. 184.
47 Carlin thesis, p. 184.
48 Carlin thesis, p. 185.
49 Carlin thesis, p. 185, n. 180.
50 CLRO, MS 39 C., fol. 138^v.
51 CLRO, MS 39 C., fol. 138.
52 PRO, SC6/Hen. 8/6026, m. 6^r (Original Ministers' Accounts).
53 PRO, E36/169, fol. 159^r (rental, 1549); Carlin thesis, pp. 50–1.
54 Shapiro, 'Bankside Theatres', plate 5.
55 The main unnamed Southwark streets were: the 'hyghe streate' (modern Borough High Street), so called in the 1540s, also known in the 1550s and 1560s as 'Long(e) Sowthewarke' (CLRO, MS 39 C., fols. 74^r, 92^r, 95^r; Southwark Local Studies Library, St Olave's Churchwardens' Accounts, 1546–1610, p. 208); 'St Margaretes Hill' (the high street running south from St Margaret's Church to Blackman Street), so called in the 1540s and 1560s (CLRO, MS 39 C., fols. 12^r, 270^r); 'saynt olaves streate' (modern Tooley Street), so called by 1550 (*ibid.*, fol. 79^r); and Bermondsey Street, so called since at least 1379 (BL, Stowe MS 942, fol. 74^r; *Place-Names of Surrey*, ed. J. E. B. Gover *et al.* [English Place-Name Society, vol. 11, 1934], p. 30), but also known from the late twelfth to the mid fifteenth century as 'the causeway leading (from Battle Bridge) to

Bermondsey' (Huntington Library, San Marino, California, MS BA 29, fol. 230^r [late twelfth century]; Magdalen College, Oxford, Southwark deeds 5C, 136 [*c.* 1240×1250, 1449]; CLRO, Bridge House deed E.31 [1381]).

56 Shapiro, 'Bankside Theatres', p. 35, n. 5, quoting the Appendix to the *Thirtieth Report of the Deputy Keeper of Public Records* (1869), p. 39.
57 Carlin thesis, pp. 143–4.
58 Carlin thesis, p. 144 and n. 63; PRO, SC6/Hen. 8/3468, fol. 43^r (Original Ministers' Accounts, 1545–6).
59 Carlin thesis, pp. 128–30.
60 Carlin thesis, p. 140.
61 Carlin thesis, p. 140.
62 The river stairs there were referred to as Pepper Alley Gate in 1547; in 1549 the stairs there were said to be very decayed and dangerous. CLRO, MS 39 C., fols. 58^v, 72^r.
63 Carlin thesis, pp. 132–44, 385–96.
64 Carlin thesis, p. 145.
65 Carlin thesis, pp. 147–8, 553–4.
66 Carlin thesis, pp. 145–6.
67 Carlin thesis, pp. 79–119; Martha Carlin, 'The Reconstruction of Winchester Palace, Southwark', *London Topographical Record*, 25 (1985), pp. 33–57; Brian Yule, 'Excavations at Winchester Palace, Southwark', *London Archaeologist*, vol. 6, no. 2 (Spring 1989), 31–9.
68 Carlin thesis, pp. 82 and n. 8, 160–1.
69 Carlin thesis, p. 77.
70 Carlin thesis, p. 149.
71 Carlin thesis, p. 151.
72 CLRO, MS 39C., fols. 1^r, 10^r, 16, 73^v, and *passim*; Carlin thesis, pp. 150–1.
73 Carlin thesis, pp. 153–6, 421–8; CLRO, MS 39C., fol. 9^r.
74 Carlin thesis, pp. 156–60.
75 Carlin thesis, pp. 156–60; CLRO, MS 39C., fols. 75^r, 226^r; *Hugh Alley's Caveat*, pp. 77, 96–7.
76 L.C.C., *Survey of London*, vol. 22, *Bankside* (1950), pp. 9, 12.
77 Carlin thesis, pp. 271–3, 364–8, 415–21.
78 Carlin thesis, pp. 354–61.
79 Carlin thesis, pp. 352–4.
80 Southwark Local Studies Library, St Olave's Churchwardens' Accounts, 1546–1610, p. 27.
81 Carlin thesis, pp. 282–3.
82 Carlin thesis, pp. 284–5.
83 Carlin thesis, p. 286.
84 Carlin thesis, pp. 345–52.
85 Carlin thesis, p. 299.
86 Carlin thesis, pp. 286–8.

87 Carlin thesis, p. 249.
88 Carlin thesis, p. 248.
89 Carlin thesis, pp. 156–7, 238–48, 396–414, 433–5; Martha Carlin, 'Medieval English Hospitals', in *Hospitals and History*, ed. Roy Porter and Lindsay Granshaw (London: Routledge, 1989), 21–39, *passim*.
90 Carlin thesis, p. 241.
91 Carlin thesis, pp. 233–4.
92 Carlin thesis, pp. 229–31.
93 Carlin thesis, p. 228.
94 Carlin thesis, pp. 224–7.
95 Carlin thesis, pp. 224–7, 554–5; Philip Norman, 'The Tabard Inn, Southwark, the Queen's Head, William Rutter, and St. Margaret's Church', *Surrey Archaeological Collections*, 13 (1897), 29–32.
96 Carlin thesis, pp. 223–4.
97 Carlin, p. 222.
98 Carlin thesis, pp. 221–2.
99 Carlin thesis, pp. 220–1.
100 Carlin thesis, pp. 213–19.
101 Carlin thesis, p. 211.
102 Carlin thesis, p. 212.
103 Carlin thesis, p. 212.
104 Carlin thesis, pp. 203–9.
105 Carlin thesis, pp. 197–8, 429–30.
106 Carlin thesis, p. 160.
107 Carlin thesis, p. 173. For the Bankside bull- and bear-rings, see C. L. Kingsford, 'Paris Garden and the Bear-Baiting', *Archaeologia*, 70 (1920), 155–78; and L. C. C., *Survey of London*, vol. 22, *Bankside*, pp. 66–77.
108 CLRO, Misc. MSS 169.6(1), fol. 10^v.
109 1279 *Kentstrate*: CLRO, Bridge House Deed C.22.
110 *Place-Names of Surrey*, p. 31.
111 David Knowles and R. Neville Hadcock, *Medieval Religious Houses: England and Wales* (London, New York, Toronto: Longmans, Green and Co., 1953), p. 95; *VCH, Surrey*, vol. 4 (1912), pp. 18, 20–1; vol. 2 (1905), pp. 64–77; William Rendle, *Old Southwark and its People* (London: privately printed, 1878), pp. 292–305; W. F. Grimes, *The Excavation of Roman and Mediaeval London* (London: Routledge and Kegan Paul, 1968), pp. 210–17; David Beard, 'The infirmary of Bermondsey Priory', *London Archaeologist*, vol. 5, no. 7 (Summer 1986), pp. 186–91, and notes on ongoing excavation in no. 10 (Spring 1987), p. 276; no. 15 (Summer 1988), pp. 413–14; Ann Saunders, *The Art and Architecture of London* (London: Phaidon, 1984), p. 415.
112 *VCH, Surrey*, vol. 4 (1912), p. 23; John Strype, *Survey of London* (1720), Book 4, p. 25; Saunders, p. 415.

113 CLRO, MS 39C., fol. 142^r.
114 Carlin thesis, p. 308; Rendle, *Old Southwark and its People*, 281–2.
115 Carlin thesis, pp. 75–6.
116 L.C.C., *Survey of London*, vol. 22, *Bankside*, plate 1.
117 Carlin thesis, pp. 176–81.
118 Carlin thesis, p. 165, 169.
119 Carlin thesis, p. 165.
120 Carlin thesis, p. 163.
121 Carlin thesis, p. 77.

III. WHITEHALL PALACE AND KING STREET, WESTMINSTER: THE URBAN COST OF PRINCELY MAGNIFICENCE

By GERVASE ROSSER and SIMON THURLEY

> Why come ye nat to court
> To whyche court?
> To the kynges court?
> Or to Hampton Court?
>   Nay, to the kynges court!
> The kynges court
> Shulde have the excellence;
> But Hampton Court
> Hath the preemynence!
> And Yorkes Place,
> With 'My lordes grace',
> To whose magnifycence
> Is all the confluence,
> Sutys, and supplycacyouns,
> Embassades of all nacyons.

JOHN Skelton's satire evokes the familiar story of the king who finally turned upon a trusted servant whose ambition had over-reached itself.[1] The rivalry between Henry VIII and his only too magnificent chief minister, Wolsey, is conventionally seen as the isolated clash of political giants. The subjects of this article, however, are the ordinary citizens and subjects whose lives were shaken beyond recognition by the self-aggrandising architectural designs of these two tyrants. At the time Skelton wrote, in 1522, Westminster, the capital of the kingdom, was in the clear possession not of the king but of the cardinal. The youthful Henry's passion for rural hunting-parties, compounded by the devastation of the medieval royal palace of Westminster by fire in 1512, had led the king to spend relatively little time in what, by the end of the Middle Ages, had become the chief *locus* of monarchical power. Instead, Westminster was dominated by the red-robed presence of the Ipswich grazier's son whom ambition had elevated to a status hardly inferior to that of the king himself.

Wolsey's residence in the capital was York Place, the great house beside the Thames at Westminster which had served as the London base of the archbishops of York since the thirteenth century. Wolsey inherited York Place on his accession to the archbishopric in 1514. The house had been largely rebuilt at the end of the previous century; nevertheless Wolsey undertook a programme of substantial enlargement. Unlike his country properties of The More in Hertfordshire and Hampton Court, where space for new building was not at a premium, York Place stood in an urban context on a constricted site (Fig. 2). The house was bounded to the east by the river Thames, to the west by King Street, to the south by tenements in the possession of Westminster Abbey, and to the north by a piece of land called 'Scotland', which belonged to the crown. To increase the scale of the house Wolsey was therefore compelled to buy up adjacent properties. He first acquired 'Scotland' and then, in 1519-20, induced his neighbours to the south, beside Endive Lane, to part with their leases;[2] on this southern side of the existing mansion Wolsey constructed a long gallery and an orchard.[3] The cardinal's methods provided a model which would be far more extensively employed by his successor.

In October 1529 Wolsey was forced to surrender all his property to the crown and King Henry at last gained, in York Place, a worthy residence at Westminster. The cardinal's palatial design was seized for his own by the king, by whom in turn it was vastly enlarged. Great architectural enterprises such as the building of Whitehall Palace are all too often studied in a vacuum, independently of their human environment.[4] Yet the royal building works of the mid-sixteenth century transformed the small town of Westminster, profoundly affecting the lives of its inhabitants. The

KING STREET, WESTMINSTER

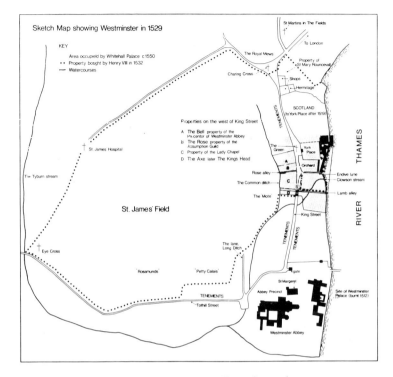

Fig. 1. Sketch map showing Westminster in 1529.

following account is a case-study in the local human impact of a grandiose building scheme.

The site of the enlarged palace was the northern half of the road between the old Westminster Palace and Charing Cross: then known as King Street, in modern times as Whitehall (Fig. 2). Until the late Middle Ages this district of Westminster remained partially undeveloped with buildings. Houses, including York Place, lined the east side of the road, extending to wharves on the river frontage; but until the later fifteenth century the view to the west of the street was open to fields belonging variously to the monks of Westminster Abbey and to St James's hospital. At that period, however, this quarter shared with the rest of the town of

Westminster the experience of a dramatic increase in population, following upon several decades of stagnation. This new growth, which must have been due both to an increase in fertility and to an influx of migrants from the countryside, was reflected in a rash of building developments in the town. Typical of the urbanisation of the period was the erection of houses on Steynour's Croft, a field beside King Street directly opposite York Place. Like the area to the west of the road in general, the land was marshy: willow trees here had produced an economic crop of osier rods. But for John Millyng, the London investor who developed the ground after 1466, the incentive to build outweighed the difficulties of the site. In that year Millyng took the vacant plot from Westminster Abbey on a cheap lease at 5 s. *per annum*, on the explicit understanding that he would build within sixteen years; by 1490 he had put up no less than seven cottages and a barn. The profits to be made from such urban development in this period are shown by the fact that in 1530 (when the annual charge on the head lessee remained only 5 s.) the total value of rents from subtenants in the cottages was over £8 *per annum*.[5]

In the same quarter of the town, Endive Lane became more densely built up at this period, again with developments of small cottages (Fig. 2). Around 1425 eleven cottages had been built here as the endowment of a chantry in Westminster Abbey. To these in the later fifteenth and early sixteenth centuries were added eight more cottages, erected by leaseholders who evidently anticipated a buoyant demand for accommodation in the district. The chief landlords, the Westminster monks, collected just over £6 *per annum* from this part of their estate; but to the lay lessees the combined rents from subtenants were in the early sixteenth century worth annually some £25.[6] Evidently urban property speculators in early Tudor Westminster might earn a significant income from rents, although this would typically supplement revenue from other sources. A chandler, William Russell, in 1530 held a 'wax house' in the Wool Staple yard near King Street which he described as his 'working house', while living at another address beside the bars in King

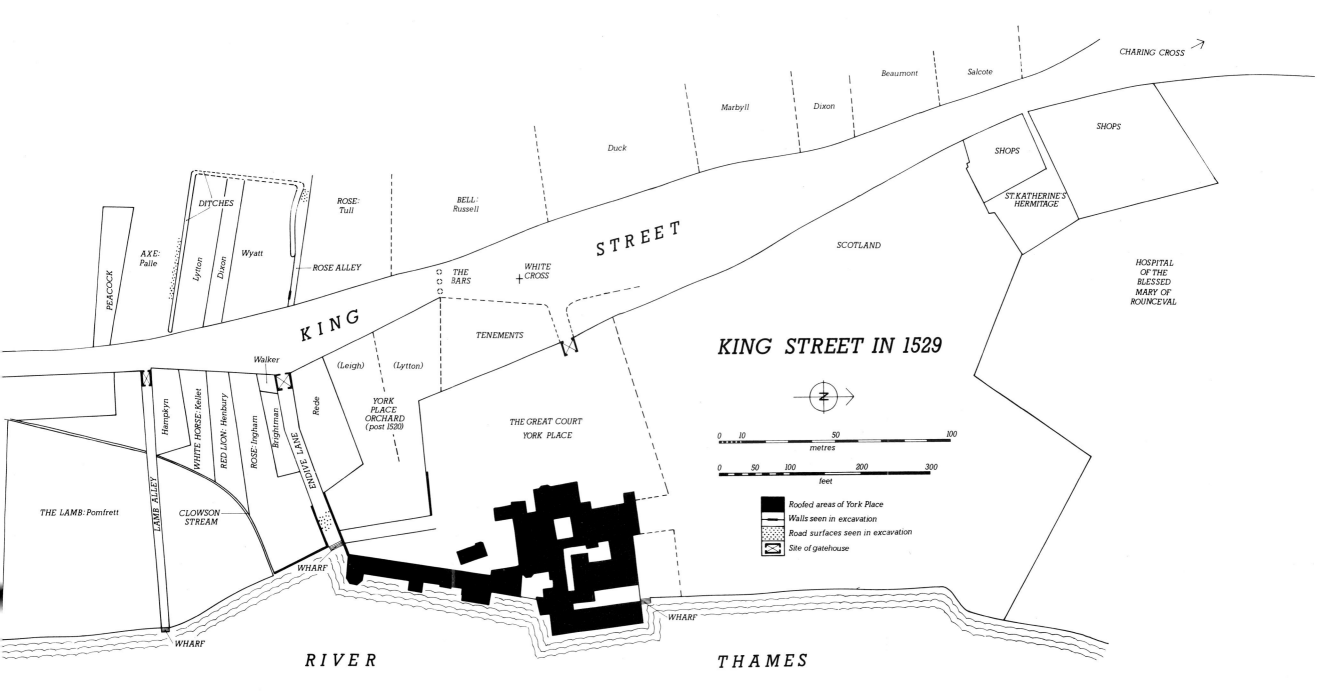

Fig. 2.

Fig. 3. Westminster in the mid-sixteenth century, after the construction of Whitehall Palace. Detail from the contemporary 'Agas' woodcut map of London.

Street, called the Bell (Fig. 3). As a sideline, he leased and sublet nine cottages next to the Bell, which provided him with a yearly clear profit of £6 15s.[7] Such evidence shows that the local population of King Street around York Place was increasing rapidly in the first third of the sixteenth century.

The society of the neighbourhood was distinctive in its composition. Westminster's status as royal capital and seat of government lent it, from a distance, the aura of an enclave of the very rich. In fact, however, the wealthy élite spent only a small proportion of their time here, contributing for the most part only indirectly to the character of the resident population. More continuously in occupation than royalty or aristocracy were members of their households, such as Dr Richard Duck, Wolsey's treasurer, who lived on the west side of King Street, opposite the cardinal's palace.[8] Those more prominent officials in the bureaucracy of central government who lived locally might manifest pretensions to grandeur, like John Manfeld, a clerk of the privy seal, who around 1480 contracted with a carpenter for extensions to his house at Charing Cross called the Swan, to include a new hall and a gatehouse, with a bay window above the entrance.[9] But the tone of the society of resident townspeople was largely determined by the overwhelming majority of representatives of the various service trades which supplied the needs of the court and the offices of government when they were present and in session.

It was characteristic of Westminster in general and of King Street in particular that a number of locally influential figures grew powerful through the drink trade. One of these was John Pomfrett, the holder until his death in 1531 of two brewhouses, the White Lion at Charing Cross and the Lamb, a short distance to the south of York Place in King Street (Fig. 2). Pomfrett's standing in the community is reflected in his election to local office: he was in 1516–18 a churchwarden of the parish church of Westminster, and in 1522–4 a warden of the leading guild of the town, that of the Assumption of the Virgin Mary. The high stakes involved in the Westminster brewing business were shown by a violent conflict between Pomfrett and another brewer, John Henbury of the Red Lion. The Red Lion lay close to the Lamb, and in fact Pomfrett leased the Lamb Alley, which ran down from King Street to the Thames-side quay, from Henbury (Fig. 2). The lease was granted on condition Pomfrett made a gate at the street end of the lane, the profits from which were to be reserved to Henbury; evidently those who controlled the points of access to the commercially vital Thames wharves could charge private tolls. But Pomfrett appropriated the lane to himself; directed his servants to drive Henbury's men away from the wharf; and maliciously damaged Henbury's property. Henbury himself was no mean rival; he could retain as his attorney no less an advocate than Thomas Cromwell. Events, however, were to overtake this tycoons' vendetta, when the enlargement of Whitehall, in the aftermath of the loftier struggle between king and cardinal, swallowed Lamb and Lion together.[10]

Wealth and prestige accrued also to couturiers in this resort of the fashion-conscious. The Westminster tailors were said to charge twenty times the cost of their material; and the waste of a cobbler's workshop found a few steps from York Place charts the change in modes of footwear among the fops, from the fifteenth century's narrow points to the bulbous style set by King Henry VIII.[11] The proprietor of a nearby bookshop, Elias Snethe, was one of the five wealthiest inhabitants of Westminster in the 1520s; and a taste for luxuries was also encouraged at this period by the clockmaker of King Street, whose stock included gilded cuckoo-clocks.[12] But while these smart establishments fronted on to the highway, the alleys and courts behind teemed with humbler artisans and labourers in the service trades. The occupants of the cottages in Endive Lane about 1508 included a representative cross-section: a butcher, a baker, a weaver, a pin-maker, a spurrier, a cobbler, a tiler, and watermen and porters employed on the quayside; also in the lane was Cardinal Wolsey's laundry.[13] Throughout the medieval period, a particular prominence had been enjoyed in Westminster by the building trades; both royal and monastic patrons had guaranteed employment for quantities of masons and carpenters. One dynasty of carpenters, the Russell family, was established near the King Street bars in the early sixteenth century. Richard Russell, master carpenter both to the crown and to the monks of Westminster Abbey, held the house called the Bell in the years 1512–17; in 1531 Russell's two sons, one a carpenter and the other a chandler, were living here (Fig. 2).[14] The abbey's impending dissolution might have spelt disaster for such building craftsmen, as the removal of the great monasteries did elsewhere, had not the royal works at Whitehall from the 1530s provided alternative employment. It was ironical that, for those living in the building zone itself, the new works brought a cataclysm of another kind; for the Bell too fell victim to the expanding monster-palace.

Within the micro-society of King Street, a wide range of personal prosperity was enjoyed by the inhabitants. Wealth was principally concentrated on the high street frontage; relative poverty in the back alleys and courts. An assessment for taxation taken in 1544 of the parts of King Street which remained after the redevelopment of Whitehall revealed that, whereas over a third of households in King Street itself were assessed at valuations of £10 and above, in the alleys the proportion was barely one-tenth; here four-fifths of households were assessed at less than £5.[15] Yet even the fine mesh of the tax-collector's net failed to trap the poorest transients, of whom King Street probably housed an unusually large number. A security search of the street in October 1519 lifted the curtain fleetingly upon the marginal world of unofficial lodgers. Among these were unemployed or impoverished servants, such as the two who 'lay in Lewis Griffith's chamber in a bed next to him and to his wife in a poor house'. In a house below this were 'two women, an old drab and a young wench, upon a sheet case upon the ground'. Elsewhere in the street the searchers picked up a foreigner, Faux Vyncent, with his mistress, and Philip Umfrey, 'servant to the king, as he saith, taken in a house by himself in a chamber, and a woman in a chamber underneath, without shutting of doors'.[16]

Diverse as it was, a number of forces operated to give this local society a degree of identity and cohesion. In the first place, all residents shared an involvement in Westminster's dominant economy of services. From goldsmith to laundress, all sought the patronage of the consuming classes which periodically frequented the town. The development of a sense of neighbourhood among the permanent inhabitants of the upper part of King Street can be observed in such personal behaviour as the prolonged residence here, in close proximity to one another, of different branches of the Russell family, or the marriage of a buckler-maker of Lamb Alley to a neighbour, the widow of Cardinal Wolsey's cook.[17] More formally, York Place itself was a focus of collective activity among the townspeople of Westminster, notably when (by the archbishop's gracious permission) the local guild of the Assumption held its triennial feast here on 15 August. The two hundred or so members of this guild were drawn from almost all ranks of local society; their joint festivity in the archbishop's hall, which they filled with music and decorated with flowers, celebrated a periodically-felt spirit of community among the resident townspeople of Westminster.[18] Another local centre of neighbourhood activity was the Rounceval hospital beside Charing Cross (Figs. 1, 2). This hostel for down-and-outs was supported by another of the Westminster guilds, whose membership, while recruited from throughout the town, included a noticeable group of close neighbours of the hospital in the northern part of King Street. Among these members and benefactors were the local bookseller, Elias Snethe, the carpenter Richard Russell, and Russell's widow, Constance, who provided bedlinen for the inmates. A barber of the neighbourhood was among those who chose to be buried in the chapel of the hospital; his widow, who during her life continued to reside close by, asked in her will to be interred beside him, and in 1504 bequeathed to the institution a feather bed, bolster, sheets, blankets and a coverlet. The brewer, John Henbury of the Red Lion, and Thurston a Mere, a local baker of King Street, held office as wardens of the Rounceval guild. Like the meetings of the Assumption guild, the Rounceval hospital was in addition the focus of regular festivities among the townspeople of Westminster.[19]

When York Place fell into King Henry's hands, the fate of this neighbourhood was sealed. At Christmas 1529 the king, in the words of the Milanese ambassador,

> designed new lodgings and a park adjoining York House which belonged to the late Cardinal Wolsey. The plan is on so large a scale that many hundreds of houses will be levelled, well nigh all of which belong to great personages.[20]

Chapuys, the Imperial ambassador, likewise reported in May 1530 that the king

> is having a great park made in front of the house which once belonged to the cardinal, and in order to go to it across the street has had a very long covered gallery built, for which purpose a number of houses have been pulled down, to the damage and discomfort of the proprietors, without there having yet been any question of indemnifying them for their losses.[21]

Neither witness had fully understood the circumstances, but both recorded what were the bare facts. Henry was planning to extend York Place by the same means as had been employed by Wolsey, but on a gargantuan scale. First, the site of Wolsey's orchard was to be extended southwards, engulfing all the properties between Endive Lane and Lamb Alley (Figs. 1, 2). Thereafter, the king's more ambitious intention was to demolish all the properties on the west side of King Street for a considerable distance to either side of the bars (Figs. 1, 2). On this western site he was to build a resort for courtly recreation, to include tennis courts, bowling alleys, a tiltyard and a cockpit. This complex was to be linked with the eastern palace buildings by a gatehouse bridging the street (Fig. 3).[22]

The royal plan necessitated the compulsory purchase of large tracts of the town of Westminster and its neighbouring fields. The acquisition of the land was a complicated operation; as his commissioners for this task the king appointed Thomas Cromwell and the abbot of Westminster, William Boston. As has been shown, King Street was populous, and some of its inhabitants were both rich and influential; the royal officers had to tread carefully. Their first step, taken at the beginning of 1530, was to survey all the properties which the king wished to appropriate. Three records of this survey survive. One lists seven properties on the west side of King Street, citing the owner and rent-value of each; a second relates to six tenements in Endive Lane; the third, much more wide-ranging, lists the yearly value of many of the properties bought both for the palace extension and for the creation of the king's new park.[23] On the basis of these surveys, and of others no longer extant, the king's commissioners assessed the delicate matter of compensation.

The first group of properties had been surveyed and valued by 23 May 1531. A surviving list of this date with annotations in Cromwell's hand gives details of moneys paid out: on that day the king spent, on the purchase of leases, more than £1,120 (see Table 1).[24] William Russell, the leaseholder of the Bell inn, received £128 in compensation for his loss; the Rose, next door, cost the king £82. A little to the north lay land in the possession of Hugh Marbill, a goldsmith; Thomas Rawlins, a tenant of Marbill, received £20 for his interest. Henry Heyse, a chandler, was paid £60 for his portion of Richard Hampkyn's property at Lamb Alley.[25] These and the other grants represented a very substantial outlay by the crown: in general they were calculated on approximately ten times the annual value of each property to its leaseholder. The expenditure indicates that the king was forced to take seriously the claims of the principal tenants of King Street. On the other hand, the poor and very much larger class of tenants at will was cynically ignored. Thus William Russell's nine sub-tenants, none of whom held leases, received nothing; like all others in their case, they were simply evicted.

With the chief landlords of the district – the owners of the ground on which the various houses stood – the king negotiated a direct exchange of properties which, although theoretically fair, was in practice inequitable. The monks of Westminster Abbey derived an annual income of some £33 in rents from their possessions in this part of the town. In

Table 1. *Principal Holders of the Properties in King Street Acquired by Henry VIII for the Extension of Whitehall Palace**

Name of lessee	Occupation	Property	Date of Abbey lease	Price paid by the crown	Date of sale to king
		West side			
Elizabeth Palle	Carpenter's widow	The Axe	6.11.1522 Sub-lease E40/1564	£80	18.5.1531 E40/12837
William Dixon	Brewer		31.1.1488 Inherited E40/13449		Missing
Henry Wyatt	Knight, soldier	4 t	1524 WAM 18019	£20	Missing
Geoffrey Tull	Tiler	The Rose 1 s 22 t	Inherited property	£53 6s 8d (SP1/67 fol. 82) £50 (E40/1563)	18.5.1531 E40/1563
William Russell	Yeoman and waxchandler	The Bell 13 t	25.9.1524 WAM 18026	£128	23.5.1531 E40/13086
Guy Gaskyn	Carpenter	1 t of the Bell	24.9.1524 WAM 18023	£33 6s 8d	23.5.1531 E40/13447
John Russell	Carpenter	1 t of the Bell	25.9.1524 WAM 18025	£20	23.5.1531 E40/13406
John Garlonde	Yeoman	1 t of the Bell	24.9.1524 WAM 18024	£13 6s 8d	18.5.1531 E40/13446

Name	Occupation	Property	Date/Source	Value	Date/Source
Dr Richard Duck	Dean of Wolsey's chapel	7 c gs	3.6.1528 WAM 18038		Missing
Hugh Marbill	Goldsmith	The Rose 7 t & g 13 acres	inherited	£92	2.8.1532 E40/1543
Thomas Rawlyns	Gentleman	6 t g & m	Sub-lease from Marbill 30.5.1526 E40/13077	£20	12.5.1531 E40/13077
William Salcote	Carpenter	cs at Ch X & m	1524 WAM 17193	£176 13s 4d	18.5.1531 E40/6071
John Bennet	Citizen and grocer of London	ts & g at Ch X The Lamb	15.11.1525 Inherited from John Pomfrett	£66 13s 4d	18.5.1531 E40/1560
		East side			
John Stephens	Yeoman of the guard	The Lamb	16.6.38 LP xiii, no. 1519		Missing
Phillip Lentall	Cutler		Sub-tenant of Lamb		18.12.1542 E315/250 f47
Richard Hampkyn	Buckler-maker	3 t	Inherited from Currer	£30	18.5.1531 E40/1536
Henry Heyse	Chandler	House	12.4.1527 Sub-tenant of Hampkyn E40/1552	£60	Missing

Table 1. (cont.)

Name of lessee	Occupation	Property	Date of Abbey lease	Price paid by the crown	Date of sale to king
John Kellet		White Horse	17.7.1528 WAM 18039	£48	18.5.1531 E40/1565
John Henbury	Brewer	Red Lion, cs at Ch X	25.7.1508 E40/1559	£160	18.5.1531 E40/1559
Robert Pennythorne	Carpenter	House and wharf	Sub-lease from Red Lion E40/1553	£7	Missing
Edward Ingham	Gentleman of the king's household	The Rose 1s 2t & g	29.10.1528 WAM 18041	£15	8.5.1531 E40/12383
Richard Walker, *alias* Hampstead.	Smith	t on W of KS & t in Endive La.	20.3.1524 WAM	£20	Missing
Thomas Brightman	Yeoman	Endive Brewhouse	1520 WAM 18004	£106 10s	18.5.1531 E40/1526
John Rede	Gentleman	3t at Endive La.	17.10.1516 E40/13448	£82 13s 4d	18.5.1531 E40/13448

* This table is based upon a list of properties bought by the king in 1532 (PRO SP1/67, fols. 82–7) supplemented by deeds and leases held at Westminster Abbey and the PRO. Each side of the road is taken separately and the properties are listed from south to north (Fig. 3). Abbreviations: c, cottage(s); ChX, Charing cross; g, garden(s); KS, King Street; m, meadow(s), s, shop(s), t, tenement(s).

September 1531 the king took the required monastic lands in Westminster in return for the estate of a dissolved religious house at Poughley in Buckinghamshire. That the urban estate was both close at hand and currently increasing in value seems to have been left out of account by Abbot William Boston, whose compliance eased the king's path.[26]

The purchase of the relevant leases presented occasional problems in disputes over the level of compensation. One such wrangle concerned the property known as 'Petty Calais', the possession of Lord Berners, the deputy of Calais (Fig. 1). This estate was to form part of the new park; Cromwell, demanding its transfer, suggested that Berners' property was of little value. Berners complained to the king, and in a subsequent letter to Cromwell outlined his grievance:

You have shown the king my demand for my interest in the lease of Petty Caleys... the king supposed I would make no such price of so small a thing as ten acres... you esteem the house but little, alleging that I could never spend £100 on it.

Berners went on to list improvements made by himself to the property to the value of over £350, and proceeded to offer it to the king for £400: 'No other man living should have it for that price, [for] when I come to London I have no other place to put my head'. Though Berners later repeated his request, the outcome of the dispute is unknown.[27] It shows, however, that the grander inhabitants of Westminster were capable, as humbler residents were not, of fighting for a fair settlement.

Even before the land had been legally acquired, building work had begun on the site. The royal works accounts record the removal of tiles from the roofs of tenements in the way of the construction, the dismantling of their timber frames and the storage of the materials for re-use.[28] Excavation on the west side of Whitehall has revealed the mutilated foundations of the King Street tenements, sealed beneath the level of the palace; also found were rubbish pits containing the discarded contents of the demolished houses from the period of their occupation.[29] Some of the larger houses were temporarily retained by the king. The Bell (Fig. 1) was used by the royal carpenters,[30] while the neighbouring former property of Dr Duck was fenced off and used as a joinery yard for the construction of the king's new gallery.[31]

It was not only those who lost their homes who were affected by the sudden expansion of York Place. Those in neighbouring properties suffered attendant inconvenience. The occupants of Lamb Alley were particularly troubled. The alley was a busy route for the transport of building materials for the palace, and the royal accounts note the purchase, for example, of 'hinges made for the hosier's stall in Lamb Alley which were broken by a cart which carried stone to the foundations'.[32] A tenant at the bottom of Lamb Alley complained that one of the king's barges employed on the works had damaged his landing stage; compensation was forthcoming only after an acrimonious dispute.[33]

Some survived the first wave of evictions only to be visited a little later by the demand for their departure. The Lamb inn initially escaped the king's grasp. In June 1531 the monks of Westminster granted a new lease on this property to one John Bennet.[34] Bennet had just been dispossessed of his lands at Charing Cross, for which he had been paid £66 13s 4d;[35] but the unfortunate was not destined to enjoy his new property for long. An undated royal valuation of the Lamb, estimating its worth as £22 13s 4d, records Bennet's successful petition to retain a garden and tenement within the property.[36] By 1538 the Lamb was in the possession of a yeoman of the king's guard called John Stephens.[37] But in the early 1540s both Stephens and his under-tenants, who included John Bennet, were finally removed, as the boundary of the king's orchard was extended yet further to the south.[38]

In addition to the properties south of the King Street bars which have been discussed, Henry VIII secured all the properties to the north of the bars on either side of the street as far north as Charing Cross (Fig. 2).

On the west side were further tenements held on lease from Westminster Abbey. On the east side were tenements at the gate of York Place, some shops further north, the hermitage of St Katherine and the guild hospital and chapel of St Mary Rounceval. All of these were acquired by the king in 1531–2. The houses at the palace gate were demolished to create a more magnificent entrance.[39] Since 'Scotland' was now the base of the office of works, the Rounceval hospital was appropriated for use as the payhouse for the Whitehall building operation.[40]

The row of shops between Scotland and St Mary Rounceval having also been demolished, the site was redeveloped with new tenements designed specifically to house servants of the crown. These were erected in two phases: those between the hermitage and St Mary Rounceval, which were built by the king's carpenter, James Needham, at a cost of £51 7s;[41] and those built directly before the hermitage, which were separated from Scotland by a newly planted hedge.[42] The surveyor Ralph Treswell drew the tenements in 1610 as evidence in a law suit; his plan shows what were at that date sixteen shops.[43] Their original function, however, had been to accommodate some of those employed at Westminster in the king's service.[44] The records of the keeper of the palace show that the first occupants included prominent figures in the office of works, such as Thomas Heritage, clerk of the works at Whitehall,[45] the painter Luke Hornebolte,[46] and the French glazier, William de la Hay.[47] In addition, current and former members of the royal household occur: John Payne, a retired groom of the stables; Edward Millet 'of the household'; Robert Hayward, a yeoman of the guard; and Robert Gardener, a sergeant of the wine cellar.[48]

Among the recorded residents in what remained of King Street after the 1530s, only three names are recognisable from the pre-existing neighbourhood. The baker named Thurstan a Mere, formerly a tenant of the Ram on the east side of King Street, still worked locally at the White Horse beside Charing Cross. Thomas Swallow, a supplier of bricks and other materials to the royal works in the 1530s, continued to rent a tenement and stable here until his death in 1540. Land at Charing Cross was rented by Thomas Arnold, whose former house had been bought, dismantled, and re-erected as a builders' workshop for the palace.[49]

Thus within a short period after 1530 a complex network of solidarities operating within the urban community of Westminster was heedlessly shattered in the name of royal ambition. The private houses in the vicinity of York Place had been compulsorily purchased and destroyed; the palace itself closed to local inhabitants; and the Rounceval hospital forcibly shut down, to be turned into a payhouse for the king's works. The Rounceval guild limped on until 1544, but with a membership understandably depleted by the removal of its chief *raison d'être*. The arrogance evinced by the king in regard to the building of Whitehall surpassed even the late cardinal's characteristic high-handedness. Without consideration for the neighbourhood, for patterns of social intercourse which had evolved over generations, Henry merely directed his agent, Thomas Cromwell, to evict the inhabitants of King Street, and by his grandiose architectural design to change at a stroke the very physical and social shape of the town of Westminster. While the holders of head leases – powerful figures like the heir of the brewer John Pomfrett – received financial compensation, poorer sub-tenants – the pin-makers and labourers of Endive Lane, and their kind – were simply turned out of their homes. Yet this was the town, and these the townspeople, on whom the monarch depended (whether or not he ever reflected on the fact) for his and his courtiers' maintenance whenever he stayed at his principal base of power.

The arrogance of attitude is epitomised in Henry's arbitrary redrawing of the boundaries of the town. In 1524 the boundary of St Margaret's parish was realigned, to spare the royal sensibilities the odour of deceased Westminster parishioners being carried along King Street to their ancient resting-ground at St Margaret's church: instead the families living in the north part of King Street and at Charing were now to look in the opposite direction, to the church and cemetery of St Martin-in-the-Fields. Two years later an even more fundamental transformation was worked upon the town, when by Act of Parliament the precinct of the king's palace of Westminster was redefined to include all of the territory now occupied by the new Whitehall. The Act did not confine the royal enclave to the palace alone, but enlarged the precinct to incorporate the entire length of King Street, from Charing Cross to the old palace of Westminster, with 'all the houses, buildings, lands, and tenements on both sides of the same street or way from the said cross unto Westminster Hall'.[50] The greater part of the town of Westminster was thereby denied such limited yet significant rights of self-determination as had been evolved during the later Middle Ages.[51] Such crude interference with the pattern of human settlement and community betrays a profound disregard, on the part of the king, for those very subjects who, of all his many thousands, lived closest to him.

1 'Why come ye nat to court?', in J. Scattergood, ed., *John Skelton : The Complete English Poems* (Harmondsworth, 1983), p. 289.
2 Public Record Office (hereafter PRO) C82/482; E40/1304.
3 S. Thurley, 'The Domestic Building Works of Cardinal Wolsey', in S. Gunn and P. Lindley, eds, *Cardinal Wolsey : Church, State and Art* (Cambridge, 1990).
4 This is a fair criticism, not only of such older works as W. J. Loftie, 'Whitehall: Historical and Architectural Notes', *The Portfolio*, xvi (April, 1895), but also of the recent *History of the King's Works*, ed. H. M. Colvin, 6 vols. (HMSO, London, 1963–82). References to some of the documents which form the basis of the present article were published, without interpretation, in *Survey of London*, xiii, pp. 3–18; xiv, pp. 3–9; xvi, pp. 1–4, 158–62.
5 Westminster Abbey Muniments (hereafter WAM) 18519–22; 17792; 23115 *et seq.*; 18049A.
6 G. Rosser, *Medieval Westminster 1200–1540* (Oxford, 1989), pp. 89–92.
7 Rosser, *Medieval Westminster*, p. 397; WAM 18049A.
8 WAM 18049A; WAM Register Book i, fol. 121.
9 PRO C1/63/213–15; J. Otway-Ruthven, *The King's Secretary and the Signet Office in the Fifteenth Century* (Cambridge, 1939), pp. 158, 183; WAM 19066–78.
10 PRO C1/517/57; Rosser, *Medieval Westminster*, pp. 127–8.
11 G. Mathew, *The Court of Richard II* (London, 1963), pp. 25–7; Whitehall excavation reports forthcoming in *Transactions of the London and Middlesex Archaeological Society*.
12 Rosser, *Medieval Westminster*, pp. 215 and n., 163 and n.
13 WAM 18599.
14 D. R. Ransome, 'Artisan Dynasties in London and Westminster in the Sixteenth Century', *Guildhall Miscellany*, ii (1964), pp. 236–47; WAM 18049A.

15 PRO E179/141/139; tabulated in Rosser, *Medieval Westminster*, table 5, p. 223.
16 Calendar of Letters and Papers, Foreign and Domestic, Henry VIII (hereafter *LP*), iii (1), no. 365 (11).
17 For the buckler-maker, Richard Hampkyn, see PRO E40/1536; E40/1552; WAM 18033, 19800–4.
18 Accounts of the guild of the Assumption, Westminster, unnumbered volume in Westminster Abbey Muniment Room.
19 Accounts of the Rounceval guild for a few years in this period are bound up with those of the guild of the Assumption cited in n. 18 above. See also Rosser, *Medieval Westminster*, pp. 313–20; and the Act 28 Hen. VIII c. 32.
20 *Calendar of State Papers Venetian*, iv (1527–33), no. 664.
21 *Calendar of State Papers Spanish*, iv (1531–3), p. 154.
22 See further H. J. M. Green and S. J. Thurley, 'Excavations on the West Side of Whitehall 1960–2. Part 1: From the Building of the Tudor Palace to the Construction of the Modern Offices of State', *Trans. London and Middlesex Arch. Soc.* (1989).
23 WAM 18048, 18049A–B; PRO SC12/3/13.
24 PRO SP1/67, fols. 82–7.
25 Ibid.; WAM 18049A.
26 *LP* v, nos. 404, 627 (23); 23 Hen. VIII c. 33. A portion of the land on the west side of King Street was simultaneously acquired from Eton College. See also *Survey of London*, xiii, pp. 11–12.
27 *LP* v, nos. 857, 1219.
28 PRO E36/251, pp. 11, 35, 95, 116, 132, 161, 352.
29 Green and Thurley, op. cit.
30 PRO E36/251, pp. 171, 233; E36/252, p. 402.
31 PRO E36/251, pp. 100, 167; E36/252, p. 402.
32 PRO E36/251, p. 102.
33 PRO E351/3322; E36/252, p. 590.
34 PRO E40/1551.
35 PRO SP1/67, fol. 85.
36 PRO SC12/21/21.
37 PRO SC12/3/13, fol. 4v; *LP* xiii (1), no. 1519 (8); PRO SC6/HenVIII/2101.
38 PRO SC6/HenVIII/2103.
39 PRO E36/252, p. 588.
40 St Mary Rounceval was finally dissolved in 1544. *LP* viii (2), no. 590; see also PRO C66/706; E322/138; E36/251, p. 166; E36/252, p. 413.
41 WAM 12257, p. 2.
42 PRO E36/252, pp. 432, 435, 620, 649.
43 See *Survey of London*, xvi, pp. 241–2; and J. Schofield, ed., *The London Surveys of Ralph Treswell*, London Topographical Society Publication No. 135 (1987).
44 See S. Thurley, 'English Royal Palaces, 1450–1550', unpublished Ph.D. thesis, Univ. of London (1989), pp. 311–20.

45 PRO SC6/HenVIII/2101.
46 PRO SC6/HenVIII/2103; see also E315/236, fol. 37.
47 PRO SC6/HenVIII/2101/3.
48 *LP* xix, no. 1036, fols. 4v, 60v; PRO E315/218, fol. 105.
49 See PRO E40/1560; E315/218, fol. 105; *London Topographical Record*, xix (1947), p. 104; PRO SC6/HenVIII/2101.
50 28 Hen. VIII c. 12. We are grateful to George Bernard for discussing this Act with us.
51 Rosser, *Medieval Westminster*, ch. 7.

IV. MOORFIELDS, FINSBURY AND THE CITY OF LONDON IN THE SIXTEENTH CENTURY*

By ELEANOR LEVY

DURING the fifteenth and sixteenth centuries both the citizenry and the corporation of London began to take a much greater interest in the largely uncultivated spaces which lay to the north of their wall. At different times both sections of this area, the smaller, marshier Moorfields in the east and the manor of Finsbury to its west extending towards Cripplegate, eventually came under the city's jurisdiction and the different rationales which lay behind the decisions to adopt the extra land explains the markedly contrasting ways in which Moorfields and Finsbury came to be part of the city. The diversity of their management and utilization helps illuminate some of the problems which confronted the city's corporation when faced with the reality of London's need to extend her perimeters in order to accommodate both a growing populace and intensified urban pressures.

During the turbulent civil war years of the twelfth century it appears that ownership of land lying to the north of the city's wall became the subject of dispute.[1] By the fourteenth century, Londoners had secured the Moorfield – as the larger expanse of Finsbury was a prebendal manor of the Dean and Chapter of St Paul's Cathedral – although the exact boundary between the sections is not easy to define in that period as the word 'Moor' was often used in official documents to describe either part or all of the fields. For example in 1309 the Mayor and Aldermen leased the (generalised) 'More' without Cripplegate to Alderman Nicholas Pikot for an annual rent of two pounds.[2] By

* I am very grateful to Dr Caroline Barron for all her advice and assistance with this article.

1319–20 this scheme is obviously being taken quite seriously as the Chamber of the Guildhall was demanding surety against the year's rent of four marks[3] and in 1374 Thomas atte Ram – 'a brewer of the Moor' – was permitted to keep the Walbrook for seven years for the annual fee of twelve pence provided that he safeguarded the Moor and kept the watercourse clean.[4] (The Walbrook, flowing southwards through the Moorfields made its way through London to meet, eventually, with the Thames.) Although culverts in the London Wall still allowed free flow through the city, to the north this channel became increasingly silted up and clogged with rubbish so that the Moorfield ground became marshy,[5] and according to Fitzstephen in the twelfth century, when frozen, 'great throngs of youth go forth to disport themselves...upon a vast spread of ice'.[6] The cleansing of the city ditch outside the wall in 1213[7] may have helped drain this lagoon – although Stow describes the still nominally undivided fen of the early fourteenth century as a 'vast and unprofitable ground'[8] – presumably referring to documents of that date which he had seen. Despite the demands made upon Thomas atte Ram (the brewer) filth continued to accumulate in the brook for in 1378 the London authorities were still discussing means of improving conditions both inside and outside the city.[9] It may have been the Moorfield's obviously problematical and inhospitable nature which initially deterred the city from automatically encompassing the tract of waste ground – but by 1411 this piece of land was both of increasing interest to local inhabitants and a continuing headache to the city's governors. On 12 January 1412 the Mayor, accompanied by a number of Aldermen, crossed the ditch in order to view for themselves the Moorfields and found it 'covered with gardens, trees, hedges as well as rubbish and filth'.[10] The problems of hygiene, administration and supervision, which were to plague the corporation for almost two hundred years were about to begin.

Three years later due to lack of management and 'a common latrine there situate which by reason thereof very many cellars and dwelling houses were overflowe in the streets to the said Moor next and adjoining, many sicknesses and other intolerable maladies arising and the horrible and corrupt and infected atmosphere proceeding from the latrine',[11] the Mayor and his advisers took the momentous decision that the Moor should be divided into different plots. These parcels could then be let out – not only would supervision be maintained but the City would in this way also receive extra revenue from the rentals charged.

Unfortunately the Chamberlain's account books for the period have not survived and it is therefore impossible to assess how much money came into the City from this scheme. Betty Masters' work on the City's Chamber Accounts for the sixteenth century includes some scattered references to the maintenance paid out by the City. In 1535 Will Palmer was paid 6s 8d for 'keeping the Moorgate and the Posterns'[12] and in 1563 Thomas Norwall was paid £2 13s 4d for making a new boat and mending the old Moor Ditch.[13] In the same year Mr Waterbailie's men and others were paid 10s for taking pains over the fishing at Moor Ditch. As the account books of the Chamber were designed to show whether the Chamberlain owed the Corporation money rather than to relate expenditure to revenue – despite the desire to capitalize on the leasing out of the Moor – it is possible that maintenance costs may have exceeded income.[14] However the fact that in 1415 Thomas Falconer, the Mayor of London, built the Moorgate 'for the ease of citizens that way to pass upon'[15] and also caused the ditch to be 'new cast and cleansed by means whereof the said Fen or More was greatly drained and dried', strongly suggests increased interest and usage of the ground. Although no fifteenth century leasing records exist and only occasional leases are documented in the sixteenth century repertories – a mere fourteen in all – the impression gained is that fairly regular use of the field was taking place. During the course of the sixteenth century, however, much criticism was levelled at those responsible for the management of the City lands, both in general terms and with regard to specific aspects of their financial management.[17]

Of the fourteen surviving sixteenth century leases recorded for Moorfield, four of them specify land use for gardens – these carry no suggestion of market gardening so that their use may have been recreational. Another lease of 1534 describes a parcel of common ground between Moorgate and Cripplegate whereupon 'tentors have been set' to be leased at £50 4s 5d p.a. but unfortunately the exact location is not specified.[18] Another lease of 1585 suggests the same use of land 'for the ease and benefit of sundry clothworkers within the City'[19] (this group of workers feature prominently in both Moorfields and Finsbury sixteenth century records). Other leases refer to the use of the land for pasture,[20] and the doghouse[21] of the City's common hunt is also mentioned.[22] The exact location of the doghouse in Moorfields is not specified in the records but in 1560[23] there were clearly local pressures to have the doghouse removed from its Moorfields site. The doghouse location as depicted on the Copperplate Map confirms the idea that it was located close to cultivated plots – and petitions are recorded from those with gardens nearby who wish to have it removed. The final word comes from the Corporation who demand that the complainants contribute the equivalent of one year's rent towards a new relocated building. In 1569[24] and 1570[25] sites in Hogg lane (a small road off the west side of Bishopsgate Street) and in Whitecross Street (to the West of Finsbury) were both suggested but no final decision was recorded.

The Copperplate Map of about 1557 provides a vivid picture of the appearance of Moorfield and this picture is further substantiated by the details of a lease later granted to John Acheley. He was to allow the citizens to dry their clothes on his leased land freely, and have other 'free' pleasures 'as heretofore they have been accustomed'. He was however allowed to charge a penny for every cart 'desiring to pass that way with brick…the said brick, timber, earth, dung and other like not to have continuance of lying there above three days without further agreement with him or them for the same'.[26] Those who wanted to bleach clothes could make individual arrangements with Acheley who, for his part, was to clear his land by an agreement not to have

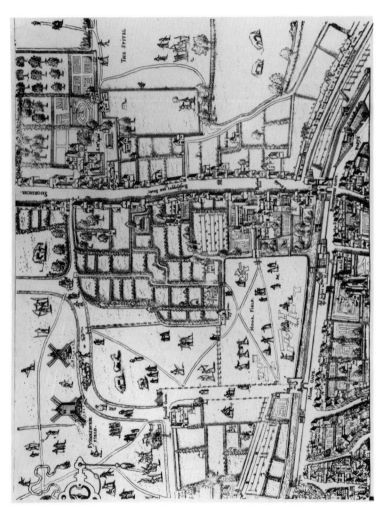

Detail from the 'Copperplate' map (north section) by courtesy of the Museum of London.

any laystalls (dungheaps) on the ground, and to pull down an existing cow house. The entry prompts several questions – were the citizens who were to be allowed to dry the clothes 'professional washerwomen' of the city? The streams on the field would have been ideal for bulk washing and the two figures to the south west of the doghouse appear to be carrying a basket which may have contained wet clothes on their way to be soaked in lye for bleaching and to be pegged out afterwards to 'white' as the lease says. With regard to the toll on carts, there is no indication in the lease that this is an innovation brought into force by a recent increase in traffic either to or from the city or crosswise from Finsbury to Bishopsgate or vice versa. The cartways themselves were not new for as early as 1513 there was a motion 'for a cartway from the Moorgate unto Finsbury Field for laystalls'. However the reference in Acheley's lease to movement, especially of 'brick, timber and earth', suggests that the districts around the Moorfield may have been used as brickfields, which implies that the surrounding neighbourhood was being built up. The mention of a cow house taken in conjunction with 'pasturage' reflects London's need for grazing land close to the city. Both the map and Acheley's lease reveal that the Moorfields by the sixteenth century was being much more intensively used for a variety of industrial and domestic purposes. The terms of the lease also show that at least some control of the area was being arbitrarily placed in the hands of lessees. It may be that the Court of Aldermen delegated responsibility for the area to those who leased the land whilst their system of government readjusted to the new demands made by a 'greater London'. A further example is the case of John Hillyard who in 1567 was only granted a lease on condition that he assumed responsibility for building a brick wall enclosing part of his rented land.[27] The lack of on-the-spot administrative machinery is further highlighted by the injunction against those who may have had designs to obstruct a ditch-cleansing operation which involved laying mud on its north-east bankside – those 'who should refuse or withstand the operation' would be 'brought before the Lord Mayor to be dealt with accordingly'.[28]

As already noted, in 1415 there was constant fear of 'intolerable maladies' arising from the Moor area and from the 1513 scheme mentioned earlier it is clear that London dung and filth were carried in the form of laystalls outside the city. Apart from industrial waste dumpage as perpetrated, for example, by the Curriers' Company who in 1526 were served with the injunction to 'forthwith cleanse all such filthe and ugly thing by them used and laid in the ditches surrounding the Moorfield',[29] the conditions of the sewers and ditches of the field continued to give rise to growing concern throughout the sixteenth century. There are frequent references to the need for these to be 'cleansed and scoured'.[30] In 1584 the keeper of the sluice at Moor Lane and the postern called Moorgate was paid 13s 4d and the same year John Kinge was paid the same amount to keep the ditch clean.[31] By the early seventeenth century the Moorfield was obviously something of a malodorous mess,[32] and a group of Aldermen and others were appointed to 'consider of some means for removyng and avoiding of the sewers and ditches in Moorfields.[33] In 1605 a seven man committee was appointed to confer with those of 'the King Majesty's works to view the springs etc.'[34] It is not clear whether the main impetus for the formation of this committee came from the City or the Privy Council. The management of the Moorfields demonstrates the pressures on the City governors to utilize the space and also the lack of an administrative network capable of managing the land as it became more intensively occupied. Yet despite this difficulty and the vexed issue of drainage, James I was able to write to the Corporation in glowing terms that 'we have been informed by some persons of great inwardness and trust about us and have also perceived by our own observations that you have of late bestowed no small cost in things that due concern the ornament of that our cittie – as namely in the Walkes of Morefields, a matter both of grace and greate use for the recreation of our people', a 'pleasurable place of sweet ayres for citizens to walk in'.[36] Thus it was that the packed populace of early Stuart London was given the chance to experience preserved leisure space within easy walk of the city. The 'park' was eventually extended to encompass the ground from Lower through to Upper Moorfields – that is in modern terms from London Wall to the northern edge of Finsbury Square. From the survey describing the boundaries of Finsbury manor it is clear that Upper Moorfields was in fact the mallow field of Finsbury, so that in the seventeenth century the two areas overlapped, once more blurring their boundaries.

In the sixteenth and early seventeenth centuries there was scarcely a year when plague was absent from the capital. Paul Slack in his essay on metropolitan government in crisis[37] comments that national authority again and again charged the rulers of the City that in matters of disease they always did too little too late. The City's need to isolate the sick was paramount and in 1583 a committee was appointed to find out some convenient place where 'such persons as shall be infected or visited with the plague may be conveyed' and 'with all convenient speed measure out a convenient plot near unto the little lane betwixt Moorgate and Cripplegate without the walls of the citie'.[38] Despite the need for 'speed' it was only after two further outbreaks of plague that building finally began in 1594. By then the site was even further out of the city, close to a path leading to Canonbury – on three acres of land belonging to Saint Bartholomew's Hospital and not in the vicinity of Moorfields as had originally been planned. (The building was still unfinished in 1603.) In 1412 when the Corporation of London intervened in the Moorfield, the motivation seems to have been primarily to contain it as an extra-mural problem spot and perhaps incidentally make some money for the City. As has been seen, by the seventeenth century the City's need had helped to shape Moorfields into a free recreational area – whereas in neighbouring Finsbury from the outset more economic and industrial considerations prevailed.

Negotiations between the Corporation of London and the Dean and Chapter of St Paul's Cathedral for the leasing of the manor of Finsbury were concluded in 1514. Whereas a hundred years earlier the Mayor and Aldermen seem simply to have taken over Moorfields, in the early sixteenth century they had to negotiate to acquire the Finsbury manor. Repertory entries describe a series of tentative meetings between numbers of aldermen and the Prebendary of Finsbury which culminated in the signing of a lease when 'the Indentures of the Farm of Finsbury, with an obligation thereto belonging were sealed'.[39] It is not clear exactly how much the City paid although the lease may have cost 300 marks.[40] This was probably the purchase price with an annual rent of 100 marks.[41] The lease was renewed again in 1544 with no apparent difficult but the city ran into problems when the lease came up for renewal ten years later in 1554. Negotiations at this date were beset by the prolonged and obstinate reluctance on the part of Edmund Bonner, then Bishop of London, to ratify the agreement. Accordingly, it was ordered that 'Mr Chamberlain shall forthwith deliver to Mr Alderman Dobner 40 marks to be paid to Mr Grymstead of St Pauls for his goodwill and assent to be obtained and given for the sealing of the City lease.'[42] When this brought nothing, a bribe was offered to the Bishop's steward upon condition that he 'make means to the said Bishop and also bring to pass that his lordship seal be set to the City's Writing for Finsbury, he shal be freely admitted into the liberties of the same city into a company without paying for the same, and further it was agreed that my Lord Mayor and others shall repair tomorrow to my said Lord of London for his lawful favour for the sealing of the said writing' in return for 'the granting of certain quantities of the city's water unto him'[43] (possibly for the Bishop of London's palace). Although this set of bribes proved to be abortive, the City was finally rewarded, after another year's bargaining, with a new lease for 90 years. As these negotiations coincided with the Marian period it is possible that delay was caused by the wider debate about the settlement of Church lands. What is notable about the negotiations is the sense of determination on the part of the City. Indeed only 12 years later, in 1567,[44] the City made further approaches to secure the manor for another 200 years after the expiration of the 90-year lease (140 years was finally agreed upon). The study of these negotiations clearly reveals how important Finsbury manor

THE CITY OF LONDON   87

had become to the City and how determined the Mayor and
Aldermen were to secure control of the area by a long lease.
Although the medieval boundary between Moorfields and
Finsbury manor is impossible to determine, yet two
sixteenth century surveys survive which delineate the
prebendal manor of Finsbury in 1567[45] and 1586[46]. These
may well have been commissioned by the City to determine
a more exact value of their property – the first survey
coincides in date with the signing of the very long lease in
1567 and was probably drawn up in that connection. Such
surveys were often made by taking evidence from a sworn
body of local jurors although, as seems likely here, 'outside'
surveyors were sometimes brought in. In fact these very
detailed written descriptions which covered all the lands of
the manor were an end in themselves – a property inventory
and they were not formulated for the final production of a
map.

From the signing of the first lease the City government
appeared anxious to exercise effective control over Finsbury.
For regulatory and judicial purposes possibly the traditional
pattern of the manor was followed, whereby a local court
continued to function on a small scale, supervised by the
City's Court of Aldermen acting as overlords of the manor.
A record book from the local court survives and notes annual
meetings from 1581 onwards.[47] This court was apparently
concerned in the main with the collection of local quit rents
and inheritance issues as they touched on the streets of
Finsbury manor. It is hard to determine from the sparse
records a clear picture of the exact functions of the Finsbury
manorial court but it seems to have combined some of the
characteristics of both a court Baron and a court Leet.[48] The
former was not a court of record.[49] Its principal concern, as
a private court of the lord (forming part of his estate and
property) was to enforce ancient customs, maintain the
rights of the lord and privileges of the tenants, settle mutual
differences and organize community affairs. It was only
attended by tenants of the manor and these occupying
'owners' of local property made up the Homage or Jury
whose members' (a minimum of three) judgements were

embodied in the form of presentments. The principal interest of the court centred around incidents of land tenure and the use of fields and waste ground. The business of the Homage or Jury was to consider and make statements with regard to property which fell to the Lord due to the lack of an heir, the surrender of tenements and the death of any tenant. Other functions were the hearing of plaints, deciding disputes and it was also ordained to 'determine injuries, trespasses, debts, and other actions'.[50] Such a court was usually presided over either by the Lord or his steward. It could be summoned at the discretion of either of these two men and was usually held every three weeks. By contrast a court Leet was a court of Record, and in theory had to be attended by every male who had lived in the manor for a year and a day. Although a court of the lord of the manor, it was in fact a local criminal court in so far as it dealt with infringements of the law. This monarch's court held by the lord had as its jury 24–6 local residents who dealt with such issues as delinquencies, consumer matters, regulation of markets and local nuisances, 'petty treasons and felonies' were enquirable in a Leet, but not punishable there[51] – all matters of indictment were transferred to the assizes. Surprisingly the Leet court was held only once a year, traditionally in the autumn. The Finsbury manor court appears to have conformed mainly to the structure of a court Baron – that is attended solely by tenants of the lordship, only meeting once a year (this may have increased towards the end of the century), containing an all male jury of between eight to ten homages sworn in by the steward, and concerning itself primarily with the issue of local quit rents (these were small annual rents and in return for their payment the occupier was 'quit' of all other tenurial obligations).[52] On occasion the court also considered other matters – it heard information concerning the death of tenants and dealt with the subsequent issues of inheritance and entrance fines. Very occasionally the court dealt with boundary disputes and the fixing of tavern signs. No other local regulatory or judicial decisions are recorded. The steward of the court – appointed by the Corporation – who may have also been the scribe, appears as an integral part of the court which met in late autumn. Martha Carlin's research on the administrative aspects of Southwark overall from 1200 to 1500[53] reveals other differences. Although 'court procedures varied from manor to manor and from century to century', by the sixteenth century the Southwark Leet court normally presented on all breaches of assize, ordinance, statute and public nuisance. In addition persons presented for immoral or anti-social behaviour could be sentenced by the jurors to the ducking stool and/or expelled from the manor. The government of the manors therefore rested almost entirely with their residents but reflected all interests. Another possible contrast with Finsbury is revealed by Dr Boulton's work on the court at Boroughside[54] (an administrative district of St Saviour's parish, Southwark) where, he suggests, local meetings lent structure and coherence to the neighbourhood. It is impossible from the 'clipped' accounts in the Finsbury manor records to assess to what degree this was true for that area, particularly as no parish records survive which might have helped to throw light on a possible network of involved local people active in both manorial and parochial affairs.

Perhaps the Court of Aldermen realized there was a need in Finsbury for more frequent meetings, if only on a trial basis, as from 1566 onwards the London repertories mention the idea of arranging extra meetings to deal with specifically Finsbury matters.[55] On one occasion the venue was the Armourers Hall,[56] whilst in 1583 'yt ys ordered that the dynner for the lordship and manor of Finsbury shall be theare holden and kept at Tuesdaye next insuinge'.[57] It is possible that the location, vaguely referred to, is the building known as Finsbury Court, which is specifically mentioned on both the surveys and shown on sixteenth century maps. It is not clear whether the new court was to be an additional manorial one or as seems most likely an off-shoot court of London. Sixteen years later the matter was still under review. 'Mr Dale and Mr Collyn, aldermen, shall have present consideration whether two leets may yearlie be kept for the manor of Finsbury or not'.[58] None of these entries clarifies the nature of the proposed court nor its scope of jurisdiction – although a ducking stool for the Lordship of Finsbury was ordered in 1600, at the charge of the City Chamber.[59] The manorial record books give some idea of the number of freeholding families who continued to hold property in the manor over a long period. Forty-four family groups can be traced. For example Richard Roper's family lived in Golden Lane in 1550 and kept property until 1595, and the Hilliard family likewise for thirty five years. Both Richard Roper and John Hilliard served continuously on the local court jury which may indicate that they and others like them did in fact constitute a backbone of stability within the manor. In contrast data collected from the manor book, the two surveys and a list of Freeholders which starts in 1550 and concludes in 1600, show that at least 220 tenements changed hands between 1550 and 1600.[60] There are eighty four references to tenements in Golden Lane, thirty in Whitecross Street and eight other property references include Grub Street and Beech Lane. The most common tenants were clothworker craftsmen who were attracted by the advantages offered by the open spaces close to London. It is not possible to tell if the people who lived in Finsbury were new immigrants to London, or City residents who had moved out. As is often the case, it is Stow who provides some flesh for the bones. He tells us that in Golden Lane there were almshouses,[61] Beech Lane was replenished with beautiful houses of stone, brick and timber,[62] that on the north side of Beech Lane the Drapers 'have lately builded 8 almshouses[63] and Grub Street of late years was inhabited by...bowyers, fletchers, bowstring makers'.[64] Fifty one leases granted by the Corporation between 1518 and 1596 were recorded in the Repertories, the bulk of the leases were issued after the ninety-year lease agreement, i.e. from mid 1550s onwards. Thirty were for gardens and another ten simply for pieces of ground. The largest group of lessees were, once again, clothworkers who numbered fourteen. As the field areas were suitable for tenter grounds no doubt difficult to accommodate within the city, the large number of clothworker leases is not surprising, indeed seven of them refer specifically to tenter facilities. The second largest group of tenants were widows; unfortunately their former husbands' occupations are mostly unrecorded. These women may have been attracted to the garden plots, possibly to help augment the family budget by cultivation of fruit or vegetables, the growing of medicinal herbs or for drying extra washing. Twenty five other leases also mention garden plots whilst Alderman Lionel Duckett was granted a lease of 'some gardens' enabling him to enlarge his own plot by the acquisition of three others adjoining his own.[65] Whether it was Duckett himself and his colleagues who fell foul of Stow's censure is not certain, but it is interesting to read the antiquarian's acid description of suburban garden owners who 'built many fair summer houses...not so much for use or profit as to show...betraying the vanity of mens minds'.[66] There was very little similarity between Moorfields and Finsbury in terms of how the City administered the areas. In Moorfields, as we have seen, problems of management were wished away rather than tackled in a systematic and sustained way. By contrast there were four officials appointed by the Corporation to deal with Finsbury matters. The 'Farmer of Finsbury' is first mentioned in 1515[67] immediately after the City acquired the manor. In return for unspecified quarterly payments to the City, the farmer could apparently collect and keep the Corporation lease revenues for himself. The bailiff's work was that of a general overseer. When Lawrence Nashe was bailiff in 1553 he had blue crosses delivered to him by the court, in order to set them up at Finsbury, on the doors of those houses whose occupants were plague victims, and was also ordered to remove a dung heap which was lying on a highway near Finsbury Court.[68] The third officer, already mentioned, was the steward of the local manorial court who was appointed by the Corporation. Richard Hawks, gentleman, was granted the post with the signing of the first lease and as recognition of his 'good diligence and labour' he was officially given a stripped gown – in accordance with the livery of the Chamber – which he could retain whilst in office, in addition to the sum of 13 s 4 d.[69] In the Chamber Accounts for 1535–6, the steward appears to have been paid an annual salary of twenty shillings.[70] The fourth officer was the rent gatherer who worked with the steward, presumably involved with the collection of quit rents. As with the Farmer he paid rent to the Corporation himself,[71] though this was not a full-time job as William Veer who held the post in 1535 was also Clerk of the City's works.[72] In the Chamber Accounts for 1584–5 the Finsbury Steward, Rentgatherer and Bailiff are all given twenty-four shillings for cloth designated for winter liveries.[73] It is important to remember that Finsbury, unlike Moorfields, had to be paid for and from the start of the negotiations with St Paul's, the manor was envisaged as a well regulated metropolitan extension. Coincidental with the signing of the first lease, a group of Aldermen were assigned in 1514 to requisition carts and to set vagabonds to work levelling old laystalls – in order to prepare the area for metropolitan use.[74] At another Court of Aldermen meeting in 1515, one member was designated to 'oversee Finsbury both East and West...for the cleansing of the streets'.[75] In the same year nine Aldermen were appointed to raise 'volunteer' money within the City in order to offset the expenses already incurred there and in anticipation of further costs.[76] Whilst the acquisition of Finsbury was obviously expensive both in terms of money and effort, the Corporation had gained an asset in broader respects than a mere rent potential. In all the ages of patronage it was extremely useful for the City to obtain a new source of help with regard to the boost of pay of worthy city officials and the provision of extra income for their dependants, as well as to furnish Aldermen and common councillors with advantages and opportunities to rent extra land. In 1538 Thomas Welles, late Clerk of the Wool Beam at Calais, was licensed to erect a windmill in Finsbury after the court agreed to show him as much favour as they reasonably could.[77] This was to be in Finsbury Field on the east side of the mill recently set up by George Barne, an Alderman, a goldsmith and a gentleman.[78] In 1575 'The Quenes majestie dyrected her most generous letters...in favour of William Bowle one of the ordinary yeomen of her highes chamber for a lease to him be granted by this courte of the manor of Finsbury.' This was granted after the 'determynacions of the right and interest alreadye granted by this courte unto William Dalby gentleman one of the clerkes in the Lord Mayors court'.[79] In 1569 Sir Walter Mildmay, a privy councillor to Queen Elizabeth, was granted a licence to 'make certain bricks' in Finsbury Fields'... as it be to his owne use provided always that this shall be no precedent to grant the like hereafter'.[80] The windmills are familiar enough from depictions of Finsbury and are evidence of the industry and entrepreneurial interest in the area. Moreover the potential for brickmaking was well known to Londoners from at least 1477 when Mayor Ralph Joceline 'for repairing the wall of the city, caused the More to be searched for clay and brick to be brent there'.[81]

In Finsbury in the seventeenth century, defenders of medievalism, embodied by citizen archers, were destined to do increasing battle with the growing forces of industrial expansion, as represented by local brickmakers, over the right to utilize the open spaces to the north of the City. This contrast with the recreational development of Moorfields must be a reflection of the irresistible and diverse pressures from within the City, which in their turn helped to fashion the adjacent areas into very different areas of development, aspects of a newly enlarged and enlarging London.

1 William Page, ed., Victoria County History, *London* (1909), vol. I, 556.
2 R. R. Sharpe, ed., *Calendar of Letter Book D* (London, 1902), 210–11.
3 R. R. Sharpe, ed., *Calendar of Letter Book E* (London, 1903), 118.
4 R. R. Sharpe, ed., *Calendar of Letter Book G* (London, 1905), 324.
5 *Report on the Royal Commission on Historical Monuments* (London, 1927), vol. III, 17.
6 C. L. Kingsford, ed., John Stow, *Survey of London* (Oxford, 1908), vol. I, 93, hereafter Stow, *Survey*.
7 Stow, *Survey*, I, 19.
8 Stow, *Survey*, II, 76.
9 E. L. Sabine, 'Cleaning in Medieval London', *Speculum*, vol. 12, 1937, 333–4.
10 R. R. Sharpe, ed., *Calendar of Letter Book I* (London, 1909), 101.
11 ibid., 137.
12 Betty Masters, *Chambers Accounts of the Sixteenth Century* (London Record Society 1984), vol. XX, 107.

13 ibid., 125.
14 ibid., xxxviii
15 Stow, *Survey*, I, 32.
16 Stow, *Survey*, II, 76.
17 Neal Shipley, 'The City Lands Committee 1592–1642', in *Guildhall Studies* in London History, vol. II, 1977, 164. With regard to the registration of leases in 1589 Robert Smith was voted £40 in 'consideracion of the travayle and paynes heretoforetaken...in searching and sertyninge of a great part of the evidences of the Cytes Lands found in a room where they of longe time remaned unknown and where they have receved some hurt by wet and otherwise and in placing therein several boxes in cupboards in Mr Chamberlynes custody with the other cityes evidences of their lands', Corporation of London Records Office, Rep. 22 f. 18r. Sustained neglect appears to have characterized this aspect of corporate record-keeping so that the documented number of leases for Moorfields (and for Finsbury which was also part of the City Lands Estate after 1514) may well be under-representative of their commercial activity in the sixteenth century.
18 C.L.R.O., Rep. 9, f. 60r.
19 C.L.R.O., Rep. 21, f. 222r.
20 C.L.R.O., Rep. 1, f. 105v.
21 C.L.R.O., Rep. 1, f. 105v.
22 C.L.R.O., Rep. 2, f. 126v.
23 C.L.R.O., Rep. 14f, 318r.
24 C.L.R.O., Rep. 16, f. 492r.
25 C.L.R.O., Rep. 17, f. 22r. See also Betty Masters, 'The Mayors' Household before 1600', in W. W. Kellaway and A. J. Hollander, eds, *Studies in London History* (1969), 100–1.
26 C.L.R.O., Rep. 21, f. 222r.
27 C.L.R.O., Rep. 16, f. 252r.
28 C.L.R.O., Rep. 23, f. 390v.
29 C.L.R.O., Rep. 8, f. 230v.
30 C.L.R.O., Rep. 7, f. 202v.
31 Betty Masters, *op. cit.*, 17.
32 See earlier references to problems on the Moor.
33 C.L.R.O., Rep. 26ii, f. 403v.
34 C.L.R.O., Rep. 26ii, f. 553.
35 Analytical Index to Remembrancia 1579–1664 (Corporation of London, 1878), 46.
36 Richard Johnson, *The Pleasant Walkes of Moore-fields* (1607).
37 Paul Slack, 'Metropolitan Government in Crisis', in Beier and Finlay, eds, *London 1500–1700* (London, 1986), 65.
38 C.L.R.O., Rep. 20, f. 415v.
39 C.L.R.O., Rep. 3, f. 72r.
40 Betty Masters, *op. cit.*, 112, 114.
41 ibid.

42 C.L.R.O., Rep. 13i, f. 254v.
43 C.L.R.O., Rep. 13i, f. 293r.
44 C.L.R.O., Rep. 16, f. 242v.
45 Manor of Finsbury, Copy of Survey of the Manor, C.L.R.O. Misc. MSS 42.5.
46 Manor of Finsbury Court Baron Minute Book, C.L.R.O. Shelf 39.B.1.f.
47 Regular records of Court meetings start in 1581 but the book does include a selection of entries from the previous thirty years' inquests. All appear to have been written up afterwards.
48 Webb S. and B., *English Local Government From the Revolution to the Municipal Corporations Act. The Manor and the Borough* (London, 1908), vol. II, ch. I.
49 Namely a court whose acts and proceedings are permanently recorded, and which has the authority to fine or imprison for contempt.
50 John Kitchin, *Jurisdictions, or the Lawful Authority of the Court Leet etc.* (1598). (John Kitchin, Gent, was granted the lease of a garden in Finsbury for forty years.)
51 ibid.
52 These were small annual rents. In return for their payment the occupier was 'quit' of all other tenurial obligations. As these payments were fixed no sharp picture of property values for sixteenth-century Finsbury emerges from the records. In addition the entries themselves are general in description. 'A small tenement at the upper end of Golding Lane' carried a quit rent of 12 pence (C.L.R.O., Manor of Finsbury Court Baron Minute Book, f. 29). The Wardens of the Parish Clerks of London paid 24 shillings for 'certain tenements in Whitecross Street (ibid., f. 3). Other 'certain tenements at the Antelope in Barbican' were charged at sixteen pence (ibid., f. 9).
53 Martha Carlin, *The Urban Development of Southwark c. 1200–1550*. Unpublished University of Toronto thesis, 1983, ch. 13.
54 Jeremy Boulton,
55 C.L.R.O., Rep. 16, f. 114r; Rep. 20, f. 462r; Rep. 21, f. 214r; Rep. 23, ff. 109v, 291r.
56 C.L.R.O., Rep. 22, f. 436r.
57 C.L.R.O., Rep. 24, f. 471v.
58 C.L.R.O., Rep. 24, f. 471r. 'It is ordered that the Leete Court Baron for the Lordship of Finsbury shalbe there kept on Tuesday 30th October.' C.L.R.O., Rep. 26i, f. 26v.
59 C.L.R.O., Rep. 24, f. iv.
60 Alphabetical extract of the Admissions of the Freeholders of the Manor of Finsbury 1550–1760. C.L.R.O., Misc. MSS, 43.4.
61 Stow, *Survey*, I, 302.
62 ibid., 301–2.
63 ibid., 302.
64 Stow, *Survey*, II, 79.

65 C.L.R.O., Rep. 17, f. 394v. William Denham goldsmith also secured a lease of a garden and a lodge for twenty one years (C.L.R.O., Rep. 19, f. 329v.)
66 Stow, *Survey*, II, 78.
67 C.L.R.O., Rep. 3, f. 71v.
68 C.L.R.O., Rep. 15f, 263v.
69 C.L.R.O., Rep. 3, f. 52r.
70 Betty Masters, *op. cit.*, 115.
71 ibid. 105.
72 ibid., 105.
73 ibid., 49.
74 C.L.R.O., Rep. 2, f. 173v.
75 C.L.R.O., Rep. 2, f. 180v.
76 C.L.R.O., Rep. 2, f. 192r. (These may have included the laying of water pipes ordered by the Court.)
77 C.L.R.O., Rep. 14, f. 9r.
78 C.L.R.O., Rep. 14, f. 13v.
79 C.L.R.O., Rep. 19, f. 5r.
80 C.L.R.O., Rep. 16, f. 447v.
81 Stow, *Survey*, II, 76.

V. THE BROADWAY CHAPEL, WESTMINSTER: A FORGOTTEN EXEMPLAR

By PETER GUILLERY

ON the north side of Victoria Street opposite Strutton Ground there is a small open space enclosed by dwarf brick walls. This was formerly the southern part of the burial ground to a church known as the Broadway Chapel. The remainder of the churchyard extended north to what is now the junction of Westminster Broadway and Caxton Street. The church was not always called the Broadway Chapel. While building in the 1630s it was the Tothill Fields Chapel, and from its opening in 1642 until well into the eighteenth century it was simply the New Chapel (Fig. 1). As a result of the timing of its completion it received no dedication, nor was it ever consecrated. The Broadway Chapel was demolished in 1841–3 and Poynter's Christ Church was built in its place.[1]

The Broadway Chapel was a remarkable building that reflected some of the contending forces at work on church architecture in the 1630s. Occupying a position close to the centre of London, it had an important and hitherto unacknowledged influence on post-Restoration church architecture. Before setting out the building history a description of the chapel is necessary to adumbrate this architectural interest.

There are several illustrations of the exterior. In 1817 Robert Blemmell Schnebbelie drew the chapel from two vantage points (Figs 2, 3).[2] It was a low brick building, with only slightly projecting transepts to disrupt a rectangular plan. The angles were quoined and the eaves had large shaped brackets. The round-headed doors at both ends of each of the long elevations had rusticated and pedimented surrounds.[3] As thus far described the exterior was simply

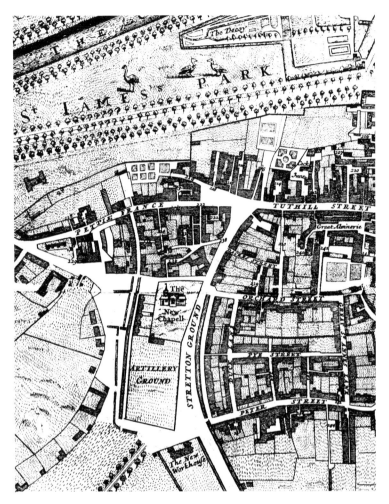

Fig. 1. 'The New Chapel', Westminster (William Morgan, *London etc. Actually Survey'd*, 1682).

but purely classical in its features. The windows were very obviously different. The larger windows at the centre of each elevation, despite unambiguously round heads, housed five lights of sub-Perpendicular tracery. The smaller, square-headed three-light windows could almost pass as conven-

A FORGOTTEN EXEMPLAR 99

Fig. 2. Broadway Chapel, view from south-west
(CWVL, Print Box 51, no. 23).

Fig. 3. Broadway Chapel, view from north-east
(CWVL, Print Box 51, no. 25 B).

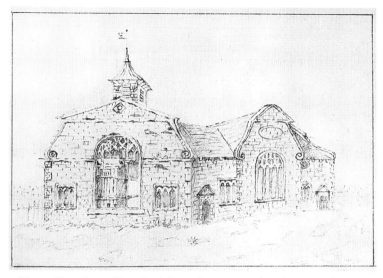

Fig. 4. Broadway Chapel, eighteenth-century view from south-west (CWVL, Print Box 51, no. 25A).

tionally Gothic. However, all of the windows seem to have had classical architraves. To the east of the transepts there were oval windows just below the eaves.[4] The roof of the building was tripartite with aisle ridges at a lower level than those of the nave and transepts. At the west end there was a timber turret.

There is another view of the chapel, a crude pencil sketch, captioned in ink, 'N.W. View of Broadway Chapel, Westmr.', in an eighteenth-century hand (Fig. 4). This shows the building with scrolled and pedimented 'Holborn' gables and a smaller turret. The existence of these gables can be confirmed by reference to the elevation of the chapel on Morgan's map (Fig. 1), and to an early-nineteenth-century account which relates that, 'The ancient finish to the gable is destroyed, and a modern coping substituted'.[5] The cutting down of the gables and rebuilding of the turret probably followed a 1792 recommendation that the chapel 'undergo a thorough Repair'.[6]

Following the jumbled and humble impression created by

A FORGOTTEN EXEMPLAR

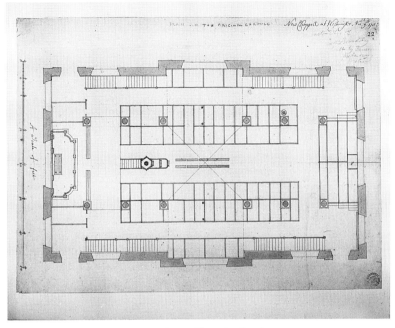

Fig. 5. Broadway Chapel, plan in 1711
(CWVL, Print Box 51, no. 26 A).

the exterior it comes as a surprise to find that the chapel, as recorded in the early eighteenth century, had an utterly coherent centralized plan (Fig. 5).[7] The dominant axis was east–west, but there was no chancel, nor even an eastwards emphasis in the architecture. Only the early-eighteenth-century fittings imply orientation, allowing the assumption that east is to the left on this plan. Twelve columns divided the building into a nave and aisles of seven bays, of which the midmost is widened to correspond to the transepts, and at the same time form a central square space. Crossed vaults are indicated. Within the rectangle of approximately 96 × 56 ft the central space was about 24 ft square; the other nave bays were 24 × 12 ft. The windows were about 8 ft and 12 ft wide and were arranged to fit the rhythm of the columniation. The internal arrangements and the shell were clearly conceived as one. There are no inherent reasons to doubt the authenticity of this as a Carolean plan. A case in Chancery in 1721 sets out the history of the building in some detail, but refers to no significant alterations between 1642 and 1711, the date of the plan.

It is less clear to what date the arrangement of the fittings in 1711 should be ascribed. The north and south galleries were evidently mid-to-late-seventeenth-century additions.[8] In the early nineteenth century the hexagonal pulpit immediately east of the crossing was described as original, but said to have been moved from the south-east crossing column. The reredos and the pews were said to be post-Restoration, perhaps datable to 1664 when, reportedly, a new communion table and rails were provided. In 1706 £100 was spent on 'adorning and beautifying' the chapel.[9]

The 1711 plan is a good indication of the interior of the chapel, but it is not wholly satisfactory as evidence. In 1829 Thomas Allen described the Broadway Chapel in some detail and his account is of great value as no internal views have come to light: 'The interior is not remarkable for decoration; it is made in breadth into a centre and side aisles by two rows of columns of an order between the Doric and Tuscan, six being disposed on each side of the central aisle, the intercolumniation in the middle answering to the transept being wider than the others; the columns sustain an entablature, which is broken at the transepts, and the cornice returned to the side walls. The ceiling of the central aisle and transepts is elliptically arched and groined at the intersection; the side aisles have plain horizontal ceilings'.[10]

There is a surviving London church to which Allen's internal description fits almost perfectly. This is the Church of St Matthias in Poplar, formerly known simply as the Poplar Chapel.[11] Long assumed to have been rebuilt in the late eighteenth century, there are, in fact, strong reasons to believe that this church retains its original form of 1642–54, when it was built by and for the inhabitants of Poplar and Blackwall. The foundations of the Poplar Chapel were apparently laid in 1642, the same year the Broadway Chapel was opened, but further work was not undertaken until 1652.

A FORGOTTEN EXEMPLAR

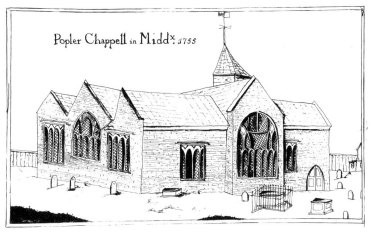

Fig. 6. Church of St Matthias, Poplar, view from north-east in 1755 (Oxford, Bodleian Library, Gough Maps 17, fol. 26^r [top]).

There is no documentary evidence for a rebuilding and the surviving fabric points to a seventeenth-century date. The walls are of two-inch bricks laid in English bond. The oak roof timbers of large scantling have carpenter's marks in undisrupted sequences. Internal columns are tenoned into, and must be integral with, this roof structure. There are suspended king-post trusses, a constructional type apparently introduced to England by Inigo Jones, yet the crossing timbers show a tentative approach to the problem of roofing a vaulted interior.[12]

Another crude mid-eighteenth-century sketch shows the exterior of the Poplar Chapel before various alterations (Fig. 6). The essential similarity to the Broadway Chapel is apparent. In certain respects the resemblance was stronger than this view suggests. The window openings at Poplar have been altered more than once, but straight joints show that the larger windows were originally round-headed. The unaccomplished eighteenth-century 'artist', understandably perhaps, imagined pointed arches over the Gothic tracery. The doors were probably round-headed and there were certainly angle quoins and eaves brackets similar to those shown at Westminster. These last two features survive.

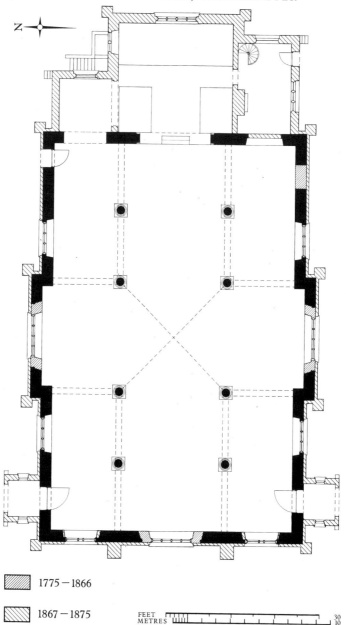

1775–1866

1867–1875

Fig. 7. Church of St Matthias, Poplar, plan (1642–54 building in solid black).

A FORGOTTEN EXEMPLAR 105

Fig. 8. Church of St Matthias, Poplar, interior,
view to west end in 1976.

The plan of the Poplar Chapel provides much more compelling evidence of the close relationship this building has to the Broadway Chapel (Fig. 7). Only the reduction in the number of columns from twelve to eight at Poplar distinguishes the two plans. The Poplar crossing is another 24 ft square.[13] The other nave bays are not double square

Fig. 9. Church of St Matthias, Poplar, interior, view to east end in 1987.

rectangles as at Westminster, but squares and a half of 24 × 16 ft.

There is every reason to believe that the Poplar Chapel was virtually a copy of the Broadway Chapel. Once this is accepted it becomes reasonable to examine the interior of the surviving building as a substitute for a depiction of the lost interior (Figs 8, 9). The impression of harmony given by the plans is borne out by the beautifully simple interior at Poplar, even in its present state of advanced dilapidation. As Allen observed at Westminster, the order is neither Doric nor Tuscan. The entablature indicates that the intended order was Tuscan. The use of Attic mouldings on Tuscan columns was widespread in the seventeenth century. The columns are on low plinths, not, as in post-Restoration churches, raised above pew level. With the exception of the central ceiling boss, which holds a form of the arms of the East India Company not in use after 1709,[14] and which was probably inserted sometime shortly after 1656, there are no traces of any seventeenth-century ornament or decoration.[15] This is a strikingly pure design which in its proportions and its simplicity is neither Mannerist nor Baroque. It merits the label 'Palladian'.

The Broadway Chapel was thus a centrally planned church with, by inference, a 'Palladian' Tuscan interior. At the same time it featured apparently incongruous Gothic tracery and 'Holborn' gables. Interpretation of this architectural curiosity must follow an account of the building history.

In the early seventeenth century the area south of St James's Park and west of Westminster Abbey was being developed at a rapid rate.[16] The land to the west of the Abbey was owned and leased by the College of Westminster. In 1626, following the plague of the previous year, the inhabitants of the parish of St Margaret's, Westminster, petitioned the Dean and Chapter for a new burial ground.[17] On 13 December 1626 a deed was signed conveying a plot of land at the north end of 'Tuttle Fields' from the College to the parishioners, for use as a churchyard for burial.[18] The ground had apparently already been enclosed by a brick wall, presumably erected in anticipation of the grant.[19] The day after the deed was signed the ground was consecrated by Dean John Williams.[20] At this stage there is no evidence of any intention to build a chapel on the site.

One of the Prebendaries at Westminster signatory to the 1626 deed was George Darell. He died in October 1631 and his will set out arrangements for property he held in Westminster.[21] Buildings that he believed would bring in about £40 per annum were to have their rents set aside for eight years, together with money accruing from the sale of his undisposed goods, to raise £400 towards building a chapel on 'Tuttle Fields' for 'the inhabitants of those parts of the Towne now utterly unprovided in the necessary groundes of religion'. This was apparently an intention of long standing, but Darell stipulated that if the project did not become a reality within ten years of his decease the money was to revert to his daughters. Sir Robert Pye, William Man, William Ireland and 'Mr. Tufton of the bowlinge alleye' were named as trustees for the £400.[22]

Sir Robert Pye (1584/5–1662) rose to prominence as a client of the Marquis of Buckingham.[23] From 1618 until 1655 he was an important civil servant, holding the appointment of Auditor of the Receipt of the Exchequer, which gave him a central role in the management of Government finance and, more particularly, in some of the administrative measures promoted by Archbishop Laud in the 1630s. Pye seems to have been a man with some independence of mind. In 1637–8 he clashed with Laud over the reorganization of Irish revenues. Despite his position at Court he was a Parliamentarian, a member of the Long Parliament from December 1640 until 1648.[24] In 1642 Pye was elected to the vestry of St Margaret's and became Chairman of the 'Committee for the College of Westminster', formed to stand in place of the Dean and Chapter.[25] His eldest son, also Robert, married the daughter of John Hampden, became a prominent Parliamentarian, and served as a Colonel under Essex.[26] The older Pye may have been inclined towards moderate support for Puritanism, even during the years of Laud's ascendancy.

Pye evidently had an interest in the arts; Ben Jonson said that he 'loved the Muses'.[27] With Philip Herbert, the Earl of Pembroke, Pye acted as a trustee for the young Duke of Buckingham in the 1630s. In 1635 the two men were joint signatories to an inventory of the Duke's collection of paintings.[28] In the following year (when Pembroke began rebuilding the south front of Wilton), Pye was employing Nicholas Stone to make two marble chimneypieces for his house in St Stephen's Court, Westminster,[29] occupied from 1618, *ex officio*.[30] More crucially for the subject in hand, in 1634 Pye was appointed to the sub-committee formed to oversee the repair of St Paul's Cathedral.[31] From the date of Darell's will right up to his death, Pye was closely involved with the Broadway Chapel. If any one was likely to have co-ordinated the building of the chapel it was Pye, for he clearly had the standing to do so effectively. Pye's youngest son Thomas set up in business as a merchant in the Levant and died in Syria in 1660.[32] Poplar was used as a London base by merchants travelling overseas and a Thomas Pye was a churchwarden for Poplar from 1641 to 1643.[33] This might well have been the same man. If so he would be a convenient medium for the transmission of the design of the Broadway Chapel to Poplar in 1642.

Two of the other trustees named by Darell are closely linked.[34] William Man was Surveyor to the Fabric of Westminster Abbey from about 1600, probably until his death in December 1635.[35] He was charged with rebuilding a prebendal house in 1631 and would certainly have carried out numerous repairs in and around the Abbey, but no evidence seems to survive to illustrate his accomplishments as an architect.[36] There is no reason to suspect that he was endowed with more than pedestrian skills. The same holds true for William Ireland, who succeeded Man as Surveyor to the Fabric, holding the post only until his own death in November 1638.[37] It would be premature to consider how these men might have contributed to the building of the Broadway Chapel before further explanation of the circumstances.

No steps towards building the chapel were taken immediately following Darell's bequest. The evidence for the next stage in the story comes from Chancery. In 1721 the inhabitants of the parish of St Margaret's, Westminster, took the Dean and Chapter of Westminster to law over rights to the site of the Broadway Chapel and to the nomination of the minister. The inhabitants' Bill of Complaint, in setting out the history of the building of the chapel, relates that in 1634 their predecessors applied to the Dean and Chapter for permission to build a church on the plot granted in 1626. The Dean and Chapter apparently considered that the proximity of the site to the Royal Palace obliged them to refer the matter to the Crown. The inhabitants petitioned the King citing the lack of space in St Margaret's owing to the increasing population and the fact that Darell's bequest would be lost if the chapel were not built before 1641. The question was then said to have been considered by the Archbishop of Canterbury, William Laud, and the Lord Privy Seal, Henry Montagu, first Earl of Manchester. They apparently examined witnesses before issuing a report which approved the building of the chapel, allegedly stating that, 'His Majesty was pleased therewith and would have meanes provided for him that should serve the place and to have the Church or Chaple erected and required to have the Work go then presently in hand'.[38] In support of this version of events the 1721 inhabitants produced as evidence the petition to the King and an endorsed reference signed by Laud and Manchester carrying the date 12 May, without record of the year.[39] As the process of petitioning commenced in 1634 (old style), the endorsement is probably datable to 1635.

The Dean and Chapter in 1721 claimed no knowledge of these events, saying that they were not recorded in any of their books or papers.[40] They invoked the excuse that during the Interregnum 'many of the Evidences and Writings relating to the Estates and Rights of the Dean and Chapter were lost or otherwise destroyed'.[41] Of course, this suited them well as acknowledgement of the inhabitants' right to build a chapel through Royal Licence would have weakened their case.[42]

Despite the denials of the Dean and Chapter and the

absence of an entry in the Privy Council Registers it is improbable that a new church so near the Palace would have been built without some form of Royal Licence.[43] In the 1630s the Privy Council was much exercised over the question of building control, particularly around the Palace. All new building in London was actively discouraged and Commissioners, including Inigo Jones, oversaw permitted buildings. The Broadway Chapel must have been approved by the Privy Council or it would not have been built.

Laud and Manchester's encouragement was heeded and work on the chapel apparently began within the year 1635.[44] According to Newcourt the shell of the building was complete in 1636.[45] Following the official prompting money was evidently found to enhance Darell's benefaction. Sir Robert Pye, as a particularly well-connected trustee for the original bequest, must have had a crucial role in raising money from amongst the wealthier inhabitants of Westminster, no doubt contributing something himself.[46]

Leaving the chapel in 1636 as a shell, a digression into the politics of Westminster Abbey will help to illuminate subsequent developments. In July 1637 Laud finally managed to have his opponent of long standing, Dean Williams, condemned for treason and imprisoned in the Tower, where he remained until November 1640.[47] Laud's supporters seized control of the College of Westminster during Williams's imprisonment. In November 1637 the Sub-Dean, Robert Newell, was given nominal charge of the College's affairs, but the real authority during the three-year interregnum lay with Prebendary Peter Heylyn, who was to become Laud's biographer.

John Williams's censure in 1637 stemmed from his anonymous publication of 'The Holy Table, Name and Thing', a seminal document in the altar wars of the 1630s. Williams was strongly opposed to Laud's innovations in liturgical practice, believing that during services the communion table should be positioned in such a way as to allow the minister to be audible. He believed that, 'if you mark the most part of the olde Churches in England, you shall plainly see, that the Chancells are but additions' and, ' a number of our old Churches have their Iles of such a perfect Crosse, that they cannot possibly see either high Altar or so much as the Chancell'.[48] There is no reason to believe that Williams's disagreements with Laud extended to contemporary church architecture. Lincoln College Chapel, Oxford, of 1629–31, is sufficient witness to this. Paid for by Williams, it is an entirely Gothic building that has generally passed as 'Laudian'. Nothing suggests that Williams was in any sense a patron of the Broadway Chapel, and it does seem clear that the College of Westminster had washed their hands of any formal responsibility for the building. As an advocate of toleration and conciliation Williams is unlikely to have had any desire to interfere with the building of the Broadway Chapel. Its centralized arrangement, well suited to the requirements of auditory worship, may even have consorted with his own beliefs. Laud's followers may have been less well disposed to such a plan, preferring something 'east and west, without tricks'.[49] They were certainly not averse to interference in church building.[50] It is thus not surprising to find that after 1637 there is evidence of Laudian meddling with work on the Broadway Chapel.

At his trial in 1644 Laud was charged with paying for a stained-glass window in the new chapel at Westminster. This apparently represented Pentecost.[51] Laud denied the charge, claiming to know nothing of the window. Adam Browne, a joiner whose career had a special attachment to Laud, was called as a witness, he having been employed in setting up the window, together with a Mr Sutton, the glazier.[52] Browne and Sutton explained that Robert Newell, not Laud, had given the order for the window, but at whose cost they knew not.[53] At some point between 1637 and 1640, the Broadway Chapel was adorned with at least one 'Laudian' window. The incongruous Gothic tracery can be explained by assuming that in 1637, with the walls up and architraved openings formed, plain glazing was intended but not yet put in place. Newell and Heylyn might then have interposed Browne to see to it that the building was given what they would have considered more comely windows, with full-blown Gothic tracery.

Adam Browne benefited from the Laudian interregnum at Westminster in so much as he was appointed to succeed William Ireland as Surveyor to the Fabric in December 1638, which post he retained until his death in 1655.[54] It is clear that Browne was working at the chapel prior to this appointment. In 1638 St Margaret's vestry paid out £200 for work on the chapel.[55] Adam Browne, joiner, received £75 8s 0d for 116 yards of 'Cornish', twelve 'Capitalls' at 4/- a piece, work towards the pews, and further unspecified work. Other payments were made to Thomas Hamond, carpenter (for 'the Compasse Cealing and Turrett'), to Phillip Lilly, plumber (for leading the turret), to Thomas Galloway, carpenter (for the plates for the pews), and to James Kitsyn, plasterer. This bill shows that work had progressed to the interior and roof by 1638. It is also further evidence that the interior described by Allen and reflected in Poplar was indeed original. The references to a 'compasse cealing' and to twelve capitals surely correspond to the building as described.[56]

As work on the pews had commenced by 1638 the building must have been approaching completion in 1639. The inhabitants had laid out £1,600 by this date, yet finishing was seen to require another £610.[57] On 10 July 1639 the inhabitants applied to the Dean and Chapter for power to levy a rate for the residue.[58] A month later Sir Robert Pye and others were granted permission to collect the money.[59] In 1641 a small part of this money was spent when Nicholas Stone was paid £10 for a black and white marble font.[60] This font now stands in St Margaret's, Westminster, at the west end of the south aisle, having been moved to the mother church when the Broadway Chapel was demolished. It is an elegant classical font and it must have been set off well by the interior of the chapel; it is a pity that it is all that survives.

The chapel was doubtless ready for use by the time the font was installed. However, it was not opened until after 9 December 1642 when Sir Robert Pye, now Chairman of the Committee for the College of Westminster, was ordered by the House of Commons to, 'cause the Doors of the new Church in Tothill Fields to be forthwith opened; to the End that People may resort thither to hear Prayers read, and Sermons preached, as to other Churches they may'.[61] The reason for the long delay, four years since the finishing touches were in hand, must have been the absence of provision for a minister.[62] On 18 December 1642 William Gouge, a prominent Puritan, gave the first sermon in the chapel.[63] The first minister was Herbert Palmer, another active Puritan, installed by Pye's Committee. His post was no doubt funded through the College estate.[64] Pye's personal connection with the chapel continued through the Interregnum.[65] In 1652 he gave eight houses in Petty France in trust to maintain the minister on the condition that he and his heirs held the right of nomination.[66] After this benefaction Pye had a gallery and several pews built to accommodate the tenants of his property.[67] By the Restoration the building had all but evolved into the proprietary chapel of a public servant. Pye was buried in the chapel in 1662.[68]

The initiative for the building of the Broadway Chapel lay with the inhabitants of Westminster. Archbishop Laud and Dean Williams both figure in the story, but neither in a demonstrably primary role. Sir Robert Pye appears to have acted as the principal agent for the inhabitants, though it is impossible to know how much he was representing a collective desire or how much indulging in personal patronage. Window tracery notwithstanding, the Broadway Chapel was conceived on a far higher plane than the numerous peculiar examples of architectural compromise and confusion typical of early-seventeenth-century 'Churchwarden's Gothic'. Even 'Laudian' churches in and around London like St Katharine Cree, old St Paul's, Hammersmith, and St John the Evangelist, Great Stanmore, do not seem as clearly expressive of a relationship between architecture and worship. In churches of this period the meaning of 'Italian' as opposed to 'old-fashioned' forms was not clearly defined. It would be quite wrong to imagine a battle of styles. Laud did not abjure classical forms, but in emphasising a return to axiality and pre-Reformation liturgical practice he seems to have found the revival of Gothic forms helpful. Very few of his contemporaries would have been more fastidious. Yet the Broadway Chapel does have a scent of the programmatic. Without wandering too far into baseless assertions as to how ideology or churchmanship might have found architectural expression in the 1630s, it can be suggested that the Broadway Chapel may have been designed for the inhabitants of Westminster along the lines of an imagined ideal form for a new Protestant church. In this respect St Paul, Covent Garden, is the only contemporary church to compare usefully. This introduces Inigo Jones, in terms of whom any sophisticated work of architecture in Carolean London must, in any case, be considered, whether or not he was the architect.

The characteristics of the Broadway Chapel that mark it out as more 'architectural' than most contemporary churches are the plan and an apparently systematic use of the Tuscan order. Sir John Summerson has discussed the conscious primitivism behind Jones's use of the Tuscan order at St Paul, Covent Garden, between 1631 and 1633, and in his restoration of St Paul's Cathedral from 1633 to 1641.[69] For Jones the Tuscan order was 'a plain, grave and humble manner of Building', 'retaining in it a Shew (as it were) of that first Face of Antiquity'.[70] Summerson's suggestion that the Tuscan order at St Paul, Covent Garden, reflects the fundamentalist views of the Puritanically inclined patron, the Earl of Bedford, has been sustained.[71] There is good reason to believe that the simple classicism of the Earl's church was intended to express Protestant fundamentalism and 'the "natural" beginnings of architecture, unalloyed by association with the corrupt mutations of the Roman church'.[72] Sir Robert Pye and the inhabitants of Westminster, perhaps holding St Paul, Covent Garden, in their minds, might well have wished their church to be similarly expressive.

Jones was not, of course, the only architect to use the Tuscan order in the early seventeenth century. It was employed widely, in many cases, as in the chapel at Lincoln's Inn, as if because its profile and mouldings resembled the familiar forms of the Perpendicular. The mouldings of the columns at Poplar are solecisms that Jones would not have committed, nor did he at Covent Garden, but no other English architect would have been so careful. The mouldings at Poplar compare very closely with those of the 'Tuscan' arcade of the Canterbury Quadrangle at St John's College, Oxford, where Adam Browne was employed, as well as with those of the lower 'Tuscan' order on the screen at Castle Ashby, attributed to Edward Carter, Jones's deputy at St Paul, Covent Garden, and St Paul's Cathedral. If we accept that the intention at Poplar was Tuscan, and continue to assume that the interior, with its trabeation and lack of ornament, is an accurate reflection of that at Westminster, then the Broadway Chapel represents an exceptionally pure and thorough attempt to use the Tuscan order as more than decoration.

On closer examination the exterior of the Broadway Chapel can be read as having been conceived, though not built, in much the same spirit as the interior. That is with the same drive towards Tuscan simplicity exemplified much more obviously by the exterior, and in particular the portico, of St Paul, Covent Garden. In both buildings large expanses of plain brick wall are broken up by angle quoining. At the Broadway Chapel the symmetrical equilibrium of the whole is clearly marked by the placement of the window and door openings. The early-nineteenth-century drawings show all the windows in architraves without impost mouldings. Jones used this sort of architrave at Covent Garden as well as in the 'Tuscan' part of his restoration of St Paul's Cathedral.[73] The door surrounds of the Broadway Chapel were also very simple. The well-known doorway of 1633 at St Helen, Bishopsgate, gives a measure of the distance between the classicism of the Broadway Chapel and contemporary mannerist design. Much closer in type were the gateways that flanked the portico of St Paul, Covent Garden (Fig. 10).[74] The Broadway Chapel door surrounds are even simpler, approaching the very basic form of the rusticated gateway designed for an entrance to the Palace Yard from Cannon Row, Westminster, in 1634.[75]

A FORGOTTEN EXEMPLAR 117

Fig. 10. Eastern gateway to Churchyard of St Paul, Covent Garden, demolished.

The parts of the Broadway Chapel that do not fit neatly into the 'Tuscan' schema cannot be overlooked. The elliptically vaulted ceiling is a feature that has nothing to do with St Paul, Covent Garden. It must relate to the ceiling of similar section at the Queen's Chapel at St James's Palace, the first truly classical church in England, which antedates the Broadway Chapel by just a decade. The carpenter for the ceiling of the Broadway Chapel, Thomas Hamond, worked at St James's Palace. In 1632-3 Hamond was making oak pedestals for Charles I's new sculpture gallery in Jones's Tuscan colonnade.[76] Perhaps he was also employed at the Queen's Chapel in the 1620s.[77] As he made the ceiling, Hamond might also have been responsible for erecting the roof trusses of the Broadway Chapel. These will very probably have had a form analogous to those of suspended king-post type at Poplar. The Queen's Chapel also utilizes this previously foreign constructional form.[78]

Other features of the Broadway Chapel cannot be explained by such direct reference to Jones. The popularity of 'Holborn' gables in the 1620s and 1630s, as at Raynham Hall and Swakeleys, may derive from Jones's work at Sir Fulke Greville's house of 1617-19. However, these gables are not a form that Jones himself would have used in the 1630s. The finishing of the chapel gables would probably have had to await completion of the roof in 1638. Their form might then have been dictated by Adam Browne, an architect-craftsman, like any other practising at this date, not attuned to Jones's fully developed classical style. The gables might have replaced an earlier intention or solved an as yet unsolved problem. It is difficult to see how projecting eaves *à la* Covent Garden might have been suitably adapted to the triple pitch of the end elevations. On the long elevations, in place of the severe mutules used at Covent Garden, there were shaped brackets. These have a profile that Jones did use, for example to support the internal gallery at the Banqueting House, but they are inappropriate in terms of 'Tuscan' purity and should more probably be associated with mannerist usage, as at Baulms House or Balls Park. Placement of the eaves brackets would have followed construction of the roof, so they too can be interpreted as part of Browne's contribution. The window tracery is the most obvious irregularity in terms of a coherent classical programme. Unless it is read as a remarkably persistent case of Gothic Survival it can, as already suggested, be attributed to Laudian meddling not dissimilar to that which forced alterations to the eastern entrance at St Paul, Covent Garden.[79] Externally the 'Tuscan' and 'non-Tuscan' work can be divided as that complete by 1636 and that done later. However, it would be unduly mechanistic to insist that the Tuscan features imply that the Gothic tracery and 'Holborn' gables must be alterations to a pure original scheme. The possibility that the building was stylistically incoherent from the start cannot be ruled out. Even so the strength of the Tuscan element is remarkable. However imperfectly realized, the Broadway Chapel does appear to have been conceived with a Tuscan, and, by implication, fundamentalist, programme.

Galleries are often considered an essential part of a Protestant church. However, neither the Broadway and Poplar Chapels, nor St Paul, Covent Garden, were built with galleries. These churches all had galleries inserted before the end of the seventeenth century by which time new churches were being built with galleries.[80] The mid-seventeenth-century congregations were presumably provided with adequate seating within earshot of the pulpit, giving the architects freedom to design uncluttered interior spaces.

The plan of the Broadway Chapel is its most astonishing feature. It was centralized in so much as it had bilateral symmetry on both north–south and east–west axes. It was not a proper quincunx, or cross-in-square plan, but rather a cross-in-rectangle. The symbolic resonance of the Tuscan order aside, the arrangement does seem to reflect the requirement of Reformed worship that a church should serve as a lecture room. With the exception of St Paul, Covent Garden, there were no churches in London built for Protestant worship, and there were effectively no useful models in the rest of the Kingdom. The Earl of Leicester's great but uncompleted Puritan church at Denbigh of 1578–84 had Tuscan columns, and its rectangular plan may have had a proportional base, but it has no meaningful relationship to the Broadway Chapel.[81] The kirk at Burntisland in Fife of 1592 seems to be the only earlier British centrally planned Protestant church.[82] Again it cannot be sensibly related to the Broadway Chapel.

Comparable plans can be sought in the Netherlands. The Dutch built new churches in the early seventeenth century that clearly reflected the requirements of a Calvinist society. These were buildings of a chaste and solemn character made up of pure, geometrical forms, with pulpits that were clearly visible and audible, where possible without resort to galleries. Hendrick de Keyser, Master Mason and Sculptor to the City of Amsterdam, had his works engraved and published in 1631 by Salomon de Bray in *Architectura Moderna*. This was available in England and was probably used as a pattern book by educated builders/surveyors. Nicholas Stone had been an apprentice to de Keyser and married his daughter. De Keyser's sons worked in England from time to time as artisans.[83] The plans of three churches illustrated in *Architectura Moderna* can be usefully compared to the Broadway Chapel. The Zuiderkerk of 1603-11 is a simple six-bay rectangle of nave and aisles with Tuscan arcades and parallel barrel vaults. Some lateral emphasis is given in bays two and five by cross vaults in the aisles, taller windows and external gables. The Westerkerk of 1621-31 is similar but the lateral emphasis of bays two and five is more pronounced as separately vaulted 'transepts' cause cross vaults to interrupt the barrel vault of the nave. The Noorderkerk of 1620-3 has a truly centralized plan, a Greek cross with barrel vaults crossed at the centre on complex 'Doric' piers. None of de Keyser's churches provides a good model for the plan of the Broadway Chapel. They have in common a general centralizing tendency with crossed nave and 'transept' vaults. As to elevational detail de Keyser's churches are more Mannerist than Palladian. The Dutch interior that is most like the Broadway Chapel's is in Jacob van Campen's Nieuwe Kerk at Haarlem of 1645-9. This has a quincunx plan and crossed vaults with flat ceiled corners. The date of this building rules it out as a model, but since it is often quoted as an influence on Wren the similarity is worth noting.

If the architect of the Broadway Chapel was looking at

A FORGOTTEN EXEMPLAR

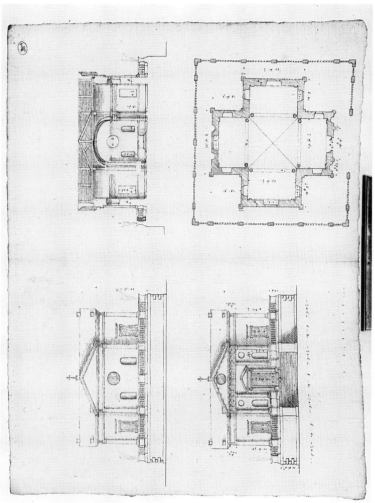

Fig. 11. Plan, elevations, and section of a small centralized church, drawings attributed to John Webb (Worcester College, Oxford, Drawing nos. 146A–D) [Gotch 1/38A]).

Architectura Moderna it was with an eye not for detail but for planning, and even then not with strictly imitative intentions. The Dutch connection could be a chimera. The plan may derive its architectural authority from Inigo Jones's own quest for antique authenticity. Amongst John Webb's drawings at Worcester College dated to *c.* 1635 to *c.* 1650 there is a design for a church that falls typologically between St Paul, Covent Garden, and the Broadway Chapel. It has a Greek cross plan, elliptical vaults crossed at the centre, and Tuscan detailing that owes nothing to the Dutch (Fig. 11).[84] St Paul, Covent Garden, is, of course, evidence that Jones was able, even eager, to design in a mode suited to Puritan requirements. It is not necessary to believe the description of Jones as 'Puritanissimo Fiero', quoted by Wittkower, to accept that he designed what is as near as we know to a 'Puritan' church.[85] Italian Renaissance church planning would have been tainted in Reformed eyes. If the spirit of primitivism that may have determined the choice of the Tuscan order at the Broadway Chapel prevailed, the plan may represent an attempt to look beyond Serlio and Palladio, back to early Christian antiquity. It might, however indirectly, reflect Jones's ideas about ancient church or temple plans. Perhaps the centralization owes something to Jones's study of Stonehenge, a monument that he believed to be a Roman temple of the Tuscan order built on a carefully proportioned circular plan.[86]

The Broadway Chapel was a remarkable architectural achievement for Carolean England. It offered an elegant solution to one of the most vexing problems for seventeenth-century European church architects. It retained a clear east–west axis within a boldly centralized plan. This was done with modest means and, as perhaps initially intended, with a restrained and purely classical vocabulary. The clarity and proportions of the plan, with the use of the Tuscan order, have pointed irresistibly to Inigo Jones. Jones cannot, of course, be held responsible for the chapel as built. Even taking into account the inferred departures from the first design, the building as a whole, most notably in its external elevations, is simply not 'Italian' enough to be attributed to

Jones. The Broadway Chapel may well fall into the category of buildings postulated by John Newman as, 'a significant intermediate class to some extent designed or remotely controlled by [Jones]'.[87] As Surveyor of the King's Works Jones had much to do with the implementation of building controls, particularly near Royal residences. The Privy Council relied heavily on his advice before making decisions on building. Approval of the Broadway Chapel can have come only after consultation with Jones.[88] When the Crown 'required to have the Work go then presently in hand' a call to Jones might have been the natural course. He was, however, very busy in the mid 1630s. If he was asked to prepare plans for the chapel in 1635 he might well have been obliged to delegate control of the building, possibly equipping a subordinate with a ground plan.

Buildings of this period were often products of collaborative design. It would be fatuous to look for a single architect for the Broadway Chapel. However, it does seem likely that between Jones and the craftsmen a surveyor of considerable sophistication was involved. In looking for a substitute for Jones the obvious candidate is John Webb. In 1635 Webb was still in his early twenties and unlikely to have had much independence, let alone experience, as a surveyor. He was Jones's Clerk Engrosser, but probably no more, for the work at St Paul's Cathedral after 1633.[89] His theoretical drawings that do relate loosely to the Broadway Chapel are too late in date to suggest more in this context than that Webb might have been responding to existing work. If it is impossible to credit the Broadway Chapel to Jones it is equally impossible to credit Webb with such an early departure from Jones. Nicholas Stone had an intimate knowledge of de Keyser's work and he did work for Pye in 1636. However, there is nothing in Stone's Account Book or Note Book to indicate that he was involved with the Broadway Chapel in any way other than as sculptor of the font. If the building were attributed to Stone the interior and most of the external features would stand apart in his known architectural work as showing an unlooked for fluency in Jonesian classicism.

The same is true, but with more force, for Adam Browne,

William Man and William Ireland. As Browne was setting up the windows as well as carrying out the internal joinery perhaps he was acting as surveyor from 1637–8. There is no absolute reason to exclude him from involvement at an earlier date but Laud's and Heylyn's manoeuvres do provide an explanation for his arrival at this point. As a craftsman Browne practised in a busy mannerist style. His limitations have been lucidly defined in the context of his work at the Canterbury Quadrangle in the early 1630s.[90] The fundamentalist character of the Tuscan interior which, as Allen put it, was 'not remarkable for ornament', cannot plausibly be ascribed to Laud's joiner, even if it were accepted that he had made an extraordinary leap in his understanding of classicism. In the normal course of building affairs William Man, a trustee for the original bequest and the Abbey Surveyor, might well have expected to supervise the work. However, a new church was a rare opportunity. It is difficult to believe that Man or William Ireland, who succeeded him in 1635, conceived a building as architecturally radical as the Broadway Chapel. Perhaps they were responsible for managing the building work prior to 1637.[91]

Another candidate for the role of surveyor from 1635–7 presents himself in the person of Edward Carter, the man who in 1643 succeeded Jones as Surveyor of the King's Works. Carter was older than Webb, and less dependent on Jones. As Jones's deputy at St Paul, Covent Garden, and at St Paul's Cathedral Carter had a central role in matters of church architecture in London in the mid-1630s. It is also clear that, in later life at least, he had Puritan sympathies.[92] A further tenuous link comes through the East India Company which seems to have employed Carter at Poplar in 1627, remodelling the house that fifteen years later gave part of its grounds for the Poplar Chapel.[93] Perhaps Carter is more likely to have been an architect of the Broadway Chapel than Webb or Stone, but all three are similar in that, as architects practising in the 1630s, so little is known of them that they can, in any case, only be described as shadows of Jones.

London's church architecture of the 1630s has been seen as a flowering without fruit. Inigo Jones's two great projects, St Paul, Covent Garden, and the restoration of St Paul's Cathedral, have quite properly been hailed as *tours de force* amongst otherwise inchoate attempts to redefine English church architecture following the Reformation. The enforced stop to almost all new building through the 1640s and 1650s, together with the artistic distance between Jones and his contemporaries, left the promise of these works unfulfilled. With the exception of the inspiration given to Wren by Jones's great west portico to St Paul's Cathedral, the work of the 1630s has seemed to bear little direct relationship to church building after the Restoration. The identifiable influence of the Broadway Chapel makes it necessary to modify this perception.

St Paul, Covent Garden, is remarkable for the unwavering strength of its classicism. The Broadway Chapel as built was less severe, and perhaps for that reason more acceptable as a model. Pye family connections as much as fashion or ideological concerns may explain the replication of the Broadway Chapel at Poplar. After the Restoration the Broadway and Poplar Chapels had a distinct impact on church architecture in London. There are two churches in east London that are clearly derivative. St Mary Magdalene, Bermondsey, of 1675–9 is attributed to Charles Stanton of Southwark (Fig. 12).[94] The exterior, now altered, was plain, with round-headed windows and quoins. There is a cross-in-rectangle plan with projecting transepts and a chancel that is an addition. Eight Doric/Tuscan columns support a trabeated and incorrect entablature. Elliptical vaults cross at the centre and the aisles are flat-ceiled. This seems to be an artisan's interpretation of the familiar interior. The second east London church is St Nicholas, Deptford. This is also attributed to Stanton and is dated to 1696–7.[95] It too is a brick church with round-headed windows and quoins to imply transepts with 'Holborn' gables. Without its medieval tower it has a cross-in-rectangle plan with a very shallow chancel. Although extensively rebuilt after bomb damage the

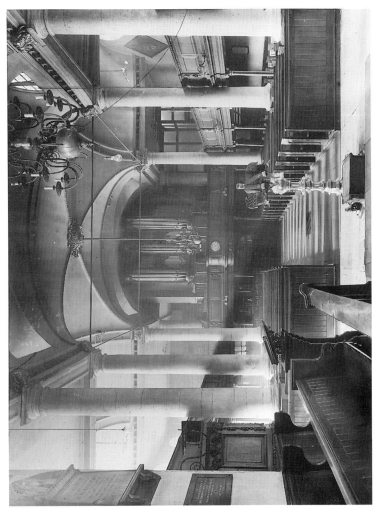

Fig. 12. Church of St Mary Magdalene, Bermondsey, interior, view to west end.

interior is of the same type with six Doric/Tuscan columns, trabeation and elliptical vaults crossed to the west of centre. The architecture of the Bermondsey and Deptford churches has been ascribed to the influence of Wren's post-Fire City churches, although there is no single church of Wren's that serves as a close model. Wren did not use Doric or Tuscan orders in the manner seen in these churches and his vaulted ceilings were rarely as shallow as those at Poplar, Bermondsey and Deptford. Stanton must have been looking not at Wren, but at the Poplar and Broadway Chapels.

Much of what has traditionally been termed 'Dutch' influence in Wren's churches may have been much closer at hand. Summerson has suggested that 'the simple, vernacular terms of many of the City Churches derive from ... the quasi-Tuscan of old St Paul's'.[96] The exteriors of St Benet, Paul's Wharf, and St Mary-le-Bow, for example, illustrate this point, but more strongly with additional reference to the Broadway Chapel. The quincunx plans with crossed vaults of St Mary-at-Hill, St Martin, Ludgate, and St Anne and St Agnes do recall the Nieuwe Kerk at Haarlem, but these churches, with St James, Garlickhythe, are also reminiscent of the Broadway Chapel, not precisely, but one does not expect imitation from Wren. It is unnecessary to have Wren looking to Holland when there was such a good model in London.

Some of Hawksmoor's early-eighteenth-century work contains echoes of the mid-seventeenth-century churches. His unexecuted project for a 'Basilica after the Primitive Christians' proposed a church square on plan with shallow north and south projections.[97] The plan of the main body of St George-in-the-East, particularly the positioning of the doorways, is also comparable. More strikingly, the original interior of this church was an elliptically vaulted cross-in-rectangle on Doric columns. St Anne, Limehouse, has a plan and ceiling that imply a cross-in-rectangle. Even Archer's church plans are similarly based on intersecting axes, and follow the Broadway Chapel in harmoniously combining centralization with the axiality required by the Church. Through the Broadway Chapel, the Jonesian Renaissance provided an exemplar for much of London's best church planning of the late seventeenth and early eighteenth centuries.

This paper is the product of work done for the Survey of London (R.C.H.M.E.) arising out of research for the forthcoming Poplar volume. I am grateful to the General Editor for allowing me to pursue and publish this material, and to my colleagues at the Survey, and elsewhere in the Commission, for directing me to numerous sources, commenting on drafts of the paper, and helping me in many other ways. I am specially indebted to Stephen Croad for drawing my attention to the Broadway Chapel in the first place. I wish also to express thanks to Neil Burton, whose work on St Matthias gave me a jump start, to John Newman for his encouragement and criticism, and to Howard Colvin and Kerry Downes, who kindly read and commented on the text. The illustrations are reproduced by permission as follows: Figs 2–5, Westminster City Archives; Fig. 6, The Bodleian Library, Oxford; Figs 7, 9, 12, Royal Commission on the Historical Monuments of England; Fig. 8, Greater London Photograph Library; Fig. 10, The Marquess of Tavistock and the Trustees of the Bedford Estates; Fig. 11, The Provost and Fellows of Worcester College, Oxford, and the Conway Library, Courtauld Institute of Art.

1 Christ Church was demolished in 1954.
2 The drawings and a watercolour are in the City of Westminster Victoria Library Archives (Print Box 51, nos. 23, 24 and 25B). Another related watercolour is at the Guildhall Library (Pr. W2/BROA/way).
3 The north-east pediment was broken by a tablet said to commemorate Thomas Jekyll, d. 1698 (Westminster Abbey Muniments, 64263).
4 These windows were probably inserted later in the seventeenth century when galleries were added, and were perhaps also intended to improve the light on the pulpit.
5 Thomas Allen, *The History and Antiquities of London, Westminster, Southwark and Parts Adjacent*, iv (1829), p. 216.
6 The advice of James Wyatt, Surveyor to the Fabric at Westminster Abbey, from where the chapel was then maintained (WAM, 57090). I am grateful to Mr H. M. Colvin for this reference.

7 CWVL, Print Box 51, no. 26A, plan titled 'New Chappell at Westminster, Nov. 9, 1711'. The 1711 plan was almost certainly made in conjunction with the investigations of the Commissioners for Building Fifty New Churches, perhaps by Nicholas Hawksmoor or William Dickinson. Consideration was given to assigning the chapel its own district before the decision to build St John, Smith Square, was taken. On 16 Nov. 1711 the minister and trustees of 'the new chapel in Tothill Fields' represented that their chapel could hold 1,200 and that it was fit to be made parochial (E. G. W. Bill, *The Queen Anne Churches* [1979], p. 66).
8 WAM, 24789.
9 M. E. C. Wallcott, *Memorials of Westminster* (1849), p. 288; T. Allen, op. cit. (n. 5), p. 217.
10 T. Allen, op. cit. (n. 5), p. 217.
11 For a full and annotated discussion of this building see *The Survey of London*, xliii, forthcoming.
12 D. Yeomans, 'Inigo Jones's Roof Structures', *Architectural History*, xxix (1986), pp. 85–101.
13 In fact it is almost cubic as the height to the apex of the vault is 25 ft.
14 I am grateful to Mr R. Baldwin for this information.
15 The ribs in the vaults are a Victorian insertion.
16 It has been estimated that as many as 200 new houses were built in Westminster in the 1630s (N. G. Brett-James, *The Growth of Stuart London* [1933], ch. 6).
17 Public Record Office, C11/2687/41.
18 CWVL, E3304.
19 PRO, C11/2687/41. In 1627 Walter Hall, bricklayer, was paid £165. 6s. 7d by St Margaret's vestry for building the churchyard wall (J. E. Smith, *A Catalogue of Westminster Records* [1900], p. 55).
20 PRO, C11/2687/41.
21 PRO, PCC 135 St John.
22 The executor of Darell's will was his brother, Sir Marmaduke Darell, Cofferer of the King's Household, who died a few months later. His elaborate tomb is in the Church of St James, Fulmer, Bucks., a simple brick church which he had built in 1610.
23 DNB; G. E. Aylmer, *The King's Servants: The Civil Service of Charles I* (1961), pp. 195–6, 308–13, 338.
24 M. F. Keeler, *The Long Parliament, 1640–1641: a Biographical Study of its Members* (1954), p. 317.
25 CWVL, E2413, f. 16; H. F. Westlake, *St. Margaret's Westminster* (1914), p. 133.
26 DNB; Sir Sampson Darell, son of Sir Marmaduke, also married into the Hampden family.
27 Quoted in DNB.
28 *Calendar of State Papers, Domestic Series, 1635–1636*, p. 342; British Library Add. MS 18,914, f. 2. Pye was apparently a supporter of

Pembroke's in Parliament and later one of his executors (*Cal. S. P., Dom., 1652–1653*, p. 47).
29 W. L. Spiers, ed., 'The Note-book and Account Book of Nicholas Stone', *The Walpole Society*, vii (1918–19), pp. 104–5.
30 *Cal. S. P., Dom., 1635–1636*, p. 4. The family's status in the area is commemorated by Old Pye Street.
31 *Cal. S. P., Dom., 1634–1635*, p. 16.
32 Aylmer, op. cit. (n. 23); PRO, PCC 49 May.
33 G. W. Hill and W. H. Frere, ed., *Memorials of Stepney Parish*, (1890–91), pp. 169, 176. This Pye is said to have died in 1658 or 1659. Uncertainty about this date may indicate that it is unreliable. A death in Syria might well have been inaccurately reported/recorded.
34 Mr Tufton was probably Richard Tufton, who died in the same month as Darell, and who is said to have built the street in Westminster that still carries his name (M. E. C. Wallcott, op. cit. [n. 9], p. 322).
35 WAM, 40959–64; PRO, PCC 128 Sadler.
36 WAM, *Chapter Minutes 1609–1642*, f. 51v.
37 WAM, 9906–7 and Surveyor cards; PRO, PCC 178 Lee. Ireland seems to have been a somewhat wealthier man who leased the 'Mason's Lodge' on the north side of the Abbey from 1626.
38 PRO, C11/2687/41.
39 PRO, C33/337, f. 163v.
40 PRO, C11/2687/41. This does appear to be the case.
41 Ibid.
42 The Dean and Chapter won the case, largely on the strength of their undisputed freehold, but the right of the parishioners to use the building for worship was protected (PRO, C33/337, f. 164r).
43 In 1708 Newcourt stated that the chapel had been built with a Licence under the Privy Seal (R. Newcourt, *Repertorium Ecclesiasticum Parochiale Londinense*, i, 2 [1708], p. 923).
44 WAM, 24804, 'Memoranda relating to the building of the New Chappell', n.d.; this is probably a rehearsal of the inhabitants', 1721 Bill of Complaint.
45 R. Newcourt, op. cit. (n. 43), p. 923.
46 Newcourt credits Pye with a contribution of £500, but this may be confused with a rate levied in 1639 (ibid., p. 722). Pye was undoubtedly a wealthy man. In 1639 he lent £2,000 to the King (M. F. Keeler, op. cit. [n. 24]).
47 For the background to this antagonism see Williams's entry in the DNB and E. Carpenter, ed., *A House of Kings* (1972), ch. 9.
48 'The Holy Table, Name and Thing' (1637), pp. 20, 224.
49 From Laud's directions for the new church in Hammersmith in a letter to the Earl of Musgrave in 1629 (J. Bliss, ed., *The Works of William Laud*, vii [1857], p. 27).
50 *The Survey of London*, xxxvi (1968), pp. 66–70, 98–103.
51 J. Bliss, ed., op. cit. (n. 49), iv (1854), p. 228. This window and others

in the chapel were destroyed in 1643, or thereabouts, under the supervision of a Parliamentary Committee; 'When the Rebellion broke out, Sir Robert Harley defaced the windows, laid the painted glass in heaps upon the ground, and trod it to pieces, calling his sacrilegious antics "dancing a jig to Laud"'. The replacement south window contained the arms of Sir William Wheeler and the date 1649 (M. E. C. Wallcott, op. cit. [n. 9], p. 288). Wheeler was an associate of Sir Robert Pye's in the Exchequer and with Pye was amongst those purged from Parliament in 1648 (D. Underdown, *Pride's Purge: Politics in the Puritan Revolution* [1971], pp. 50, 168, 252).

52 J. Bliss, op. cit. (n. 49), iv, p. 228; H. M. Colvin, *Biographical Dictionary* (1978).

53 It was later stated that Laud contributed £1,000 towards the building of the chapel as well as the stained-glass windows (D. Defoe, *A Tour thro' London about the year 1725*). No evidence for this has been found and it is difficult to reconcile such a large sum with what is known about the funding of the chapel. Laud's 1635 endorsement and exhortation to raise funds together with the charge laid at his trial may have been misinterpreted as an outright contribution.

54 H. M. Colvin, op. cit. (n. 52).

55 CWVL, E21, p. 70.

56 Four shillings seems a small price for a column capital. The payment might have been for something less weighty, such as wainscot ornament. I am grateful to Mr J. Newman for this suggestion.

57 PRO, C11/2687/41. This adds up to a credible sum for the building of a church at this date. The Poplar Chapel cost just over £2,000 to build. It is possible that the total of £2,210 excludes the cost of the traceried windows and that these were indeed paid for either by Laud or by the College of Westminster.

58 Ibid.

59 PRO, C33/337, f. 163v; WAM, 24804.

60 Entry in Account Book dated 16 April 1641 (W. L. Spiers, ed., op. cit. [n. 29], pp. 78–9, 129); A. C. Fryer, 'Fonts Sculptured by Nicholas Stone', *Archaeological Journal*, lxx (1913), pp. 137–40.

61 *Journals of the House of Commons*, ii (1640–2), p. 882.

62 In 1643 a petition to the Crown for the consecration of the chapel was deferred until an endowment was provided for the incumbent (M. E. C. Wallcott, op. cit. [n. 9], p. 287).

63 R. Newcourt, op. cit. (n. 43), p. 923; DNB.

64 WAM, 24789; R. L. Greaves and R. Zaller, *Biographical Dictionary of British Radicals in the Seventeenth Century*.

65 In 1651 a grant of £30 was given to repair damage to the churchyard, where Scottish prisoners had been kept (*Cal. S. P., Dom.*, 1651–1652, pp. 18, 584).

66 PRO, C11/2687/41; R. Newcourt, op. cit. (n. 43), p. 923. This right was overturned in 1721.

67 WAM, 24789.
68 T. Wotton, *The English Baronetage*, i (1727), pp. 414–15.
69 Sir J. Summerson, 'Lecture on a Master Mind: Inigo Jones', *Proceedings of the British Academy*, L (1964), pp. 169–92.
70 Inigo Jones's notes as published by John Webb, *The most notable Antiquity of Great Britain, Vulgarly called Stone-Heng on Salisbury Plain, Restored* (1655), pp. 68, 102.
71 *The Survey of London*, xxxvi, op. cit. (n. 50).
72 Sir J. Summerson, *Architecture in Britain: 1530 to 1830* (1970), p. 136.
73 Sir J. Summerson, *Proc. British Academy*, op. cit. (n. 69); *Survey of London*, xxxvi, op. cit. (n. 50).
74 These were freely adapted from the Tuscan order at the Roman amphitheatre at Verona, as transmitted to Jones through Serlio and Palladio (Sir John Summerson, op. cit. [n. 69], p. 177).
75 J. Harris, ed., *RIBA Drawings Collection Catalogue: Inigo Jones and John Webb* (1972), no. 10; J. Harris and G. Higgott, ed., *Inigo Jones: Complete Architectural Drawings* (1989), pp. 124–43.
76 H. M. Colvin, ed., *History of the King's Work's*, iv (1982), p. 251.
77 Hamond was, it seems, apprenticed in 1581 at the age of 19 and was thus quite elderly by the 1630s (Guildhall Library, MF 451/2).
78 D. Yeomans, op. cit. (n. 12).
79 *The Survey of London*, xxxvi, op. cit. (n. 50). Another contemporary parallel for modifications to a church design comes from the rebuilding work carried out at the City church of St Michael-le-Querne from 1637 to 1640 where Jones's designs were probably altered by Laud and Juxon, the parishioners, and Peter Mills, the City Bricklayer (H. M. Colvin, 'Inigo Jones and the Church of St. Michael-le-Querne', *The London Journal*, xii, 1 [1986], pp. 36–9).
80 *The Survey of London*, xxxvi and xliii, op. cit. (n. 50 and n. 11).
81 L. Butler, 'Leicester's church, Denbigh: An Experiment in Puritan Worship', *Journal of the British Archaeological Association*, 3rd series, xxxvii (1974), pp. 40–62.
82 G. Hay, *The Architecture of Scottish Post Reformation Churches* (1957), pp. 32–4.
83 W. Kuyper, *Dutch Classicist Architecture* (1980), pp. 6–56.
84 J. Harris and A. A. Tait, ed., *Catalogue of the Drawings by Inigo Jones, John Webb and Isaac de Caus at Worcester College, Oxford*, 1979, Plate 109, Drawing nos. 146A–D (Gotch 1/38A); M. Whinney, 'Some Church Designs by John Webb', *Journal of the Warburg and Courtauld Institutes*, vi (1943), pp. 142–50. I am grateful to Dr John Bold for sharing his views on the dating of these drawings.
85 R. Wittkower, 'Inigo Jones – Puritanissimo Fiero', *The Burlington Magazine*, xc (Feb. 1948), pp. 50–1.
86 Inigo Jones's notes as published by John Webb, op. cit. (n. 70).
87 J. Newman, 'Nicholas Stone's Goldsmith's Hall', *Architectural History*, xiv (1971), pp. 30–9.

88 H. M. Colvin, ed., *The History of the King's Works*, iii, 1 (1975), pp. 139–49.
89 'Inigo Jones and St. Paul's Cathedral', *London Topographical Record*, xviii (1942), pp. 41–6.
90 H. M. Colvin, *The Canterbury Quadrangle, St. John's College, Oxford* (1988), pp. 45–6.
91 Walter Hall, the bricklayer who had built the churchyard wall in 1626, received regular payments from the Dean and Chapter for running repairs through the 1630s (WAM, 41768, 41889, etc.). He may have built the walls of the chapel.
92 H. M. Colvin, *Biographical Dictionary* (1978).
93 *The Survey of London*, xliii, op. cit. (n. 11).
94 B. Cherry and N. Pevsner, *The Buildings of England : London 2 : South* (1983), p. 599.
95 Ibid., p. 402.
96 Sir J. Summerson, *Proc. British Academy*, op. cit. (n. 69), p. 191.
97 K. Downes, *Hawksmoor* (1979), pp. 161–6.

VI. SOME NOTES ON HOLLAR'S *PROSPECT OF LONDON AND WESTMINSTER TAKEN FROM LAMBETH*

By PETER JACKSON

AT some time before the Great Fire, Hollar etched the four plates which the LTS have reproduced as publication no. 138.

It is immediately apparent that the impressions of the four plates in their original state, before alteration, are roughly printed, dirty and under-inked. They are also unfinished. The captioning is incomplete and only the first half of the word 'LONDON' appears in the sky above the City with the 'DON' yet to be added on plate 4.

These are obviously only proof impressions.

They are, however, all that appear to exist of this pre-Fire state, which suggests that the print was never published. One explanation for this is that Hollar produced the view immediately before September 1666 when the Great Fire made the City skyline obsolete. He then set it aside with the intention of bringing it up to date, but never did.

This original version was lost for many years. It was unknown to A. M. Hind when he wrote his book on Hollar's views of London in 1922.[1] Parthey[2] had presumably seen it since he included it in his standard catalogue of Hollar's works. 'This view,' Hind wrote, 'is described by Parthey as *London before the Fire* but I have only seen impressions showing the new St Paul's Cathedral.'

It was acquired by the British Museum in 1926 and was registered as 1926-6-17-10 (1–4), i.e. on 17 June 1926. It appears to have come directly from Lieut. Col. G. A. Cardew, C.M.G., of Stafford, Dorchester.[3] How it came into his possession is not known, nor do we know its provenance between 1853 when Parthey saw it and 1926 when Lt-Col. Cardew owned it.

The fate of the four copper-plates, however, is clearer. The up-dating that Hollar had failed to do after the Fire was done by some unknown hand when the plates were acquired by an eighteenth-century publisher, probably Henry Overton. There can be no doubt that the later, reworked impression dates from 1707. St Paul's Cathedral is shown with only one of its towers, the clock tower on the south west corner. It is known that the carpenters began to frame the roof of this tower soon after the 1707 building season opened and did not start work on the other tower until 1708.[4]

Altering copper-plates was not an uncommon practice. The best-known examples of this are the plates of squares by Sutton Nicholls which are such an attractive feature of the 1754 edition of Strype's Stowe. Most of these had been brought up to date since they were first engraved in about 1728 and published in *London Described* (1731). The one showing the most alteration is that of Leicester Square. In the original state, the garden in the centre is square and laid out into four lawns planted with trees and flower beds with a single tree in the middle. In the altered state, all the trees and flower beds have gone and the equestrian statue of George I appears in the centre. The square has become octagonal, gates have been altered and oil-lamps added to the railings. Savile House on the north side has been given an additional storey and a shop on the south-west corner has been removed completely.

The labour involved in alterations of this extent is enormous. First, the copper around each redundant line or area has to be scraped down to the depth to which it had been engraved. Then the scraped-away area has to be raised to plate surface level by hammering it from the back, after which the area is burnished smooth ready for re-engraving. The same method also applies to etching, of course.

In the case of Hollar's view, a crude and unsubtle hand has been at work. Apart from removing old St Paul's, there is little sign of scraping or burnishing; so delicate was Hollar's etching that the re-toucher has found it unnecessary. He has merely engraved his heavy-handed lines on top of the existing etched lines.

Most of the alterations and additions are self-explanatory when comparing the two states but some of them illustrate interesting topographical points.

Plate 1

Peterborough House is named only on the second state. The house, which had been begun by Alexander Davies and not completed until after his death in 1665, had no name when Hollar drew it.[5] Around 1673 it was leased to Lord Peterborough and was thus known as Peterborough House by the time of the second state.

Plate 2

In the second state, Highgate is identified and the hills of Hampstead make their first appearance above Westminster Hall. Some new churches have been added quite skilfully and the references 13 (Salisbury House), 14 (Worcester House) and 17 (Arundel House) in the strip below have been erased.

Plate 3

State two of this plate shows the most re-working. New St Paul's and eight Wren spires have been added but much of the new engraving is merely a crude attempt to emphasize buildings which Hollar had delicately etched already. Salisbury, Worcester and Arundel Houses are still standing unaltered though they had all been demolished by 1707 and their names had been deleted from the identification strip on Plate 2, State 2.

It has been suggested that 'Theator' indicated by the letter 'l' is the Blackfriars.[6] However, Blackfriar's Playhouse was 'pulled downe to the ground on Munday the 6 day of August 1655 and tenements built in the rome'.[7] The

'Theator' is more likely to be the Dorset Gardens or Duke's Theatre which stood until 1709.[8]

The identity number 28, which, on state one indicated 'St Martin's in Thames Street', is transferred in the second state to a new, crudely inserted, church tower on the south bank. This, although its identity is left blank in the key strip, must represent Christ Church, Lambeth.

Plate 4

In the second state, eleven churches, the Monument, the Royal Exchange and the Tower Armoury have been added quite carefully and no attempt has been made to intensify Hollar's existing etching.

1. A. M. Hind, *Wenceslaus Hollar and his Views of London and Windsor in the Seventeenth Century* (London, 1922).
2. Gustav Parthey, *Kurzes Verzeichniss der Hollarschen Kupferstiche*. Berlin (1853).
3. Information kindly supplied by Miss K. J. Wallace, Archivist, British Museum.
4. Jane Lane, *Rebuilding St Paul's after the Great Fire of London* (Oxford, 1956).
5. Charles T. Gatty, *Mary Davies and the Manor of Ebury* (London, 1921).
6. Richard Pennington, *A Descriptive Catalogue of the Etched Work of Wenceslaus Hollar* (Cambridge, 1982).
7. A note written in Sir Thomas Phillipps's copy of the 1631 edition of Stow's *Annals* quoted in Erwin Smith, *Shakespeare's Blackfriars Playhouse* (New York University Press, 1964).
8. City Corporation plaque on the site.

VII. GUIDE BOOKS TO LONDON BEFORE 1800: A SURVEY

By DAVID WEBB

As a class of literature, the guide book to London was a relative latecomer, in comparison with those for other European capitals. In his bibliography of the guide books to Rome,[1] Schudt lists no fewer than 81 editions of the *Indulgentiae* (lists of relics in churches), in six different languages, published before 1550; by 1563 the *Indulgentiae* had been joined to the *Mirabilia* (the legends and miracles attached to the historical sites) to produce the first true guide to the City of Rome. In London, the earliest guides concentrated on particular buildings – Westminster Abbey,[2] or St Paul's Cathedral[3] – though even here little attempt was made to chronicle other than the most important monuments. There are no substantial works on the Tower, nor the Royal Exchange, nor the Palace of Whitehall, to name but a few of the most obvious, before 1700; as for the City of London as a whole, the earliest true guide book was published as late as 1681. Occasional intrepid travellers from the continent recorded their impressions of London throughout the late sixteenth and seventeenth centuries;[4] beyond them loomed the gigantic shadow of John Stow.

It seems likely that the Great Fire of 1666 had an indirect influence on the publication of London's first guide books. The massive work of rebuilding thrust London into the forefront of European architecture, while the ending of the Dutch wars in the 1670s had appreciably reduced the threats to the traveller's safety.

If (the foreign traveller) is young and has some knowledge of English, he will do well to consider "The art of living in London", wherein he will find guidance how he may avoid the harpies and tricksters who prey upon his like. Of guide-books there are few. But if his visit occurs towards the close of the century, he will find information of a somewhat arbitrary kind, it is true, in Chamberlayne's *Angliae Notitia* (1669)... Moreover there are many little books of voyages in French and German which will tell him something of the life, sights, and scenery of England, not to mention Camden and Blome's large descriptive works each styled *Britannia*.[5]

When Nathaniel Crouch published his *Historical Remarques and Observations of the Ancient and Present State of London and Westminster...with an account of the most remarkable accidents...till the year 1681*, at his own expense, under the pseudonym of 'Richard Burton' in 1681, he can have had no idea that his was the first true guide book to London. The text is a reworking of Stow, with an admixture of Howell's *Londinopolis* (1657); the information is brief and scrappy (apart from an inordinately long description of the Tower); there is a total lack of order in the presentation, and the prevailing tone is heavily moralistic. John Dunton got Crouch about right[6] '...(he) prints nothing but what is very useful and very diverting', but he 'has got a habit of leering under his hat'. Crouch's real contribution to guide book history was the pocket format size of his publication which, despite its poverty-stricken pedigree, had achieved five editions by 1703, and was even updated by an unknown continuator as late as 1730 (as '*New View and Observations on the Ancient and Present State of London and Westminster*'), several years after Crouch's death. In fact, Crouch's claim to guide book prominence would probably have been disputed by Thomas de Laune, who brought out a very similar publication, also in a pocket format in the same year, 1681: *The Present State of London : or, Memorials Comprehending a Full and Succinct Account of the Ancient and Modern State thereof*, which achieved a second edition in 1690 as *Angliae Metropolis*. De Laune's little volume, however, has more of the flavour of an almanac than a guide book (he also cribs from Stow), and undoubtedly derives from Chamberlayne.

Burton/Crouch and De Laune had both managed to squeeze into their guides a few very small, poorly engraved woodcuts, some featuring the coats of arms of the livery companies. But neither of them included a map – and indeed, it would be 75 years before a guide included such an obvious concomitant; perhaps the difficulty in producing a map on such a small scale (even if folded) would have considerably increased the price. Until the middle of the eighteenth century, most guides were content to rely wholly on the text – none more so than the most remarkable of the late seventeenth-century productions, and the first to be published by a foreigner for the use of foreigners: *Le Guide de Londres pour les Estrangers. Dédié et Offert aux Voyageurs Allemands et François*, by François Colsoni, in 1693. Despite his French-sounding name, Colsoni was actually an Italian – Francesco Casparo Colsoni – probably a schoolteacher, whose first published work[7] was a kind of school textbook. The first edition is extremely rare[8], and was reprinted by the London Topographical Society in 1951, with commentary by Walter Godfrey.[9] It had reached a third edition by 1697 (the second edition is not known to exist[10]), and this was further reprinted in 1710. Colsoni was the first author to realise the value of the personal recommendation as well as the usefulness of a suggested selection of sights for successive days. Colsoni arranges his material in the shape of five tours, with further notes on the suburbs. Despite his packed itinerary (even the last edition barely reaches 64 pages), Colsoni may be seen as the ancestor of the *London in X days* guide book, today virtually the standard type. He cannot resist taking a swipe at the English language on behalf of his tourist readership: 'Throughmorten-Street, qu'on prononce Tragmarten-Strit'! Though the guide was clearly successful, three quarters of a century passed before any further attempts were made on behalf of French visitors. In the meantime, an enterprising Hamburg publisher had issued the first guide book to London in German, *Die Sehen-Wurdigkeiten der Welt Beruhmten Stadt Londen in Engelland*, in 1706, by an author identified solely as M.V., to be followed twenty years later by Johann Basilius Kuechelbecker's *Der nach Engelland Reisende Curieuse Passagier, oder Kurche Beschreibung der Stadt Londen und derer Umliegenden Verter*, which achieved a second edition in 1736. This bulky and detailed work continually quotes from the French even listing recommended reading in that language; it would seem likely to be a translation from the French, possibly by way of Guy Miège's *The New State of England* (seventeen editions, 1691–1748). There is certainly no direct evidence that either M.V. or Kuechelbecker ever set foot in London.

The English guide books to London during the first half of the eighteenth century departed little from the model set by Crouch. They consist for the most part of a history of London followed by various classified lists of buildings. Occasionally, a concession was made to possible foreign visitors with a text in English and another language on alternative pages; *The Foreigner's Guide; or a necessary and instructive companion both for the foreigner and native, in their tour through the cities of London and Westminster* first appeared in 1729, and had reached a fourth edition by 1763. Here English alternated with French; the author may have been one of the publishers, John Pote. An amusing aside in the second edition of 1730, in the section on diversions, notes 'Without money, man is a nobody at Tunbridge'! Two other guides first appeared in the 1720s: *The Antiquities of London and Westminster*, by the lexicographer, Nathan Bailey, which ran through three editions (1722–34), and in the same year, William Stow's *Remarks on London: being an Exact Survey of the Cities of London and Westminster*, which failed to progress beyond the edition of 1722. The two guides share a certain indefinable connection (they certainly shared the same publisher, H. Tracy), though Stow, who is not known to be a descendant of John Stow, claims in his preface that the 'like never before extant', a somewhat disingenuous comment especially as he goes on 'this piece is to show people how to spell and write proper their superscription on letters...'. Stow's guide has the distinct overtones of an early directory, though the earliest edition of Kent's series[11] dates only from 1732. Certainly Stow's attempts at tabulation (lists of fairs, of market towns, members of Parliament, stage coaches, etc.) lead naturally on to the use of statistics by the Company of Parish Clerks in their *New Remarks of London: or, a Survey of the Cities of*

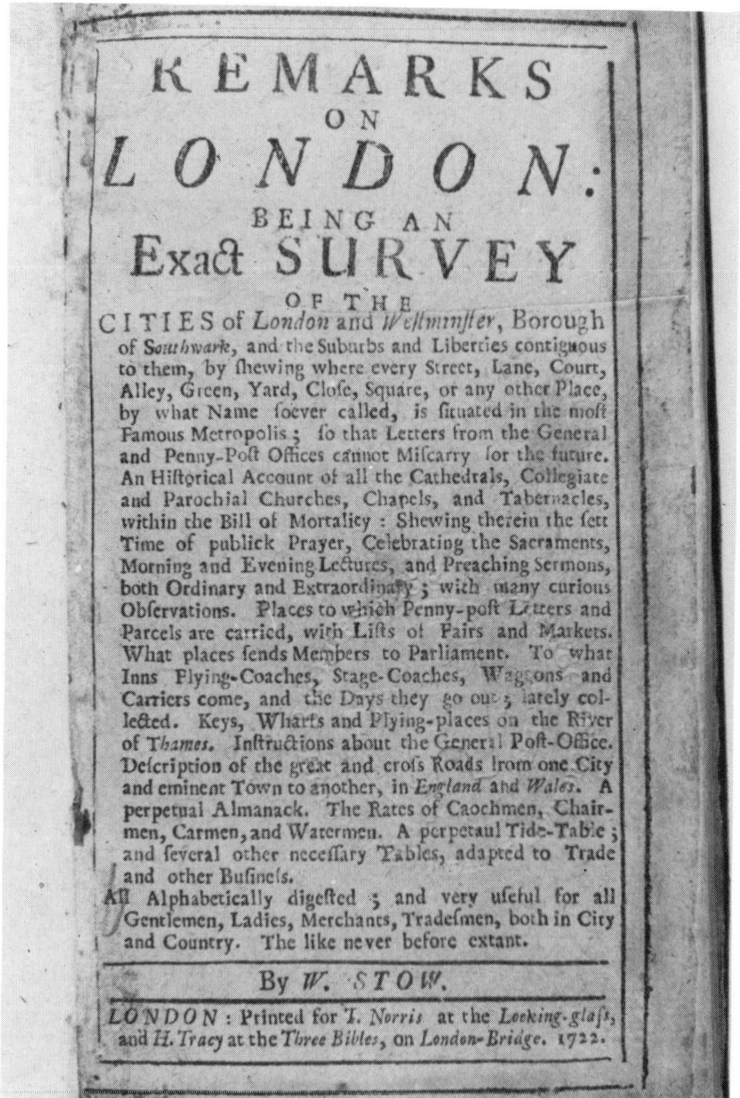

REMARKS ON LONDON:

BEING AN Exact SURVEY

OF THE
CITIES of *London* and *Westminster*, Borough of *Southwark*, and the Suburbs and Liberties contiguous to them, by shewing where every Street, Lane, Court, Alley, Green, Yard, Close, Square, or any other Place, by what Name soever called, is situated in the most Famous Metropolis; so that Letters from the General and Penny-Post Offices cannot Miscarry for the future. An Historical Account of all the Cathedrals, Collegiate and Parochial Churches, Chapels, and Tabernacles, within the Bill of Mortality: Shewing therein the sett Time of publick Prayer, Celebrating the Sacraments, Morning and Evening Lectures, and Preaching Sermons, both Ordinary and Extraordinary; with many curious Observations. Places to which Penny-post Letters and Parcels are carried, with Lists of Fairs and Markets. What places sends Members to Parliament. To what Inns Flying-Coaches, Stage-Coaches, Waggons and Carriers come, and the Days they go out; lately collected. Keys, Wharfs and Plying-places on the River of *Thames*. Instructions about the General Post-Office. Description of the great and cross Roads from one City and eminent Town to another, in *England* and *Wales*. A perpetual Almanack. The Rates of Coachmen, Chairmen, Carmen, and Watermen. A perpetaul Tide-Table; and several other necessary Tables, adapted to Trade and other Business.
All Alphabetically digested; and very useful for all Gentlemen, Ladies, Merchants, Tradesmen, both in City and Country. The like never before extant.

By *W. STOW*.

LONDON: Printed for *T. Norris* at the *Looking-glass*, and *H. Tracy* at the *Three Bibles*, on *London-Bridge*. 1722.

Fig. 1.

London and Westminster, of Southwark, and part of Middlesex and Surrey, within the Circumference of the Bills of Mortality, published in 1732. The unnamed editor notes in the preface that 'I spent above six months in making my collection', and the end result is part guide, part directory. It seems a pity that the Company never managed to follow up with any further editions. One interesting new departure is the use of a symbol to indicate a particular recurring feature to which the compiler wishes to draw attention. In this instance each street is so furnished, in order that its particular Liberty may be ascertained.

The period around the middle of the eighteenth century was rather fallow in the production of guide books. *The Pocket Remembrancer; or a Concise History of the City of London* appeared in 1741, a kind a crib for City dignitaries; and in 1755 two impressions were issued of *London in Miniature*, arranged as a series of walks up to 40 miles around 'intended as a complete guide for foreigners', as the lengthy subtitle indicates. Although several of the earlier guides had managed to issue occasional revised editions, it was not until well after 1760 that the first of the multi-edition guides began to appear. On a strict chronological basis, the first of these was published in 1767: A *Companion to Every Place of Curiosity and Entertainment in and about London and Westminster*; although several editions have not survived, and the title changed very slightly over the years, it had reached the tenth edition by 1803. The Companion included very long descriptions of the major buildings, dismissing the rest of the sights in a few pages – a trait it shared with a contemporary, David Henry's *Historical Account of the Curiosities of London and Westminster, in 3 parts* (eventually four parts), which also ran through at least ten editions between 1753 and 1805, but which concentrated almost entirely on the Tower, Westminster Abbey and St Paul's, at least in the earlier editions. The editions at the end of the century had begun to include detailed descriptions of all other major London buildings; John Harris reduced it to a single volume for the last edition in 1805. The arrangement of the Companion is rather haphazard; the preface notes 'We have avoided all useless ceremony in conducting you from place to place'. The 1783 edition included a map for the first time – though clearly an out-of-date map was used,[12] as it only showed buildings to 1769! The eighth edition, of 1795, is the first London guide to include advertisements – in this instance, for Delarue's Incomparable Liquid for cleaning and preserving all kinds of house paint, as well as for Simson's Infallible Aethereal Tincture for the Toothache. Both of these preparations were available from the publisher, John Drew, at one shilling a bottle. By the time of the tenth and last edition in 1803, it seems that the anonymous compiler was feeling worried about the sales of his production, since he added on an abridged version of Thomas Pennant's celebrated *Account of London*, already in its fourth edition. It is unlikely that the abridgement was authorised, and poor Pennant himself received no benefit, as he had died in 1798. Moreover, this was not the first occasion on which Pennant's work had been 're-arranged' as a guide; John Wallis had prepared an earlier version in 1790, and this guide's subsequent career paralleled the editions of the history.

The real trouble with multi-edition guides lay in the expense of constant revision. It was unusual for the original editor/publisher to sustain his initial enthusiasm beyond the first few editions, after which the labour of updating soon developed into a chore. New editions appeared at increasingly infrequent intervals; out-of-date information was served up to readers with bland unconcern or complete ignorance. Before the moribund guide finally expired of neglect and old age, a new editor might well be brought in – perhaps a well-known topographer, whose prominently-displayed name might be expected to produce extra sales. All of these points can be demonstrated in the career of the Companion's most successful rival at the end of the eighteenth century, the Ambulator.

The Ambulator; or, the Stranger's Companion in a Tour round London, within the Circuit of Twenty-five Miles, which first appeared in 1774 compiled by its publisher John Bew, was the earliest guide to use an almost entirely alphabetical approach. The expense of resetting each item for every new edition was enormous, and Bew resorted to corrigenda sheets, inserted either at the front or back, as well as increasing the scope of the preliminary description of London to take into account any new development. Bew even advertised[13] for the nobility and gentry to 'favour him with a communication of whatever mistakes may have been discovered in the former edition, and of what further alterations and additions it may be necessary to make in the next, in consequence of the improvements and embellishments any seats may lately have received'. Bew's death in 1793 led to an unseemly power struggle for the rights to his guide between his widow, Jane, and his successors, Scatcherd and Whitaker, both of whom brought out rival editions in that same year. Mrs Bew appears to have lost out in the end, as all later editions, up to the twelfth in 1820, were brought out by the firm of Scatcherd, under various imprints. Only four editions appeared between 1800 and 1820, this last edition being completely revised by the antiquarian Edward Wedlake Brayley, who not only changed the title (*London and its Environs; or, the General Ambulator*), but also permitted himself a lengthy moan about the iniquities of the Copyright Act, which required the deposit of eleven copies of each publication: '(this) can be viewed in no other light than as legalised robbery'. The Ambulator's last gasp was issued in 1824, under the title *The Stranger's Guide; or, New Ambulator...*; Brayley was not involved in this tail end of a once famous series.

The publisher John Wilkie mined the first gems from a vein much exploited by nineteenth-century guide publishers – the same guide published under two completely distinct titles. *A Brief Description of the Cities of London and Westminster* of 1776 is the same in all respects as *The London and Westminster Guide, through the Cities and Suburbs* of the same year, even down to the inclusion of ten pages of cautions allegedly penned by Sir John Fielding. In its latter incarnation, it had originally been published under a different imprint eight years earlier. The 'trick' was repeated at the end of the century, when *A New Picture of London; or, the Strangers' and Foreigners' Correct Guide...* of 1791, was reprinted as *Kearslys' Stranger's Guide, or Companion through London and Westminster* in 1793 completely unaltered apart from the maps, which curiously were updated. Presumably the Kearsly brothers had acquired the copyrights of the earlier publisher, John Wilson.

In the late 1780s, further efforts were made to attract foreign visitors to London with the issue of guides in two languages, modelled very much after *The Foreigner's Guide* of half a century earlier. John (i.e. Giovanni) Mazzinghi's *New and Universal Guide through the Cities of London and Westminster*...(1785, with later editions in 1792 and 1793), was paralleled almost exactly by that issued by the print dealer, Samuel Fores: *Fores' New Guide for Foreigners, containing the most Complete and Accurate Description of the Cities of London and Westminster, and their Environs*, issued ironically just as the French Revolution broke out. Like the *Foreigner's Guide*, each is in English and French on alternate pages. Only Fores' guide possesses a map – but only Mazzinghi's guide contains a dedication to Charles James Fox! For sheer novelty value, however, it would be hard to surpass the amazing guide issued by that late Georgian literary lion, the Rev. Dr John Trusler, eccentric clergyman, promoter of the Literary Society (intended to abolish publishers), successful sermon deviser and remainder publisher. Trusler's compilation of 1786, *The London Adviser and Guide : containing every Instruction and Information Useful and Necessary to Persons Living in London and Coming to Reside there* was aimed, almost for the first time, not at foreign visitors (whom he detested), but at country bumpkins at large in the wicked city, an easy prey to the villains lying in wait. Trusler dispenses with long historical descriptions and involved itineraries in favour of practical advice: 'Do not walk under a pent-house, lest persons watering flower pots or other slops, should drip upon your head'; 'If you walk with an umbrella, and meet a similar machine, lower yours in time, lest you either break it, or get entangled with the other'; 'Never stop in a crowd in the streets, to see what occasions it: if you do, it is two to one, but you either lose your watch or your pocket handkerchief'; The man who

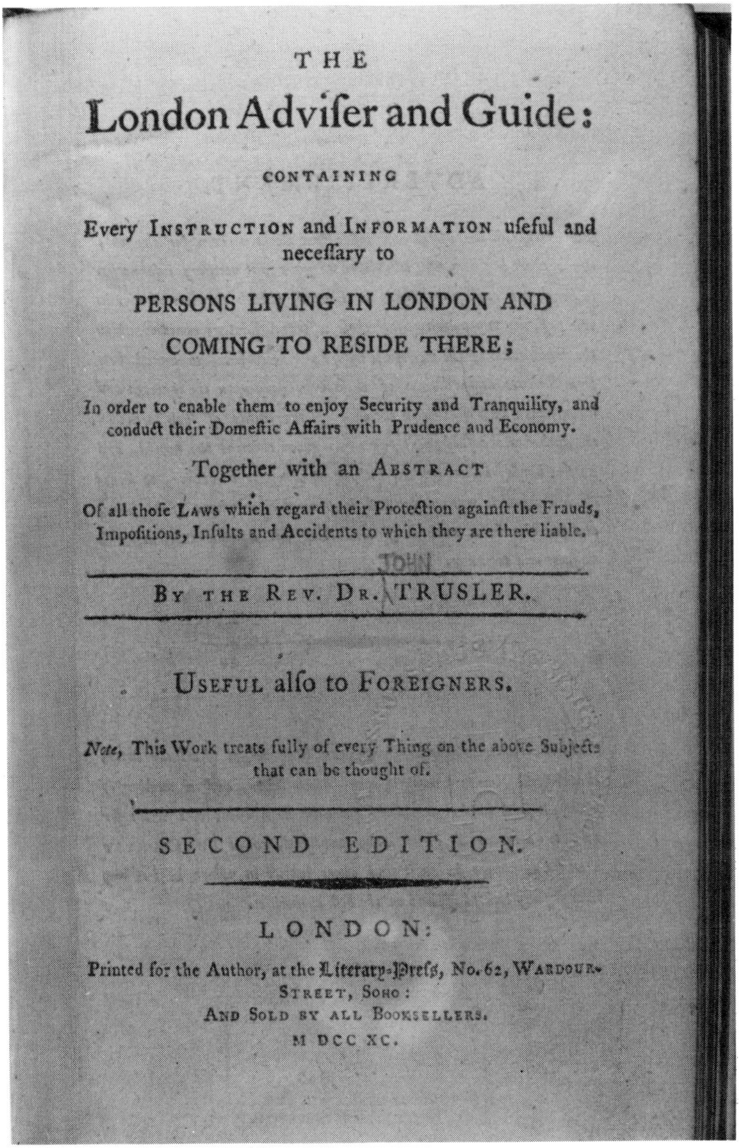

THE

London Adviser and Guide:

CONTAINING

Every INSTRUCTION and INFORMATION useful and neceſſary to

PERSONS LIVING IN LONDON AND COMING TO RESIDE THERE;

In order to enable them to enjoy Security and Tranquility, and conduct their Domeſtic Affairs with Prudence and Economy.

Together with an ABSTRACT

Of all thoſe LAWS which regard their Protection againſt the Frauds, Impoſitions, Inſults and Accidents to which they are there liable.

BY THE REV. DR. TRUSLER.

USEFUL alſo to FOREIGNERS.

Note, This Work treats fully of every Thing on the above Subjects that can be thought of.

SECOND EDITION.

LONDON:

Printed for the Author, at the Literary-Preſs, No. 62, WARDOUR-STREET, SOHO:
AND SOLD BY ALL BOOKSELLERS.
M DCC XC.

Fig. 2.

shows (the Westminster Abbey waxworks) will ask for a few halfpence for himself; but this is optional'; 'The festivity and gambols of the lower class of people: rolling down Greenwich Park Hill, White Monday and Tuesday; occasional floating through the atmosphere in balloons' (a very early reference to the last named). Though the style of these cautions sounds unbelievably stilted 200 years later, at the time Trusler's guide was popular enough to achieve a second edition in 1790.

Apart from John Wallis's abridgements of Pennant, which are referred to above, the end of the eighteenth century petered out with an early example of a class of London guide which was to become common within the next thirty years – the pseudo-guide, detailing the sordid side of London life, and culminating in quasi-pornography. John Roach was an appropriate compiler for such a publication, having been jailed for a year for publishing an immoral book. His *London Pocket Pilot, or Stranger's Guide through the Metropolis* first appeared in 1793, and managed a second edition in 1796. Roach concentrates almost exclusively on places of amusement, inns, coffee houses, pleasure gardens, etc, with lip-smacking anecdotes interspersed with the basic topographical details. Ahead lies Pierce Egan and Corinthian London; behind, the sub-literature of titles such as *Tricks of the Town Laid Open*, from the 1720s.

While the native offerings in the London guide book trade declined before the advent of Richard Phillips' *Picture of London* which was first published in 1802, and hence outside the scope of this survey, those published abroad were altogether of a higher quality. Ignoring, perhaps, Georges Louis Lerouge's *Curiosités de Londres et de l'Angleterre* (1765, with later editions in 1766 and 1770), which was merely a French translation of *The Foreigner's Guide* of 1763,[14] noted above, by far the best written is that by the police commissioner, François Lacombe, under the title *Observations sur Londre* [sic] *et ses Environs, avec un Précis de la Constitution de l'Angleterre, et de sa Décadence, par un Athéronome de Berne*, in 1777. There were at least six editions, the last appearing in 1784, under the title *Tableau*

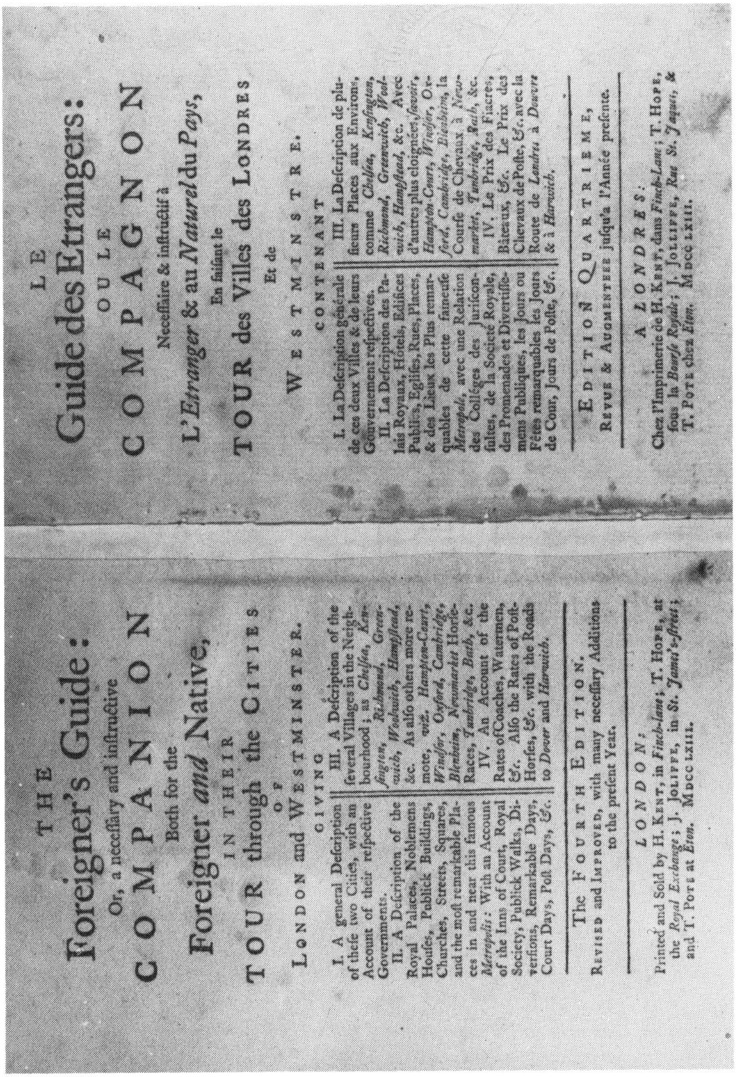

THE
Foreigner's Guide:
Or, a necessary and instructive
COMPANION
Both for the
Foreigner *and* Native,
IN THEIR
TOUR through the CITIES
OF
LONDON and WESTMINSTER.
GIVING

I. A general Description of these two Cities, with an Account of their respective Governments.
II. A Description of the Royal Palaces, Noblemens Houses, Publick Buildings, Churches, Streets, Squares, and the most remarkable Places in and near this famous Metropolis: With an Account of the Inns of Court, Royal Society, Publick Walks, Diversions, Remarkable Days, Court Days, Post Days, &c.
III. A Description of the several Villages in the Neighbourhood; as *Chelsea, Kensington, Richmond, Greenwich, Woolwich, Hampstead,* &c. As also others more remote, *viz. Hampton-Court, Windsor, Oxford, Cambridge, Blenheim, Newmarket Horse-Races, Tunbridge, Bath,* &c.
IV. An Account of the Rates of Coaches, Watermen, &c. Also the Rates of Post-Horses, &c. with the Roads to *Dover* and *Harwich*.

The FOURTH EDITION,
REVISED and IMPROVED, with many necessary Additions to the present Year.

LONDON:
Printed and Sold by H. KENT, in *Finch-Lane*; T. HOPE, at the *Royal Exchange*; J. JOLLIFFE, in *St. James's-street*; and T. POTE at *Eton*. M DCC LXIII.

LE
Guide des Etrangers:
OU LE
COMPAGNON
Necessaire & instructif à
L'*Etranger* & au *Naturel* du *Pays*,
En faisant le
TOUR des Villes des LONDRES
Et de
WESTMINSTRE.
CONTENANT

I. La Description générale de ces deux Villes & de leurs Gouvernemens respectives.
II. La Description des Palais Royaux, Hôtels, Edifices Publics, Eglises, Rues, Places, & des Lieux les Plus remarquables de cette fameuse *Métropole*, avec une Relation des Collèges des Jurisconsultes, de la Société Royale, des Promenades et Divertissemens Publiques, les Jours ou Fêtes remarquables les Jours de Cour, Jours de Poste, &c.
III. La Description de plusieurs Places aux Environs, comme *Chelsea, Kensington, Richmond, Greenwich, Woolwich, Hampstead,* &c. Avec d'autres plus eloignées, *sçavoir, Hampton-Court, Windsor, Oxford, Cambridge, Blenheim, la Course de Chevaux à Newmarket, Tunbridge, Bath,* &c.
IV. Le Prix des Fiacres, Bateaux, &c. Le Prix des Chevaux de Poste, &c. avec la Route de *Londres* à *Douvre* & à *Harwich*.

EDITION QUARTRIEME,
REVUE & AUGMENTEE jusqu'à l'Année présente.

A LONDRES:
Chez l'Imprimerie de H. KENT, dans *Finch-Lane*; T. HOPE, sous la *Bourse Royale*; J. JOLLIFFE, *Rue St. Jaques*, & T. POTE chez *Eton*. M DCC LXIII.

Fig. 3.

de Londres. The cynical Lacombe used two prefaces: 'Aux Bretons (s'il y en a encore); au lecteur impartial (si j'en ai)', and found space, in between the architectural descriptions and the history, to tweak the noses of his Londoners: 'L'usage immodéré du thé parmi le peuple et le bourgeois, me fait croire qu'il passionne beaucoup de maladies qui deviennent incurables à la longue'; 'Le peuple Anglais s'amuse peu à jouer, il préfère la bouteille et la bière'. Lacombe certainly paid a visit to London in person to compile the material for his guide, but this was probably not the case with his close contemporary, Alphonse de Serres de Latour, whose *Londres et ses Environs ou guide des Voyageurs, Curieux et Amateurs dans cette Partie de l'Angleterre* (1788, with second edition in 1790) bears all the hallmarks of library research and second-hand reportage. Each edition ran to two bulky little volumes, with extensive gazetteers. A correspondent in *Notes and Queries*[15] notes Latour's amusing spelling of the names of suburbs 'swallowed up into London' – e.g. Wenlax Barn, Wauxhall etc, on a par, perhaps, with Lacombe's riverside retreat of 'Kiou'!

The last guides to London in French before 1800 were both written by political refugees from the Revolutionary terror. Henri Decremps' *Le Parisien à Londres, ou Avis aux Français qui Vont en Angleterre* is very much the product of an involuntary exile, published in two volumes at the end of 1789. Decremps, a law professor, who lived long enough to return eventually to Paris as a science teacher, clearly hankers after the return of law and order to France, harping continually on the freedom, harmony and sweet reasonableness in London, though he does not omit to detail the street swindlers, rapacious booksellers and executions among the unpleasant practices which he encountered on his enforced visit.[16] The Abbé Tardy, an émigré priest, in his *Manuel du Voyageur à Londres ; ou Recueuil* [sic] *de Toutes les Instructions Nécessaires aux Etrangers qui Arrivent dans cette Capitale* of 1800, is both excited and heartened by the tolerance of the government to differing viewpoints, and the lack of a written constitution to enforce them.

The final item in this brief survey can only be described as a curiosity – a mid-eighteenth guide to London in Dutch. *De Leydsman der Vreemdelingen*,[17] published in Amsterdam in 1759, is a surprisingly detailed guide, with extensive historical descriptions, detailed architectural accounts, a parallel text in English – and a 120-page account of the trial and execution in 1660 of the regicides. Apart from the assumption that the anonymous author (possibly the publisher, who seems to have adopted the pseudonym of Dirk onder de Linden) was a fanatical monarchist, several decades before the Dutch had their own monarchy, it seems a singularly pointless insertion at that inordinate length, in a book of only 340 pages. Whatever the reason, the production seems to have exhausted Dutch efforts in this field – no further guide to London was issued from the Netherlands for over three quarters of a century.

I should like to thank the authorities at the libraries where the majority of the research was carried out, for their permission to examine guides in their collections. These were: the Guildhall Library; the Bishopsgate Institute; the British Library; the Greater London Record Office, and Westminster Public Libraries. This survey has been partly derived from the author's Library Association thesis: *Guide Books to London before 1900. A History and Bibliography* (1975). A further section was published as 'For Inns a Hint, for Routes a Chart: the Nineteenth-century London Guidebook', *London Journal*, vol. 6 (1980), pp. 207–14.

1 Ludwig Schudt, *Le Guide di Roma* [German] (Vienna, 1930).
2 William Camden, *Reges, Reginae, Nobiles, et Alii in Ecclesia Collegiata B. Petri Westmonasterii Sepulti* (London, 1600). Later editions in 1603 and 1606.
3 William Dugdale, *The History of St Paul's Cathedral in London. From the foundation until these times ... Beautified with sundry prospects of the church ...* (London, 1658).
4 e.g. Paul Hentzner. *A Journey into England in the Year 1598* (London, 1757; originally published Breslau, 1617); Samuel Sorbiere, *A Voyage to England* (Paris, 1664; English translation London, 1709).
5 Joan Parkes, *Travel in England in the Seventeenth Century* (Oxford, 1925).
6 John Dunton, *The Life and Errors of John Dunton, Citizen of London*

(Edited by John Nichols, 2 vols. London, 1818, originally published London, 1705).
7 *The New Trismagister, in Four Parts* (London, 1688). New Wing C5422B (wrongly as Colson).
8 British Library only in Great Britain.
9 Presented to the Society by Esmond Samuel de Beer, the author of the useful article 'The Development of the Guide-book until the Early Nineteenth Century', *British Archaeological Association Journal*, 3rd series, Vol. 15 (1952), pp. 35–46. This concentrates on the wider European field.
10 Not in New Wing, nor in Arber's *Term Catalogues*.
11 *The Directory; or list of principal traders in London* (London, Henry Kent, 1732). The earliest surviving edition is the 3rd, of 1736.
12 Not listed in Darlington & Howgego, *Printed Maps of London circa 1553–1850* (2nd edn, Folkestone, 1978).
13 In e.g. the classified advertisements columns of the *Morning Post* 9 June 1780.
14 See, for this and other instances, Madeleine Blondel, 'French and English Eighteenth-century Guide-books to London: Plagiarism and Translation', *Notes and Queries*, Vol. 230 (1985), pp. 240–1.
15 *Notes & Queries*, 1st Series, Vol. 5 (1852), p. 102.
16 Chapter 7 was translated by W. H. Quarrell in *Notes & Queries*, Vol. 189 (1945), pp. 26–31, 268–9, as 'A Frenchman in London', 1789.
17 i.e. *The Foreigner's Guide*. The title continues...'of, nodig en nuttig, med gezel beyde, voor den vreemdelingen inboorling in hunne Wandeling door de steden Londen en Westmunster'.

VIII. SOURCES FOR LONDON HISTORY AT THE INDIA OFFICE LIBRARY AND RECORDS

By MARGARET MAKEPEACE

THE collections of the India Office Library and Records are known to be a vast source of information for the history of many parts of Asia, but researchers are often unaware of the richness of material about London which is also held. There is scope for research on a number of subjects: London topography, architecture, commerce and trade, genealogy, local history. This article aims to encourage the exploitation of the sources available to students of London history at the India Office Library and Records by explaining the nature of the archive and by indicating which series are most likely to prove useful.

London was the centre of operations for both the East India Company and the India Office. The East India Company was established in 1600 by a royal charter which granted to an association of merchants the monopoly of English trade to the 'Indies'. The merchants conducted their business from Sir Thomas Smythe's house in Philpot Lane in the City of London until 1621 when they purchased a lease on Crosby House in Bishopsgate Street. The lease expired in 1638 and the Company accepted the offer of accommodation in Sir Christopher Clitherow's house in Lime Street. The Company subsequently acquired a number of properties in Lime Street and Leadenhall Street and demolished these in order to erect a new building. The first part of East India House was constructed in 1725 and, with alterations and additions, this was to remain the headquarters of the East India Company until its abolition in 1858. In that year, the Company was replaced by a government department headed by a Secretary of State which was to be known as the India Office. The new department occupied

East India House for a short period until the autumn of 1860 when the Secretary of State moved to temporary accommodation in Victoria Street whilst awaiting the completion of the new India Office in Whitehall. After the grant of independence to India and Burma in 1947 and 1948, the India Office building was used by the Commonwealth Relations Office.

In addition to East India House, the Company occupied a considerable number of warehouses in the City, at Leadenhall Street, Fenchurch Street, Crutched Friars, Billiter Lane, Seething Lane, French Ordinary Court, Jewry Street, Haydon Square, Cooper's Row, and New Street, as well as at outlying areas of London such as Ratcliff and Blackwall. The East India Company was a major employer of Londoners both in its warehouses and in its offices. In 1802, approximately 4,000 people were employed by the Company in London as warehousemen, labourers, packers, clerical staff and the like.[1] The warehouses were sold following the Charter Act of 1833 which obliged the Company to wind up its commercial activities and become a purely administrative body for Indian affairs.[2] Hundreds of Londoners were made redundant and received compensation from the Company.

The majority of East India Company shareholders were residents of London. In 1799, of the 1,824 proprietors owning stock worth £1,000 or more, nearly 1,400 lived in London or in the immediate vicinity.[3] The Court of Proprietors elected qualified members to serve on the Court of Directors. The elections for directors' posts were keenly contested, with wealthy City merchants, financiers, bankers and Members of Parliament all eager to join the Company's influential executive body. Directors might then be chosen to serve on one or more of the specialized committees which dealt with matters of business referred by the Court of Directors, for example the Committee of Buying and Warehouses, the Committee of Correspondence, or the Committee of Shipping.

The documents produced and received in London during the course of business were preserved by the East India Company, which also collected books to form a library. The officials of the India Office carried out a survey of the Company records in 1860 and decided to discard a large number of papers considered to be of little lasting importance, including many commercial documents. Some of the papers destroyed were of potential value to twentieth-century researchers interested in London history, notably the minutes of the Committee of Buying and Warehouses. Proper provision was however made thereafter for the care of both the surviving Company archives and of the materials being accumulated by the India Office, so that the India Office Library and Records now has a well-preserved collection of records, printed books and serials, official publications, prints, drawings, photographs and maps.

A number of the archive series which have survived are sources for London history from the sixteenth to the twentieth centuries. The minute books of the Court of Directors are the central record of East India Company administration and business transactions, and cover the period 1599–1858 with some gaps during the seventeenth century. Indexes are available. The minutes record the Company's dealings with the central authorities of monarch and Parliament as well as with the City of London Corporation and local parish officers, list any letters received giving a brief indication of the subject matter, and enter reports submitted to the Court by the various committees.[4] From the 1680s the minutes of the committees were normally recorded separately from the Court minutes, and the majority of those which survive are associated with the Committee of Correspondence for the years 1700–1858. This was the most important of the standing committees, responsible for supervising non-secret correspondence with India, controlling overseas administrations, and making appointments to overseas and home establishment posts.[5] Details of the staff employed in London are to be found in the records of the Accountant General's Department, which also provide details of the commercial activities of the East India Company in London from the 1660s.[6] Additional minutes and correspondence concerning the home establishments of the East India Company and the India Office

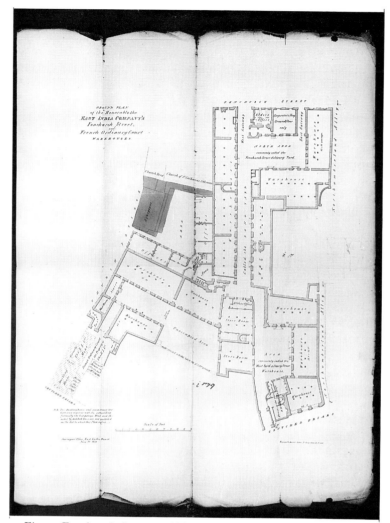

Fig. 1. Fenchurch Street and French Ordinary Court warehouses: ground plan dated 1835 [IOR: L/L/2/470].

from the early nineteenth century to 1950 are contained in the papers of the Financial Department.[7] One of the Department's duties was to deal with the financial administration of buildings in London occupied by the Company or by the India Office: expenditure on leases, repairs and parish rates, and income from the sale of properties.

Amongst the archives of the Legal Adviser's Office is a large collection of property documents dating from 1498 to the 1920s, of which Sir George Birdwood wrote in 1891 '...there are a large number of parchments (some of ancient date) in the custody of the Legal Adviser. These are, however, as far as is known, of no public interest, being merely title deeds and leases of property held at different times by the East India Company'.[8] It appears that Sir George's opinion of dusty deeds was shared by subsequent generations of officials at the India Office, for little was done to make the collection accessible to researchers. A programme of sorting and listing the many hundreds of documents is now being undertaken in order to open up this valuable source for London local history. A deeds register compiled around the year 1813 has survived and this contains chronological sequences for the different sites where the Company properties were located, giving abstracts of the title deeds involved in most of the pre-nineteenth century transactions. Brief entries for some buildings acquired after 1813 were later added at the back of the register and also entered in three slim volumes which were maintained as a record of the storage place of sets of deeds.[9] As well as title deeds, the collection includes correspondence about the properties and other related documents such as copies of reports prepared by the Committee of Buying and Warehouses, thus filling some of the gaps caused by the destruction of the original minutes.

Many of the deeds in the India Office collection pre-date the establishment of the Company in 1600, for until the Law of Property Act of 1925 limited the need to prove title to a period of fifty years, all surviving documents for earlier conveyances were generally transferred to the new owners. This system meant that many documents left the custody of the East India Company when it was obliged to sell its City premises during the nineteenth century, although the deeds for several of its former properties were left by the purchasers in Company care. Over two hundred deeds for Company sites are now held by the Guildhall Library, having been deposited there by the subsequent owners.

The property documents in the Legal Adviser's archives are complemented by the records of the Surveyor's Office, which incorporate the heavily weeded papers of the earlier Clerk of the Works Department.[10] The Surveyor's Office records consist of minutes, correspondence, contracts, plans and drawings for the years 1837–1934 with one earlier item dated 1815. They document the construction, equipping and maintenance of a large number of properties owned or leased by the East India Company and the India Office in London and the home counties.

Much information about London history is contained in two collections of notes and papers created by former members of staff at the India Office. Sir William Foster was Registrar and Superintendent of the Records from 1907 to 1923, and then served for a further five years as the official historiographer, publishing *The East India House* and *John Company*.[11] In the course of researching his books and articles he filled twenty-six notebooks and compiled a separate index volume.[12] Mr Ray Desmond, Deputy Librarian at the India Office Library and Records from 1973 to 1985, deposited two boxes of papers which he had assembled concerning the buildings, docks and warehouses belonging to the Company and the India Office 1647–1981.[13]

Foster's work contains a great deal of topographical information about London, such as the sketch maps of different East India Company properties which are scattered throughout his notebooks. Maps and plans of sites in London are found in various archive series, including a large number in the Surveyor's Office records and two compilation volumes of nineteenth-century plans amongst the Home Miscellaneous papers,[14] as well as isolated items such as the plan of the site of East India House in July 1709 in the Committee of Correspondence reports.[15] Plans are also found amongst the property records, on separate sheets or annexed to deeds. Many Company deeds dating from the eighteenth and early nineteenth century have plans drawn in the margin, as well as containing detailed descriptions of the

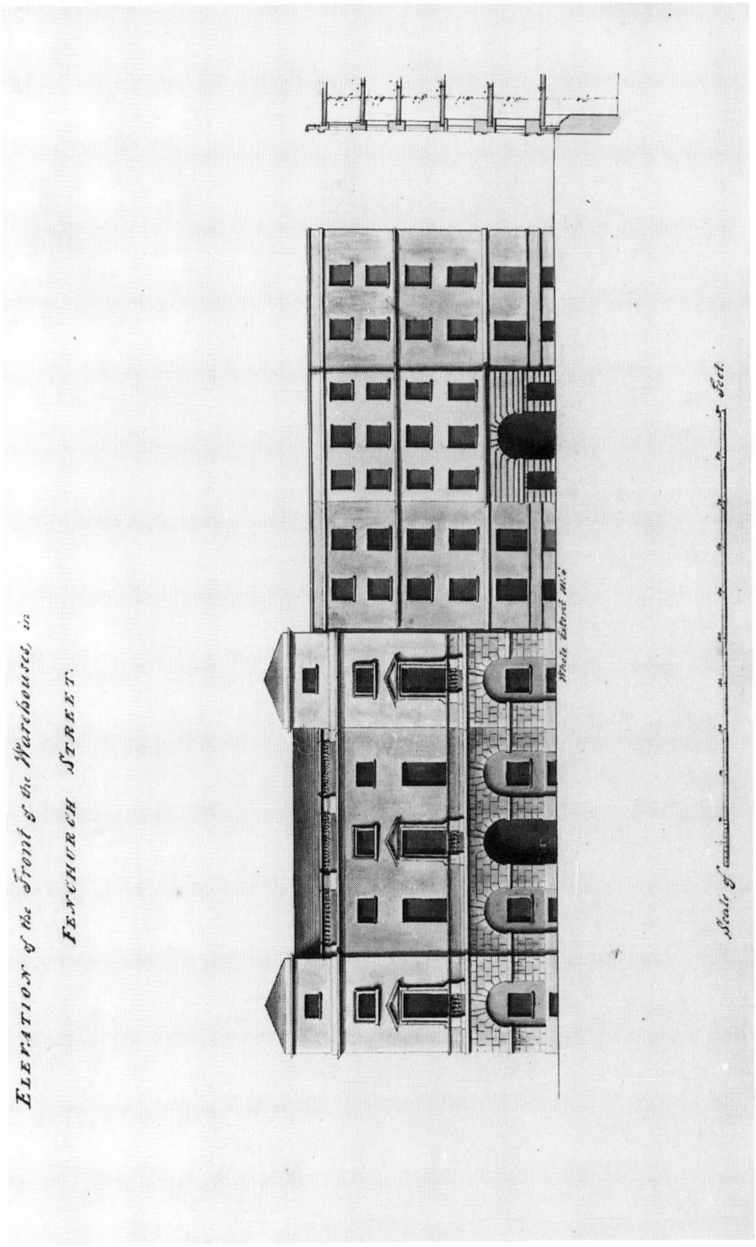

SOURCE FOR LONDON HISTORY 159

Fig. 2. Fenchurch Street warehouse: front elevation in 1806 [IOR: H/763 B f. 7].

precise location of the street where the premises stood and of the adjacent properties and their occupants.

Most of the earlier deeds in the collection are less useful for topographical research. Many early deeds for town properties do not name the streets where buildings stood but merely state the parish, or a parish and a City ward. Houses were not given street numbers until the late eighteenth century and before that time premises were often known by the names of their owners or occupiers, for example Craven House in Leadenhall Street which was leased to the East India Company by the Earl of Craven. Some early deeds do provide pieces of information needed to build up a picture of areas of London before 1700. It is possible to discover the position of lands owned by religious bodies and their dispersal into private hands at the dissolution of the monasteries. A deed of bargain and sale made between Andrew Jenour and John Smyth on 31 January 1588 describes the property being conveyed as 'All that Messuage or Tenemente with Thappurtenances scytuate Lyeing and being in Sething lane alias Syding Lane in the parishe of All Hallowes Barking nere the Tower of London sometymes parcell of the Landes And possessions of the Late dissolved Colledge of S^{ct} Trynytye of Pomfrett in the Countye of Yorke And to the sayd Colledge late Apperteyning And belonging nowe in the Tenure of the Lady Wynefride Hastinges or of hir Assignes'.[16]

A deed of partition drawn up by Nicholas Salter and John Wolstenholme in March 1608 assists in locating the site of further houses then standing in Seething Lane. Salter and Wolstenholme had converted Muscovy House and the White Horse Inn into a single messuage and were now confirming the agreed division into apartments. Compass directions are used when specifying the position of the partition walls and Lord Lumley's gardens are said to lay behind the house. This information would be of great assistance in plotting the exact site both of Salter and Wolstenholme's house and of Lumley's property.[17]

Wolstenholme's share of the property was bought in 1654 by the government of Oliver Cromwell to accommodate the

Navy Office where Samuel Pepys was an employee. Seething Lane was situated in one of the few areas of the City to escape the Fire of London in September 1666. Pepys wrote '... my confidence of finding our office on fire was such, that I durst not ask anybody how it was with us, till I came and saw it not burned. But going to the fire, I find, by the blowing up of houses and the great help given by the workmen out of the King's yards, sent up by Sir W. Penn, there is a good stop given to it, as well at Marke lane end as ours – it having only burned the Dyall of Barkeing Church, and part of the porch, and was there quenched'.[18] The East India Company described the extent of the destruction in letters to their servants in India:

it pleased God that on the 2d of this moneth, being Sunday in the morning, a most fearefull and dreadfull fire brake forth, which hath consumed the greatest part of the Citty of London, even from Tower Dock to Temple Barr, & almost all within the walls except part of Marke Lane, Bishopsgatestreete, Leadenhallstreete, part of Broadstreete & some by the Wall toward Mooregate & Criplegate, & part by Christ Church... In this sad Callamity God was pleased to bee very favourable to the Companies Intrest, having preserved most of our goods, excepting some Saltpeeter & our Pepper at the Exchange Seller.[19]

Four-fifths of the area of the City was destroyed by the Fire and a programme of rebuilding was quickly undertaken in accordance with the regulations laid down by two Acts passed by Parliament in 1667 and 1670.[20] Brick or stone was to be used for reconstruction and all main streets were to be widened to prevent the spread of fire in future. Since eighteenth-century deeds often specify the materials used in construction, it is sometimes possible to establish an approximate date for the erection of individual houses in the City by checking which building method was used. The implementation of other provisions in the legislation is shown by the title deeds for sites affected by the Fire. A lease for a site near London Wall executed in January 1671 contains the following information:

Whereas the sayd Giles Cois in his life time being seized in fee of and in one messuage or tenement... in the parish of St Michaell

Bassishaw London...And whereas the sayd messuage was utterly burnt downe in the late generall fire happening in the sayd citty of London And by an act of Parliament since made...all persons seized of any houses burned or demolished by reason of the sayd fire for life or lifes with remainders in tayle or other life estate are thereby empowered and enabled to make leases to any persons that will rebuild thereupon for a terme not exceeding Fifty yeares att the most...And whereas the sayd Thomas Phillipps hath undertaken to rebuild the premises att his owne charges...Now this Indenture witnesseth that the sayd Rebeccah Chandler...Hath demised graunted and to farme letten...to the sayd Thomas Phillipps...All that toft soyle and ground whereon the sayd messuage abovementioned formerly stood and which was in the possession of Richard Thornton att the time of the sayd late fire conteyning in the front towards London Wall nine foote of assize and in depth sixteene foote of assize and doth abutt on the north on the common way or streete and on the south east and west on certeyne ground now in the possession of the sayd Thomas Phillipps....[21]

Taking this information with the contents of other deeds for the London Wall area, one would be able to map the pattern of properties before and after the Fire, identifying the owners and occupiers at different periods.

Whereas the reconstruction in the wake of the Fire was undertaken in order to replace the houses which had been destroyed, later London development schemes involved the deliberate clearance of existing buildings to make way for new property. One such scheme was the demolition of a considerable number of houses in Westminster during the 1860s to provide a site for the new India Office near St James's Park. There is a large body of material in the India Office Records about the compulsory purchase of properties in Charles Street, Gardeners Lane and Duke Street which was sanctioned by Acts of Parliament.[22] Amongst the archives of the Legal Adviser's Department and the Surveyor's Office there are plans of the site and of the new building; documents about the passage of the enabling legislation through Parliament; papers concerning the purchase of individual properties including abstracts of title, conveyances, and correspondence with landlords and

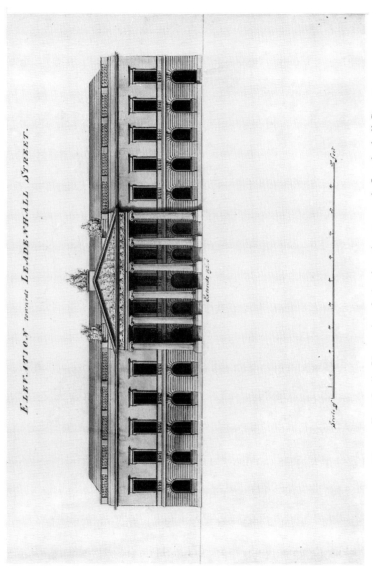

Fig. 3. East India House in 1806: elevation towards Leadenhall Street designed by Richard Jupp [IOR: H/763 A p. 23].

tenants; details of the compensation paid to owners and occupiers; tenders for contracts to build and equip the new India Office and accounts of the payments made to the appointed contractors. The records reveal that the neighbourhood which was cleared consisted of homes, small shops and businesses, with interesting features such as the barracks in Gardeners Lane and the accommodation for War Office recruits at the Royal Rendezvous public house in Charles Street. In 1881 an Act was passed to allow the India Office to sell some of the land which had been purchased in Charles Street since it was found to be surplus to requirements.[23]

Illustrations and photographs of the exterior and interior of the India Office in Westminster as well as a number of views at various dates of East India House and the adjoining buildings in Leadenhall Street are held in the Prints and Drawings Section and the European Printed Books collection. Elevations and floor plans of both East India House and the Company warehouses were prepared in the early nineteenth century providing evidence of architectural design and interior layout.[24] Researchers studying the architecture of the India Office can still see the building standing in Whitehall as part of the Foreign and Commonwealth Office. Those interested in East India House or the Company warehouses are less fortunate. East India House was sold to a syndicate in 1861 and demolished soon afterwards to make way for new offices. The only Company warehouses to survive until the twentieth century were those in the Cutler Street complex built in the area between Petticoat Lane, Houndsditch, and Bishopsgate Street in the late eighteenth century. The site was re-developed in the years 1978–82 but four of the seven original warehouses have been preserved in the Cutlers Gardens scheme.[25]

Bills of sale found amongst title deeds often provide detailed descriptions of the location, dimensions and layout of properties with the intention of attracting high bids from prospective purchasers. In July 1791 an auction was held to sell the leasehold of an estate at London Wall and Fore Street, part of which was occupied by the East India Company. The bill of sale ran thus:

Fig. 4. East India House: ground floor plan at time of sale in 1861 [IOR: H/763A p. 103].

The Warehouses occupy a valuable Plot of Ground, 182 Feet 6 Inches in Front of London Wall, 57 Feet 9 Inches at the West End, and 37 Feet 6 Inches at the East End, be the same more or less, and contain Three Floors, and One in the Roof, capable of stowing near Four Thousand Chests of Tea, a small Warehouse, formerly the Counting House, Stabling for Nine Horses, and a Yard, now occupied by the East India Company. An excellent commodious Brick Dwelling House, situate No. 45 Fore Street, in the Possession of Mr Josiah Spode, Wholesale Potter. The premises are Three Stories high, contain Three Rooms on a Floor, a genteel Dining Room, Parlour, Kitchen, roomy Warehouse, Counting House, and dry Cellar.[26]

The small number of fire insurance policies which have survived amongst the property records are useful since they describe the materials used in the building construction, and the valuation may provide an indication of the size and character of a house especially if one is able to compare a number of policies issued around the same date. A large collection of insurance company registers covering the whole of England is held by the Guildhall Library. In order to search for a particular property it is generally necessary to know the number of the policy or at least the year in which it was purchased. This may not be a problem as conveyances sometimes quote policy numbers: the document recording the mortgage of four houses in Northumberland Alley by the leaseholder John Allen to Samuel Pearce in 1788 contains the numbers of two policies issued by the Sun Fire Office, one of which has been traced in the deeds at the India Office.[27] When locating these policies in the Sun registers it proved rewarding to browse through some other entries, for on the same page was a policy taken out by one of Allen's tenants on the contents of his dwelling at 4 Northumberland Alley which included his business stock as well as household goods and clothing.[28]

Very detailed inventories or schedules of properties are sometimes annexed to title deeds. Lessors were anxious to protect their premises from the depredations of tenants, and schedules may describe the contents of each room storey by storey, stating the function of the room and mentioning elements of design such as the doors, windows, wainscot, and fireplaces. The position of rooms may be given in relation to other parts of the house with the location of staircases and corridors, thus enabling a researcher to plot a rough plan of the layout of the building.[29] One may be fortunate enough to discover an inventory of all the furniture and personal effects in a house, perhaps drawn up for probate or because of distraint of goods for debt. When Samuel Row failed to deliver a sum of over £700 which he had collected in Aldgate Ward on behalf of the Commissioners of Land Tax, his freehold house in Sugar Loaf Court and all his belongings were sold by the Commissioners in 1742 to recoup part of their loss. The inventory was meticulously prepared, listing all pieces of furniture, linen, cutlery, kitchen equipment, books and pictures, as well as the tools and materials used by Row in his trade as painter stainer.[30]

It was common practice in the eighteenth century for City tradesmen and merchants to live above their business premises. London had undergone a gradual social change, with craftsmen and merchants replacing the nobles and gentlemen who had figured prominently in City property transactions up to 1600 or thereabouts. The gentry moved west to set up homes in the more fashionable areas of London outside the City and their former residences were subdivided into tenements or transformed into commercial and government offices, for example the Earl of Craven's mansion in Leadenhall Street became the East India Company headquarters, while the dwelling in Seething Lane which had once belonged to Sir Francis Walsingham became the Navy Office. As the merchants moved in, certain trades came to predominate in particular parts of the City, and some of these clusters of businesses may be traced at the India Office from property records and from the Committee of Correspondence memoranda which contain the names, addresses and trades for a large number of Londoners standing surety for new entrants to Company service.[31] Deeds may also provide evidence of trades which might have been considered a nuisance and therefore banned by some landlords. In 1770 Mary Smith granted a lease on a house in Fenchurch Street to Stacey Till, a linen draper, on condition that he did not allow the premises to be used without her consent 'for the Art Trade or Mystery of a Brewer, Vintner,

Victualler, Tallow Chandler, Smith, Brasier, Baker, Butcher, Cook, Coffeehouse, Box-maker, Joiner, Carpenter, Pewterer, Distiller, Dyer, Flaxman, Gunpowder-man, Hemp-Dresser, Tin-worker, Poulterer'.[32]

Complicated patterns of leasing and subletting are revealed by the East India Company deeds. Much property was owned by bodies such as the Corporation of London, Christ's Hospital, and City companies such as the Coopers and Ironmongers. The long chain of leases for the London Wall estate auctioned in 1791 is well documented.[33] The City Corporation owned the whole estate and leased it to Sir Lionel Lyde for $55\frac{3}{4}$ years from Michaelmas 1766. In February 1782, Lyde sublet part of the property consisting of a house and cellar to Josiah Spode for a term of 21 years, and in November 1784 Lyde leased the entire estate including Spode's house to Thomas Wright for $18\frac{1}{2}$ years. Wright in his turn let the warehouses to the East India Company.[34]

A different system of property holding operated in East London. The manors of Stepney and Hackney administered the system of copyhold or customary tenure which had originated in grants made by the lord to his villein tenants. Once the villeins gained their freedom, they were able to sell, entail and mortgage their lands subject to certain obligations to the lord of the manor. Changes of ownership had to be effected at the manorial court baron or be reported to it. All transactions were recorded on the court roll and certified extracts were given as proof of title to the owners, who thus became known as copyholders. The East India Company was the copyhold tenant for ground at Poplar where it had established almshouses and a chapel during the seventeenth century. Some Poplar deeds which were issued by the court baron of the Manor of Stepney survive from the eighteenth and early nineteenth centuries.[35]

The manorial system of government in East London contrasted with the City administration by Aldermen elected from the ranks of wealthy merchants. Many Aldermen were also directors or major stockholders of the East India Company and it is possible to trace information about their

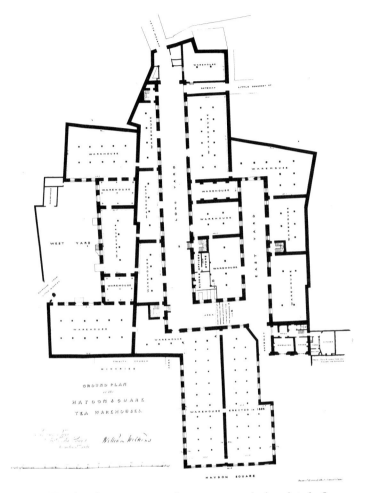

Fig. 5. Haydon Square tea warehouses: ground plan dated 1835 [IOR: L/L/2/695].

commercial activities from the Court minutes or the financial ledgers.[36] The names of some of these men indicate the presence of a foreign merchant community within London: Lethieullier, Corsellis, Hoet, Paravicini. They appear as parties to property transactions and one may thus trace the whereabouts and the changing fortunes of their families in London by studying groups of title deeds which frequently cover long periods of time.

Deeds are a valuable source for genealogists since they often describe family relationships in some detail. A will devising real estate becomes part of the title to the property and therefore a copy may be found amongst the deeds giving biographical information about a number of individuals. London women appear in the Company records not only as the beneficiaries of wills and as the wives, sisters and daughters of the men involved in property conveyances, but also in their own right as testators, freeholders, lessors, lessees, and stockholders. There are sometimes indications of the businesses they ran, such as keeping a shop or manufacturing goods.[37] The mothers, wives and widows of Company employees are named as petitioners in the minutes of the Court of Directors, asking for financial assistance or requesting that their sons and husbands be granted permission to return home from the East. Women are identified in the records of the Poplar Fund, which was established by the Company to provide pensions for former members of its maritime service and for their widows and children. Some pensioners and widows were allocated accommodation at the Poplar almshouses.[38]

Poplar was a quiet district with a small population until it was transformed in 1614 by the opening of a dock at Blackwall by the East India Company. The number of local inhabitants increased rapidly as employment for hundreds of men became available at the dock, encouraging tradesmen to move in to set up businesses catering for the needs of the labourers and their families and for the needs of the sailors belonging to ships docked at Blackwall. Although the Blackwall yard was sold by the Company in 1656, East Indiamen continued to use the dock and its later extensions until the Company ceased to trade in the 1830s. The valuable cargoes which were brought to England by the East Indiamen were unloaded in deep water at Woolwich, Long Reach and Blackwall, then taken aboard smaller vessels for transportation to the secure warehouses in the City or in the

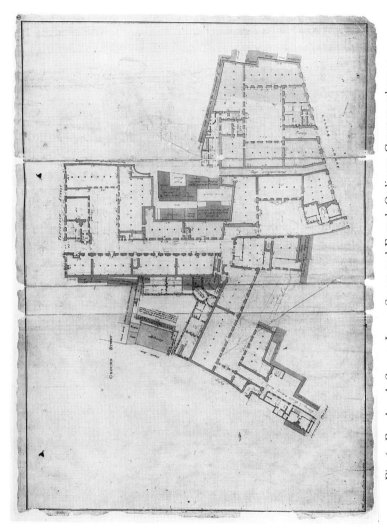

Fig. 6. Fenchurch Street, Jewry Street and French Ordinary Court warehouses, and Company's Mint in Crutched Friars: ground floor plan dated 1806 [IOR: H/763 B f. 5].

docklands. The property records and the minutes of the Court of Directors show that the Company leased, purchased and constructed wharves and warehouses along the Thames at Somers Quay, Lyons Quay, Hammond's Quay, Cumberland Wharf, Blackwall and Ratcliff. The Blackwall warehouse was built by the Company in the early nineteenth century for pepper and other spices, while the warehouses at Ratcliff were constructed in the late eighteenth century for naval stores and saltpetre. The naval store warehouse stood at Stone Stairs, Ratcliff upon copyhold ground belonging to Stepney Manor, the saltpetre warehouse upon a site to the south of Ratcliff Street leased from the Coopers' Company.

The Ratcliff warehouses were burned down in July 1794 when a fire which had started in a barge builder's premises set light to part of the Company's large stock of saltpetre. There was a great explosion and the wind spread the flames throughout the hamlet of Ratcliff, destroying 600 houses in Brook Street, Ratcliff High Street and the smaller alleys. Hundreds of people lost their homes and their working tools in the blaze and a fund was launched to provide relief for them, to which the Company subscribed 200 guineas. The Company quickly rebuilt the saltpetre warehouse and took the opportunity to enlarge it, leasing ground from Jeremiah Snow to add to the site already held from the Coopers' Company. In 1796 the City granted permission for the Company to construct an embankment on the river adjoining the warehouse, and further enlargements were made in 1801. The naval store warehouse was also rebuilt, with additional ground purchased from Susannah Freeman in 1795.[39]

During the early nineteenth century, Ratcliff became the home of a large number of Indian and Chinese seamen known as lascars. At the time of the Revolutionary and Napoleonic Wars, the British crews of East Indiamen were often subjected to impressment by Royal Navy ships in eastern waters. The Company therefore came to rely increasingly upon lascars to sail its ships home to England. The lascars had to be accommodated in London for some months until a passage back to India or China could be arranged. The Company found lodgings for them, first in Kingsland Road, Shoreditch and subsequently at Ratcliff Highway in Shadwell. Between the years 1804 and 1813 nearly 10,000 lascars arrived in London. Rumours of poor housing conditions and destitution aroused philanthropic concern about the lascars' welfare, resulting in a visit to the barracks from a Parliamentary Committee. The Committee issued a critical report which led to the passage of an Act in 1814 which obliged the Company to provide adequate food and clothing as well as accommodation for the lascars.[40] The Company's problem of coping with the influx of lascars largely disappeared after the French wars came to an end in 1814.[41]

It was in response to the wars against France that the Royal East India Volunteers were raised by the Company in 1796 to assist regular troops in the event of an invasion of England. Directors and East India House staff were eligible for the posts of field and commissioned officers, while labourers and other warehouse employees could be appointed as non-commissioned officers or privates. 1,500 men served in three regiments, bearing arms supplied by the government and wearing uniforms provided by the Company. A headquarters was established at the New Street warehouse and the Volunteers marched from there two or three times a week for training and exercise in the Company's military field situated at the northern end of Allerton Street, Hoxton. In 1801 the Company purchased the right of access through Allerton Street and secured permission to build a bridge over the ditch at the entrance to the field to enable 'their Volunteer Troops, Horse, Foot and Artillery, their servants and workmen,...with or without Horses, Oxen, etc., and with loaded or unloaded Guns and things of all sorts, at their will and pleasure, by night as well as by day, at all times to go pass and repass through Allerton Street into said Field'.[42] The Volunteers were disbanded after peace was concluded in 1814, but a new regiment of 800 men was raised in 1820 by the Company in the aftermath of the discovery of the Cato Street conspiracy against the government. The Volunteers were maintained by the Company until 1834 when the corps had to be disbanded in the period of retrenchment following the loss of the trade monopoly.

When the East India Company became a purely administrative body, its Leadenhall Street headquarters also changed in character. East India House was no longer the scene of noisy sales where buyers strove to outbid their rivals. In 1841 *Alexander's East India Magazine* wrote of East India House: '...its extent is so much beyond what is required for a mere government office that it has a very deserted appearance, especially inside; also down Lime Street, in the Market, and all round about it, except in front, where the general bustle of the busy city gives the appearance of business, along Leadenhall Street'.[43] The author of the article went on to criticize the Company directors for keeping the outside world at bay: 'The library continues virtually sealed up; it is the most valuable oriental collection in the world; but it is unvisited...The Company hides itself in Leadenhall Street, and it hide[s] its library there'. The library collections are now freely available for public consultation, together with the nine miles of official records created by the Company and its successors. The India Office Library and Records has for many years welcomed visitors researching the history of Asia and hopes in the future to welcome visitors researching the history of London.[44]

1 Cyril Northcote Parkinson, *Trade in the Eastern Seas 1793–1813* (Cambridge, 1937), pp. 26–8.
2 3 & 4 Will IV c. 85.
3 *Trade in the Eastern Seas*, p. 11.
4 IOR: B
5 IOR: D
6 IOR: L/AG
7 IOR: L/F
8 Sir George Birdwood, *Report on the Old Records of the India Office* (1891), p. 288.
9 IOR: L/L/2/1–4.
10 IOR: L/SUR.
11 William Foster, *The East India House, its History and Associations* (1924) and *John Company* (1926).
12 IOR: L/R/5/209–35.

13 IOR: MSS. Eur. D.1131.
14 IOR: H/763.
15 IOR: D/91 f. 76.
16 IOR: L/L/2/1, p. 608, no. 1582.
17 IOR: L/L/2/1, p. 315, no. 590.
18 Robert Latham (ed.), *The Shorter Pepys* (1986), p. 665.
19 IOR: E/3/87 f. 20 Letter from East India Company to Madras 14 September 1666.
20 19 Car 2 c. 3 & 22 Car 2 c. 11.
21 IOR: L/L/2 Lease 11 January 1671, Aldermanbury estate.
22 25 & 26 Vict c. 74 [1862], 27 & 28 Vict c. 51 [1864], 28 & 29 Vict c. 32 [1865].
23 44 Vict c. 7.
24 See, for example, IOL: WD 2881, WD 2465, Print box 20; IOL: X298 William Griggs, *Relics of the Honourable East India Company* (1909); IOL: X661, *Views of the East India House*.
25 See IOL: V 25210, Penelope Hunting, *Cutlers Gardens*.
26 IOR: L/L/2/1, pp. 795-7 Bill of sale, July 1791.
27 IOR: L/L/2/1, pp. 753-6.
28 Guildhall Library Ms. 11936/297, p. 397.
29 See, for example, IOR: L/L/2/403, Counterpart lease for 21 years, 6 December 1770 & L/L/2/407 Lease for 14 years, 13 June 1785.
30 IOR: L/L/2/1, p. 606 no. 1573 Bargain & sale, 20 August 1742.
31 IOR: D/91-145 Memoranda of the Committee of Correspondence 1700-1858.
32 IOR: L/L/2/1, p. 256 no. 453.
33 See above for bill of sale.
34 IOR: L/L/2/1, pp. 795-7.
35 IOR: L/L/2/1, pp. 836-43,880.
36 IOR: L/AG/1/1/2-34 General ledgers of account 1664 – 1860–index available.
37 See for example IOR: L/L/2/407 Jane Chambers, mantua maker [*dressmaker*], occupied a house in Fenchurch Street in 1785, & IOR: D/91 f. 197 Elizabeth Batson was a saddler near York Buildings in the Strand.
38 IOR: B & L/F for the Court of Directors' general policy and administrative decisions about the Poplar Fund; L/MAR/C for application papers and lists of pensioners; L/AG for payment books; L/SUR & L/F for documents on maintenance work and tenancy changes.
39 IOR: L/L/2/1, pp. 393-9, 676-88.
40 54 Geo III c. 134.
41 IOR: L/MAR/C/902 Papers relating to the care of lascars 1793-1818.
42 IOR: L/L/2/1, pp. 875-6 Grant of right of way for $96\frac{1}{2}$ years to East India Company made 4 June 1801 by John Fitzgeorge, John Reeves and Jonathan Wakefield.

43 IOL: ST 964 *Alexander's East India Magazine* (March, 1841), pp. 219–21.
44 India Office Library and Records, 197 Blackfriars Road, London SE1 8NG; telephone 01-928 9531. Opening hours, except during public holiday periods, 09.30–18.00 Monday–Friday, 09.30–13.00 Saturday.

IX. LONDON ILLUSTRATIONS IN THE *GENTLEMAN'S MAGAZINE* 1746–1863

By PETER JACKSON

THE *Gentleman's Magazine* first appeared in January 1731 and thereafter it was published every month until September 1922.*

In its heyday, from the second half of the eighteenth century until well into the nineteenth, it was the most popular and widely read of all periodicals, but by the 1850s its character began to change and its importance declined. From being a miscellany of parliamentary reports, news both domestic and foreign, obituaries, book reviews, poetry, a theatrical register, together with scholarly essays and topographical notes, it became a fiction magazine and ceased to have any significant antiquarian or news interest.

The illustrations follow the same pattern. The first four volumes contain no plates at all. Illustrations appeared gradually and were often mere wood-cuts in the text or frontispieces to the annual volumes. (The first 'London' plate, a view of Westminster Bridge, did not appear until Vol. XVI in 1746.)

By the mid-eighteenth century several plates were being published every month and they became an important feature of the paper. With the magazine's decline, fewer and fewer plates appeared until by 1877 the only pictures were story illustrations. For the purpose of this index I have chosen 1863 as the last year, since no material of any significance appeared after that date.

---

* During the last fifteen years of its existence, from 1907 to 1922, the magazine became merely a four-page number produced for the sole purpose of maintaining a continuous run and preserving the title. The British Library's holdings of these later issues were destroyed in the last war and I have been unable to trace any other copies.

The amount and quality of textual material relating to London topography contained in the *Gentleman's Magazine* can be judged by glancing through the collection of extracts published in 1904 in three volumes under the title *The Gentleman's Magazine Library: Topographical History of London*, Edited by Laurence Gomme. Many of the pieces were contributions sent in by amateur archaeologists like John Carter (1748–1817) and James Peller Malcolm (1767–1815) whose sketches were engraved and published often as small vignettes on miscellaneous plates.

Some of the illustrations present the only known views of parts of London long since swept away while others give us the earliest pictures of newly-erected buildings like the Mansion House.

In 1824 began a remarkable series on 'New Churches' which lasted until 1831 and which gave the earliest representations of many of the metropolitan churches and chapels which no longer exist.

The *Gentleman's Magazine* is notable for its contemporary maps of various parts of the World and those dealing with London are invaluable in showing proposed improvements and new street layouts.

In compiling the present index the following system has been adopted:

(1) The London location, followed by any additional information. The descriptions in quotation marks are direct quotes from the plate itself.

(2) Method of reproduction. Where no method is mentioned it is assumed that the plate is a line engraving. Wood engravings printed in the text are given the page numbers on which they appear. Lithographs and aquatints are noted when they occur.

(3) Size. No reference to size indicates that the illustration occupies a whole page, i.e. approx. 220 × 130 mm. When the plate is a larger format or a 'pull-out' its dimensions are given, the vertical followed by the horizontal, in millimetres. Where a plate shows a miscellany of unrelated items, the figure number of the catalogued subject is given. 'Half a double plate' indicates that the illustration is one of two on the same plate.

(4) Artist and engraver. These names are given with the conventional 'Del' and 'Sculp' as they occur on the plate. Location of the original art work is given where known. (The drawings in the Folger Library are catalogued in *The Nichols File of the Gentleman's Magazine* by James M. Kuist, University of Wisconsin Press, 1982.)

(5) Location. For the sake of simplicity, no volume numbers have been given. Beginning with 1783, the year was divided into two volumes, each bearing the same volume number but distinguished as Pt. I (Jan.–June) and Pt. II (July–Dec.). The complications arising from different editors regarding yearly and half-yearly volumes or parts differently, with the introduction of several 'New Series', are too devious to be examined here. All that is needed for the location of a plate is the month and year in which it was published and this is the system followed here. The exceptions are 'Supplement' plates. These are always found after the December number unless otherwise stated.

ACHILLES STATUE
J. Mills del. & sculp.
August 1822

ALDGATE
Elevation of an ancient archway at Aldgate.
Fig. (1) on a miscellaneous plate.
May 1803

ALMONRY AT WESTMINSTER
The old house called 'Caxton's House'.
Wood engraving in text p. 361
April 1846

APOTHECARIES' GARDEN
The statue of Sir Hans Sloane.
Wood engraving in text p. 388
May 1822

ARCHERY GROUNDS, FINSBURY
See Maps & Plans

BAGNOR HOUSE, SHOE LANE
May 1819

BANKSIDE
View showing the Bear Garden and the Globe Theatre. 'From the Venetian Map, 1629.'
Feb. 1816

BAYSWATER CONDUIT
Fig. (1) on a miscellaneous plate.
Robert Bolton del.
April 1798

BEAUFORT HOUSE, CHELSEA
Half a double plate.
Wood engraving. 'L. Knyff del. 1699' (Copied from Plate 13 '*Britannia Illustrata*' 1708.)
June 1829

BELL ALLEY   See Pitcher's Court

BENEVOLENT SOCIETY OF ST PATRICK
Front elevation of the Society's school in Stamford Street, Blackfriars Road.
Lithograph. B. King litho.
July 1822

BERMONDSEY ABBEY
Fragments of Saxon architectural
  ornaments.
J. Basire sculp.
Aug. 1808

5 fragments of architectural detail.
J. Basire sculp.
Nov. 1808

'Fragments found on the site...'
J. Buckler Jnr. del. G. H(ollis)
  sculp.
Dec. 1810

View showing the gate.
Sept. 1790

BETHLEM HOSPITAL
'Statues of the Maniacs...'
April 1816

BISHOPSGATE
'Statue found near Bishopsgate.'
  (Representing St Peter.)
Half a double plate.
Wood engraving. (Original pencil
  drawing by Thomas Fisher in
  the Nichols File, Folger Library.)
Sept. 1826

BISHOPSGATE STREET
Great fire in... See Maps & Plans

BLACKFRIARS
Sepulchral slabs found at
  Blackfriars.
Wood engraving in text p. 635
June 1843

BLACKFRIARS BRIDGE
'Plan & Elevation of a Bridge for
  Black Friars, humbly offer'd to
  the Consideration of the
  Committee etc. by a Bye
  Stander.'
78 × 310 mm
Dec. 1759

BOLT COURT
Medical Society, Bolt Court.
Bas-relief over the entrance.
Fig. (1) on a miscellaneous plate.
Feb. 1788

BRANDENBURGH HOUSE,
HAMMERSMITH
Front elevation across the river.
Oct. 1822

BRITISH MUSEUM
'New Gallery... N. E. Garden
  View.'
T. Prattent del. & sculp.
(Original ink & wash drawing in
  the Nichols File, Folger
  Library.)
Sept. 1810

'Interior of the King's Library...'
Drawn & Engraved by Henry
  Shaw FSA.
Jan. 1834

'The New Reading Room...'
Wood engraving p. 472
May 1859

Plan.
Wood engraving in text p. 474
May 1859

BRIXTON HOUSE OF CORRECTION
'View of the tread mill for the
  employment of prisoners...'
210 × 258 mm
July 1822

BUCKINGHAM PALACE
'Turkish Ambassador's Public
  Entry.' With a view of
  Buckingham House.
T. Prattent sculp.
April 1795

BUTTERWICK HOUSE, HAMMERSMITH
Cedar tree at the back of the house.

Wood engraving. Half a double plate.
(Reproduced from Faulkner's *History of Hammersmith*.)
Sept. 1839

CAMBERWELL
'Ancient Seat of the Bowyer Family at Camberwell.'
Wood engraving in text p. 585
(Original ink, wash & pencil drawing in the Nichols File, Folger Library.)
Supplement 1825

CANALS
'Plan of the Intended Navigable Canal from Moor Fields into the River Lee at Waltham Abbey. Surveyed by Order of London by Robt. Whitworth.'
129 × 502 mm
March 1774

CARLTON HOUSE
View of the front portico.
Wood engraving in text p. 446
An illustration from *Time's Telescope for 1829*.
Nov. 1828

'CAXTON'S HOUSE'
See Almonry Westminster

CECIL HOUSE, STRAND
Front elevation of Old Cecil House in the Strand.
Figure on a miscellaneous plate.
Dec 1786

CHARING CROSS
See Maps & Plans

CHARTERHOUSE
Coffin of Thomas Sutton.
Wood engraving in text p. 43
Jan. 1843

CHEAPSIDE CROSS
'Representation of the demolishing of the Cross in Cheapside in the year 1643.'
Supplement 1764

CHELSEA
'James Pilton's Manufactory, King's Road, Chelsea. The Interior of the Menagerie, Displaying Ornamental Works for Country Residences, and Specimens of the Invisible Fence.'
Wood engraving
April 1809

World's End Tavern
Wood engraving. Half a double plate.
(Reproduced from Faulkner's *History of Chelsea*.)
Supplement 1828

Plan of Mr Salisbury's Botanic Garden. With an elevation of the entrance from Sloane Street. With key.
160 × 345 mm
Aug. 1810

Chelsea Bun House
View of the interior.
Wood engraving in text p. 466
May 1839

Chelsea Hospital
'An East View of Chelsea Colledge & the Rotunda in Ranelagh Gardens.'
Aug. 1748

CHRIST'S HOSPITAL
'S. View of Part of the Grey Friars Monastery, Now Christ's Hospital.'
May 1820

The Grammar School
Wood engraving in text p. 129
Feb. 1821

Churches and Chapels

ALBAN (ST) WOOD STREET
Hour glass
Wood engraving in text p. 200
Sept. 1822

ALL SAINTS, POPLAR
Half a double plate
Hollis Archt. Schnebbelie del.
June 1831

ALL SOULS, LANGHAM PLACE
Half a double plate
July 1826

ANN'S (ST) LIMEHOUSE
Front elevation with details in plan and section.
F. Whishaw del.
Oct. 1828

ANN'S (ST) WANDSWORTH
Half a double plate
Supplement 1829

BARNABAS (ST) CHAPEL, KENSINGTON
Part of a double plate
Vulliamy Archt. Schnebbelie del.
(The caption gives 'St. Mary Abbot's, Kensington'.)
July 1831

BARNABAS (ST) CHAPEL, ST LUKE'S, OLD STREET
Half a double plate
Schnebbelie del.
March 1827

BARTHOLOMEW-BY-THE-EXCHANGE (ST)
(The view shows both St. Bartholomew's and St. Benet Fink before they were demolished to make way for the new Royal Exchange.)
G. Hollis del. & sc.
May 1840

A niche in the south wall
Wood engraving in text p. 153
(Original pencil drawing in the Nichols File, Folger Library.)
Feb. 1841

BARTHOLOMEW-THE-GREAT (ST)
Ruins of part of the Priory
Jan. 1790

View of Raher's tomb
Hulett del. & sculp.
Oct. 1767

Part of coffin lid
Wood engraving in text p. 521
May 1843

Detail of interior
Wood engraving in text p. 127
Feb. 1859

2 Plans. Ground floor level and triforium level
Lithograph T. Hayter Lewis del.
J. R. Jobbins.
Oct. 1863

View of interior with plan of jamb
Lithograph T. Hayter Lewis del.
J. R. Jobbins
Oct. 1863

View of ground floor looking north.
(Showing memorial to Edward Cooke before it was moved to its present position)
Oct. 1863

Capital of Apse and voussoir of Choir
Wood engravings in text p. 396. (Identical with those in text p. 609. Dec. 1859)
Oct. 1863

BEDFONT CHURCH, MIDDLESEX
Sept. 1825

BENET FINK (ST)
(The view shows both St Benet and St Bartholomew-by-the-Exchange before they were demolished to make way for the Royal Exchange.)
G. Hollis del. & sc.
May 1840

BRIDE'S (ST) FLEET STREET
Elevation of steeple
260 × 65 mm
Jan. 1825

'The West Side of St Bridget's alias St Bride's Church London.'
225 × 80 mm
1751 Supplement

Tablet in memory of Dr William Charles Wells ob. 1817 and his mother and father.
W. Jeffreys sculp.
June 1821

CAMDEN TOWN CHAPEL
Dec. 1824

CHELSEA CHURCH
Monumental column to Philip Miller, Botanist, ob. 1771 in the Churchyard.
Nov. 1815

'Capitals in Sir Thomas More's Chapel.'
Dec. 1833

CHRIST CHURCH, MARYLEBONE
Half a double plate.
1825 Supplement

CLAPHAM CHURCH
N.E. View of the Old Church
B. Howlett del. & sculpt.
Dec. 1815

CLERKENWELL CHURCH
'Ancient seat in Clerkenwell Church.'
Wood engraving from a drawing by John Carter in text p. 247
March 1846

DUNSTAN (ST) STEPNEY
Two carvings and the tomb of Sir Henry Colet, Lord Mayor of London in 1486.
Figs. 4, 5, 6 on a miscellaneous plate.
Malcolm sculp.
Aug. 1793
Three windows in the church.
Figures in a miscellaneous plate.
July 1792

View of Stepney Church.
J. P. Malcolm del. & sc.
(Etching)
May 1792

DUNSTAN-IN-THE-WEST (ST)
View of the new church with shops and houses in Fleet Street.
Oct. 1832

Ground plan with the new church superimposed on the old to show their relative positions.
Wood engraving in text p. 298
Oct. 1832

Armorial emblem of the Trinity
in the altar window.
Wood engraving in text p. 301
Oct. 1832

Altar window.
Coloured lithograph, drawn on
the stone by J. Blore.
July 1835

ENFIELD CHURCH
Monument to the Raynton family.
Whole page wood engraving
p. 209
Sept. 1823

EPPING CHURCH
Two views
H. P. Briggs del.
April 1806

GEORGE'S CHAPEL (ST) BATTERSEA
'near Nine Elms in the Parish of
Battersea.'
160 × 204 mm Edwd. Blore Arch.
Aug. 1828

GEORGE'S (ST) CAMBERWELL
With the Surrey Canal in the
foreground.
Half a double plate.
Schnebbelie del. F. Bedford Arch.
Jan. 1827

GEORGE'S (ST) SOUTHWARK
'The Old Font.'
Wood engraving in text p. 367
(Original ink & wash drawing by
Edward John Carlos in the
Nichols File, Folger Library.)
April 1840

GILES (ST) CAMBERWELL
(Original pencil drawing in the
Nichols File, Folger Library.)
April 1825

GILES (ST) CRIPPLEGATE
St Giles's Church with London
Wall and the Cripplegate
Bastion.
May 1817

GILES-IN-THE-FIELDS (ST) CHAPEL
'Chapel of St Giles in the Fields,
erected 1804.'
Half a double plate.
May 1816

GILES-IN-THE-FIELDS (ST)
'Monument of John Lord
Bellasyse in the Church-Yard
of St Giles's in the Fields.'
Longmate del. & sc.
Aug. 1817

GILES (ST) SOUTH MIMMS
F. Cary sculp.
July 1795

HACKNEY CHURCH
Effigy of Lady Latimer
Wood engraving in text p. 162
Feb. 1844

HAMMERSMITH CHURCH
Bust of Charles I erected by Sir
Nicholas Crispe.
Wood engraving, p. 558. From
Faulkner's *History* & *Antiquities
of the Parish of Hammersmith*.
June 1813

HAMPSTEAD CHURCH
'Lady Erskine's Monument.'
Conde del. C. Heath sculp.
April 1824

HANOVER CHAPEL, REGENT STREET
Half a double plate
1825 Supplement

HANWELL CHURCH
View of the church and rectory
April 1800

HARLINGTON CHURCH
Altar-tomb of Gregory Lovell
Aug. 1812

S.E. View
W. Hampner del.
May 1808

HELEN'S (ST) BISHOPSGATE
Exterior view of east end
A. Sinnot del.
May 1800

HIGHGATE CHAPEL
Interior view
April 1834

Arms of Sir Roger Cholmeley
Wood engraving in text p. 381
(Original pencil drawing by Mary
   Ann Iliffe Nichols in the
   Nichols File, Folger Library)
April 1834

HOLY TRINITY, BROMPTON
Half a double plate
Schnebbelie del.
1830 Supplement

HORNSEY CHURCH
S.W. View with details of
   sculpture and inscription.
Schnebbelie del.
July 1810

Carvings of Arms, a Niche and
   the lock on an iron chest.
Wood engraving in text
   pp. 13, 14
July 1832

ISLEWORTH CHURCH
View across the river.
Longmate del. et sculp.
Dec. 1799

JAMES (ST) BERMONDSEY
Oct. 1830

JAMES (ST) CLERKENWELL
'A Representation of a Picture on
   the Altar-piece in the Church
   of St James, Clerkenwell.'
Wood engraving p. 679
(The picture, showing the Virgin
   Mary and Child, was placed
   over the altar in 1735 but was
   considered too Papist and was
   removed.)
Nov. 1735

Bishop Burnet's Monument
Feb. 1817

A stone coffin, fragment of
   painted inscription and carving
   of shields.
Three figures in a miscellaneous
   plate.
Dec. 1788

'Ancient seat in Clerkenwell
   Church.'
Wood engraving in text p. 247
Andrew Wagner del.
March 1846

JAMES (ST) PICCADILLY
Mr Stillingfleet's Monument.
Eng. by J. Roffe
Aug. 1824

JAMES (ST) CHAPEL, TOTTENHAM
COURT ROAD
View of the Chapel of Ease to St
   James Westminster in
   Tottenham Court Rd.
Half a double plate.
May 1816

JOHN'S (ST) HACKNEY
April 1796

JOHN'S (ST) HOLLOWAY
One third of a plate.
Schnebbelie del.
Jan. 1829

JOHN'S (ST) WATERLOO ROAD, LAMBETH.
Half a double plate
F. Bedford Archt. Schnebbelie del.
May 1827

JOHN'S (ST) WESTMINSTER
Dec. 1747

JOHN'S CHAPEL (ST) BETHNAL GREEN
Half a double plate
Schnebbelie del.
Feb. 1831

JOHN'S CHAPEL (ST) HOXTON
Half a double plate
Schnebbelie del.
Mar. 1827

JOHN'S CHAPEL (ST) WALHAM GREEN
Half a double plate
Schnebbelie del.
1830 Supplement

KATHERINE-BY-THE-TOWER (ST)
N.W. View
Feb. 1826

S.W. View 'drawn by Hollar 1660'
Together with 'Representations of some curious Carvings belonging to that Ediface'.
T. Allen sculp.
March 1825

'King Edward III & Queen Philippa taken from Carvings in Wood under the Stalls in the Collegiate Church of St. Katherine near ye Tower.'
Two figures in a miscellaneous plate.
Aug. 1782

KINGSTON CHAPEL
'Inside View of St. Mary Magdalen Chapel.'
May 1809

LAMBE'S CHAPEL
'Architectural Details and Ground Plan of the Hermitage on the Wall.'
May 1825

(1) Monumental Bust of William Lambe
(2) Paintings on Glass of Saints in Lambe's Chapel.
J. Royce del. & sculp.
Jan. 1783

LUKE'S (ST) CHELSEA
Schnebbelie del.
March 1826

LUKE'S (ST) NORWOOD
Half a double plate
F. Bedford Archt.
Schnebbelie del.
May 1827

MARK'S (ST) CLERKENWELL
Half a double plate
Schnebbelie del.
1829 Supplement

MARK'S (ST) KENNINGTON
Half a double plate
D. Roper Archt.
Schnebbelie del.
Jan. 1827

MARK'S (ST) CHAPEL, NORTH AUDLEY STREET
Half a double plate
Schnebbelie del.
Nov. 1829

MARTIN-IN-THE-FIELDS (ST)
'Copied from the large engraving by Vertue.'

Wood engraving. Half a double
   plate
Aug. 1851

MARTIN LE GRAND (ST)
'N E & S E Views of the
   Remains of the Collegiate
   Church of St Martin le Grand.'
   With a plan and architectural
   details. Two plates
J. C. Buckler del. & sc.
Nov. 1818

MARTIN OUTWICH (ST)
View of the church with
   neighbouring houses & shops in
   Bishopsgate St.
J. Basire sculp.
(Original ink & wash drawing in
   the Nichols File, Folger
   Library.)
Nov. 1809

MARY ABBOTS CHAPEL (ST),
KENSINGTON
See Barnabas (St) Chapel

MARY-LE-BONE (ST) CHURCH
Half a double plate
Schnebbelie del.
July 1827

MARY-LE-BOW (ST)
230 × 70 mm
1751 Supplement

MARY OVERY (ST) SOUTHWARK
'Remains of the Old Priory of St
   Mary Overy.'
April 1791

Detail of a doorway.
Fig. (2) on a miscellaneous plate
April 1829

MARY'S (ST) BRYANSTON SQUARE
Half a double plate
Schnebbelie del.
July 1827

MARY'S (ST) GREENWICH
Half a double plate
Schnebbelie del.
Nov. 1829

MARY'S (ST) HAGGERSTON
Half a double plate
Schnebbelie del.
1827 Supplement

MARY'S (ST) ISLINGTON
Monumental tablet to William
   Hawes
J. Basire sc.
April 1811

'A View of the New Church at
   Islington...'
Feb. 1754

MARY'S (ST) PADDINGTON
Views of the old and new
   churches
Longmate del. & sculp.
1795 Supplement

MARY'S (ST) STANWELL
1   N. W. View of Stanwell
    Church
2   Monumental brass to Richard
    de Thorpe
3   Altar tomb of Thomas Windsor
Figures on a miscellaneous plate
B. Longmate del. & sculp.
Nov. 1793

Monument to Thomas Lord
   Knyvet (ob. 1622) and his Lady
Longmate del. & sculp.
April 1794

MARY'S (ST) WILLESDEN
View of the church and details of
   its font

W.T.H. del. (Original pencil
 drawing in the Nichols File,
 Folger Library.)
1822 Supplement

S. W. View of church
Sept. 1795

MARY WOOLCHURCH (ST)
'Foundation Stone of St. Mary
 Woolchurch, 1442.'
R. West del. Cook sc.
May 1788

MATTHEW'S (ST) BRIXTON
Half a double plate
Schnebbelie del.
1829 Supplement

MICHAEL'S CHAPEL (ST) ALDGATE
View of the underground Chapel
April 1789

Sketch plan of above
Fig. 6 of a miscellaneous plate
May 1790

MICHAEL'S (ST) CROOKED LANE
detail of the remains of a crypt
Wood engraving in text p. 295
April 1831

MOORFIELDS ROMAN CATHOLIC
CHAPEL
Half a double plate
Schnebbelie del.
1827 Supplement

MORAVIAN CHAPEL AT CHELSEA
Wood engraving. Half a double
 plate
From Faulkner's *History of
 Chelsea*
1828 Supplement

NICHOLAS (ST) DEPTFORD
S. W. View

Malcolm sculp.
Feb. 1795

NICHOLAS (ST) TOOTING
View of the old church which was
 demolished in 1832.
Fig. 2 on a miscellaneous plate
July 1792

PANCRAS (ST) OLD CHURCH
S. W. View
Nov. 1805

N. E. View.
Half a double plate.
D. Parkes del. (Original
 watercolour drawing by David
 Parkes in the Nichols File,
 Folger Library.)
Feb. 1806

PANCRAS (ST) NEW CHURCH
Half a double plate
Schnebbelie del.
Nov. 1827

PAUL'S (ST) BALL'S POND,
ISLINGTON
One third of a plate
Schnebbelie del.
Jan. 1829

PAUL'S (ST) CATHEDRAL
'The North East View of the
 Nave—'
March 1749

Plan. Wood engraving in text
 p. 130
March 1750

Plan. Wood engraving in text p. 79
Feb. 1753

'Plan of the structure first
 intended to be erected for St
 Paul's Church.'

(Plan of the wooden model.)
Sept. 1783

A view of the Choir from the Altar
190 × 238 mm
1750 Supplement

'Sketch of a monument for Lord Nelson.'
1805 Frontispiece

'Monumental effigy of Dr Donne in his winding-sheet in the crypt.'
J. Basire sculp.
Dec. 1808

Monument to Dean Nowell. Copied from Dugdate's *History of St Paul's*.
J. Basire sc.
May 1811

1 'East End of the North Crypt of St Paul's dedicated to St Faith.'
2 Effigy of Dr John Donne.
Malcolm
Feb. 1820

'Representation of a singular appearance on the Wall of St Paul's Cathedral.'
Two wood engravings in the text p. 99
Feb. 1822

Stone coffin lid found in St Paul's Churchyard
Wood engraving in the text p. 498
Nov. 1841

Stained glass window from Old St Paul's which was opposite the Monument of John Duke of Lancaster.
Wood engraving by Cleghorn
May 1842

PAUL'S (ST) SHADWELL
Lithograph by George Scharf
March 1823

PETER-LE-POOR (ST)
View of East end
Schnebbelie del. Mar. 9 1789
April 1789
West end
Schnebbelie del. Mar. 6 1789
May 1789

PETER (ST) CORNHILL
'A View of the Ruins of St Peter's Church, the Corner of Leadenhall Street.'
Feb. 1766

PETER'S (ST) HAMMERSMITH
Half a double plate
Schnebbelie del.
Feb. 1831

PETER'S (ST) PIMLICO
Half a double plate
Schnebbelie del.
April 1829

PETER'S (ST) WALWORTH
J. Soane, Archt. Audinet, sculp.
Sept. 1826

PHILIP'S (ST) CHAPEL, REGENT STREET
Half a double plate
July 1826

PLUMSTEAD CHURCH
Half a double plate
Nat. Orwade DD 1806
June 1807

REGENT SQUARE CHAPEL, ST PANCRAS
Schnebbelie del.
May 1828

SAVIOUR'S (ST) SOUTHWARK
South-east View showing the
  Lady Chapel restored.
Lithograph drawn by T. Faulkner
Feb. 1832

Interior of the Lady Chapel
Lithograph drawn by T. Faulkner
Feb. 1832

Gower's Monument
F. Nash del. & sc.
May 1830

Monument to the memory of
  J. R. Harris Esq. M.P.
March 1833

'Elevation of the Ancient Altar
  Screen—'
Lithograph. Drawn on Stone by
  John Blore Archt.
Feb. 1834

View of the south end of the
  Crypt with a section.
G. Buckler del. G. Hollis sc.
June 1835

Plan of the Crypt.
Wood engraving in text p. 603
(Original ink sketch in the
  Nichols File, Folger Library.)
June 1835

SEPULCHRE'S (ST) HOLBORN
Interior of the porch.
Fig. 1 on miscellaneous plate.
J. Swaine sc.
1819 Supplement

SOMERS TOWN CHAPEL,
ST PANCRAS
Half a double plate
Schnebbelie del.
Nov. 1827

SOUTHGATE CHAPEL
I. K. Miller del. F. Cary sculp.
(Original ink & wash drawing in
  the Nichols File, Folger
  Library.)
Dec. 1802

STAINES CHURCH
Nov. 1828

STEPHEN'S (ST) WALBROOK
Interior view.
175 × 265 mm
Sept. 1750

Monument to Dr Thomas Wilson
  and his wife Mary.
I. F. Moore del. C. Grignion
  sculp.
340 × 160
Nov. 1788

See also Mansion House

STEPNEY CHAPEL
Half a double plate
Supplement 1829

STEPNEY CHURCH
The ancient font.
Wood engraving in text p. 446
Illustration from *Time's Telescope
  for 1829*.
Nov. 1828

Three windows
Detail on a miscellaneous plate.
(Original ink & wash drawing in
  the Nichols File, Folger
  Library.)
July 1792

Two examples of ancient
  sculpture.
Tomb of Sir Henry Colet.

Malcolm sc.
Details on a miscellaneous plate.
Aug. 1793

TEMPLE CHURCH
'View of the porch of the Temple Church from an old shop-bill of O. Lloyd and S. Gibbons, Stationers.'
Figure (2) on a miscellaneous plate
Dec. 1784
Note. The shop-bill, or trade-card from which this was copied, is in the possession of the Society of Antiquaries; Harley Collection Vol. v. p. 30.

'External elevation of a Norman wheel window over the Western door-way —'
L. N. Cottingham del.
Jan. 1841

'View of the fine Saracenic Arch at the Entrance of the Temple Church.'
190 × 160 mm
July 1783

Twelfth century tomb.
Wood engraving.
March 1835

TOOTING CHURCH
Fig. (2) on a miscellaneous plate.
July 1792

TRINITY CHAPEL, TOTTENHAM
Part of a double plate.
Savage Archt. Schnebbelie del.
July 1831

TRINITY CHURCH, CLOUDESLEY SQUARE, ISLINGTON
One third of a plate.
Schnebbelie sc. & del.
Jan. 1829

TRINITY CHURCH, LITTLE QUEEN STREET, HOLBORN
Jan. 1832

TRINITY CHURCH, MARYLEBONE
Half a double plate
Schnebbelie del.
April 1829

TRINITY CHURCH, NEWINGTON BUTTS
Audinet sculp.
Nov. 1825

TWICKENHAM CHURCH
North west view
Dec. 1809

UPPER TOOTING CHURCH
Wood engraving in text p. 188
August 1855

WEST HACKNEY CHURCH
Half a double plate
Smirke Archt. Schnebbelie del.
June 1831

WESTMINSTER ABBEY
'Plan of the Vault in King Henry VII's Chapel...'
Wood engraving in text p. 138
Mar. 1751

Elevation of West Front
230 × 120 mm
Supplement 1751

'Monument to James Flemming Major General of all His Majesties' Forces–'
L. F. Roubiliac Inv. et fec. Ant. Walker Sculpt.
Aug. 1754

Monument to John Duke of
  Argyll
L. F. Roubiliac Inv. et fecit. Ant.
  Walker Sculp.
Dec. 1754

Monument of Cavendish Duke of
  Newcastle & his duchess.
(The plate has no caption)
I. Dowling delin. Ant. Walker
  Sculp.
Feb. 1755

Monument to John Sheffield
  Duke of Buckingham.
Ant. Walker Sculp.
April 1755

Monument to George Villiers
  Duke of Buckingham.
Ant. Walker Sc.
June 1755

Monument to John Holles Duke
  of Newcastle
Oct. 1755

Monument to Lewis Stuart Duke
  of Richmond
A. Walker Sc.
Jan. 1756

Monument to Lady Anne
  Duchess of Somerset
A. Walker Sc.
Dec. 1756

Monument to Philippa de Mohun
  Duchess of York
Jan. 1757

'The Funeral Procession of his
  late Royal Highness Edward
  Duke of York & Albany.'
130 × 310 mm
Hulett sculp.
Nov. 1767

Monument to Captains William
  Bayne, William Blair and Lord
  Robert Manners who fell in
  Rodney's engagement, April 12
  1782.
Jan. 1794

'Abbot Islip's Architectural
  Memorial as being the finish of
  the Western Part of the Abbey
  Church, Westminster – at the
  West End of the North Side of
  the Nave. Destroyed 1807.'
J.C. (John Carter) del. J.B. (J.
  Basire) Sculp.
(See Brayley & Neale's *History &
  Antiquities of the Abbey Church
  of St Peter Westminster*, vol. II,
  p. 23
April 1808

Fragments of the coffin of
  Edward I
Fig. in miscellaneous plate
Mar. 1809

Fragment of a S. E. View of
  Henry VII Chapel drawn by
  Benjamin Carter about 1747.
Together with details of the old
  parapet of the Chapel from
  various sources and the new
  parapet of 1811. etc.
J.C. (John Carter) del. J.B. (J.
  Basire) sculp.
Nov. 1811

West View – representing the
  Ceremony of his Majesty's
  Crowning.
East View – with the Archbishops
  and Bishops doing Homage.
Two wood engravings
173 × 215 mm
August 1821

Altar in the Chapel of St Blase
 embelished with a painting of
 St Faith.
Dec. 1821

The oaken enclosure of Sebert's
 Tomb.
Wood engraving in text p. 303
Oct. 1825

Painting of Edward the Confessor
 in its present (1825) state.
The same painting as it had
 appeared when Schnebbelie
 drew it in 1791.
Two plates, the first 250 × 82 mm
Oct. 1825

'View of the Choir, Theatre and
 Area.' (During the Coronation
 of William IV.)
Wood engraving 155 × 159 mm
Sept. 1831

'The Homage in the Theatre.'
 (Coronation of William IV.)
Wood engraving 180 × 140 mm
Sept. 1831

'View in the Nave. Showing the
 latter part of the Procession.'
(At the Coronation of William
 IV.)
Wood engraving 152 × 160 mm
Sept. 1831

'Spherical window –'
Wood engraving. Half plate,
 Jewitt del. & sc.
Nov. 1850

(The following Westminster Abbey illustrations, together with the text accompanying them, were published in book form as *Gleanings from Westminster Abbey* by George Gilbert Scott. The first edition was reviewed in the *Athenaeum* on 12 October 1861 and although no credit is given to the *Gentleman's Magazine* in *Gleanings*, it is clear that the material appeared in the magazine first since it pre-dates the publication of the book. I am indebted to David Webb for this information.)

Plan of Westminster Abbey and
 adjoining buildings.'
G. G. Scott Arch. J. H. LeKeux sc.
Feb. 1860

'Archway in the Dark Cloister.'
Wood engraving in text p. 128
1860 Feb.

'Chapel of the Pyx in its present
 state, 1859.'
Wood engraving p. 133
1860 Feb.

'Paving Tiles.'
Wood engraving printed in
 brown.

Two plates facing p. 356
April 1860

Architectural details.
Wood engravings in text on pages,
 131, 132, 134, 135, 136, 137,
 257, 353, 354, 356, 357, 358,
 359, 462, 463, 468.
1860 Feb. to May

'Archway now forming the
 Passage from Little Dean's
 Yard to Great Dean's Yard.'
Wood engraving p. 607
June 1861

Hall of Abbot Litlington.
'Fireplace in the Kitchen.'
Wood engraving p. 605
June 1861

Architectural details from Abbot Litlington's Hall.
Wood engravings. Pp. 603, 604, 605, 606, 608.
1861 June

Part of the Refectory.
Drawn and engraved by J. H. LeKeux
June 1861

Part of the South Walk of the Cloisters.
Drawn and engraved by J. H. LeKeux
June 1861

Jerusalem Chamber. View and plan.
Wood engraving p. 5
July 1861

Mosaic pavement on the Altar Platform.
Lithograph. Jewitt del. & sculp.
March 1863

Mosaic pavement in Chapel of Edward the Confessor.
Wood engraving in text p. 277
March 1863

'Chapter House in its present state.'
Wood engraving in text p. 465
1860 May

'Restoration of the Chapter House.'
G. G. Scott Arch. J. H. LeKeux sc.
July 1860

Chapter House.
1 Entrance from Cloister
2 Vestibule to Chapter House
3 Inner Entrance
4 Eastern Stalls
Four views on one plate. LeKeux sc.
August 1860

(The following illustrations did not appear in the first edition of *Gleanings* but were added to the 'considerably enlarged' second edition published in 1863.)

Iron Work.
Grille of the Tomb of Queen Eleanor. Wood engraving in text p. 662
Part of screen of Henry V's Chantry. Wood engraving in text p. 664
Part of Gates of Henry VII's Chapel. Wood engraving p. 667
Part of Grille of Henry VII's Tomb. Wood engraving p. 669
Early English Chests. Wood engraving p. 673

Chest in Chapel of Pyx with details. Wood engraving p. 675
Dec. 1862

[End of CHURCHES AND CHAPELS]

CLAPTON
See Hackney

CLERKENWELL
Cloister of a Nunnery on the north side of St James's Church.

Figure (1) on a miscellaneous
    plate
Dec. 1785

'Oliver Cromwell's House,
    Clerkenwell Close.'
July 1794

Carved chimney-piece at the
    Baptist's Head in St John's
    Lane.
Figure (1) on a miscellaneous
    plate
Nov. 1813

'Old Houses, St John Street.'
T. Prattent del. et sculp.
Oct. 1814

'House in St. John's Square
    formerly the Residence of
    Bishop Burnet.'
T. Prattent del. & sc.
(Original ink & wash drawing in
    the Nichols File, Folger
    Library.)
June 1817

CLOTH FAIR
Coat of arms in front of a woollen-
    drapers in Cloth Fair.
Figure (4) in a miscellaneous
    plate.
T. Prattent del. & sculp.
Oct. 1795

COLDBATH FIELDS PRISON
'The House of Correction for the
    County of Middlesex, 1794.
    Taken from Grays Inn Lane.'
Malcolm del. et sc.
Supplement 1796

CONDUIT STREET
Elevation of Trinity Chapel in
    Conduit Street.
J. Swaine sc.
June 1804

CORN EXCHANGE
'A Perspective View of the Corn
    Factors Exchange, erected in
    Mark Lane.'
150 × 166 mm
March 1753

CORNHILL
Fire in – See Maps & Plans

COVENT GARDEN
View from the church portico.
April 1747

'In the time of Charles II.'
Wood engraving. Half a double
    plate.
An illustration to *The Story of
    Nell Gwyn* by Peter
    Cunningham.
August 1851

CROSBY HALL
Portion of the exterior showing
    part of the bay window.
E. L. Blackburn del.
W. Wilkinson sc.
Sept. 1836

The new North Front in Great St
    Helen's
Wood engraving in the text p. 286
Sept. 1843

DEAF & DUMB ASYLUM, KENT
ROAD
April 1822

DEAN'S YARD, WESTMINSTER
(1) 'North-east view of the Old
    Dormitory in 1758 when the
    new buildings were begun –
    Copied from a scarce print of
    the same date.' (The 'scarce
    print', which is by W.
    Courtenay, is among the John
    Carter drawings in British

Museum Department of Manuscripts, Add. 29,942 f. 118.)
(2) 'North-east view of the remains of part of the crypt of the same Dormitory discovered in 1815 when the ground was levelled for a new arrangement of the yard.'
(3) Plans and sections of the same.
J. Carter del. J. Basire sculp.
Sept. 1815

Docks
'The West India Docks as they appeared in March 1802 – before the Bason was filled with water.'
J. P. Malcolm del. & sculp.
Oct. 1802

'View of the Proposed St Katharine's Docks.' A bird's-eye view.
Aquatint 170 × 234 mm
Jan. 1826

Drury Lane
'Nell Gwyn at her door in Drury Lane watching the Milkmaids on May-day; the Maypole in the Strand restored.'
Wood engraving in the text p. 30
An illustration to *The Story of Nell Gwyn* by Peter Cunningham
Jan. 1851

Drury Lane Theatre
'W. Front of the New Theatre –'
(Original drawing by T. Prattent in the Nichols File, Folger Library.)
Oct. 1812

Duke's Theatre, Dorset Gardens
View from the river.
155 × 202 mm
July 1814

Dunkirk House
'A Perspective View of Lord Clarendon's House in London, known by the Name of Dunkirk House.'
G. Hart fecit. Cook Sculp.
August 1789

East India House
View of old East India House from a shop-bill of William Overley, joiner of Leadenhall St.
Figure (3) on a miscellaneous plate.
Note. The trade-card from which this was copied is in the Ambrose Heal Collection. See Heal's *Signboards of Old London Shops* p. 45.
Dec. 1784

Eastcheap
'Transverse Section of Roadway in Eastcheap shewing the relative position of the Roman Way.'
Wood engraving in the text p. 422.
(From *The History & Antiquities of – St Michael, Crooked Lane* p. 22.)
Nov. 1833

Edmonton
Elevation of a Gateway.
Figure (2) on a miscellaneous plate.
May 1801

Eltham Palace
Doorway of the Hall.
Jan. 1822

Enfield
Ground Plan of Mother Wells's House and Garden.
View of Mother Wells's Kitchen. (This was the house in which Elizabeth Canning was confined.)
Wood engravings on pp. 306, 307
July 1753

Plan of the site of an Ancient Castle or Camp in Enfield.
Wood engraving in the text p. 427
May 1823

View of Forty Hall – Seat of James Meyer, Esq.
Half page wood engraving
June 1823

House of the late Richard Gough, Esq.
Half page wood engraving
June 1823

Ancient painting formerly in Enfield Church.
Whole page wood engraving. Walker cc.
Supplement 1823

Exchequer
See Westminster Palace

Exeter Hall
View of the Strand frontage.
July 1832

Fetter Lane
Ancient grave stone found in Fetter Lane.
Wood engraving in the text p. 639
Dec. 1843

'Cruelties exercised on two orphan servant girls by Elizabeth Brownrigg of Flower-de-luce Court, Fetter Lane, for which she was tried and executed Sept. 14 1767.'
Sept. 1767

Finsbury
See Maps – Archery ground

Fireworks Edifices
'View of the Fireworks exhibited on the Thames, July 7 1713, being the Thanksgiving Day for the Peace of Utrecht –' (the fireworks edifice which was 'on the Thames over against Whitehall'.)
Figure IV on a miscellaneous plate.
May 1749

'– eastern view of the edifice for the Royal fireworks, which were played off on April 27 1749.' (The fireworks edifice erected in Green Park for the Peace of Aix la Chapelle.)
Figure V on a miscellaneous plate.
May 1749

Fleet Ditch
Anchor found in Fleet Ditch.
Wood engraving in the text p. 417
Oct. 1843

Foundling Hospital
'The habits of the children of the foundling hospital, taken May 1 1747, at the breakfasting, they having baskets of flowers to present to the ladies –'
Figs. V, VI on a miscellaneous plate.
T. Jefferys Sculp.
June 1747

'South view of the Foundling Hospital.' (Key below.) (1) The Wing already built. (2) Chapel now erecting. (3) A Wing to be built.
Sept. 1748

FULHAM
Old Golden Lion. Ancient fire places therein.
June 1838

Prior's Bank, Fulham
(1) Two chairs, a suit of armour, a statue carved in oak of the Emperor Rudolph, and a painting by Zucchero of the Earl of Essex.
(2) View of Prior's Bank.
(3) Three cabinets.
(4) Three Chimney pieces.
Wood engravings on four plates.
A 'gothic' lantern in ormolu.
Wood engraving in text p. 21
'A Gothic Beaufet.' [sic]
Wood engraving in text p. 22
Jan. 1842
(All the items illustrated were auctioned at the sale of contents at Prior's Bank which commenced on 3 May 1841.)

GATEHOUSE WESTMINSTER
'Wall of the Gatehouse remaining in 1836.'
Hollil del. et sc.
'The Gatehouse, Westminster, looking towards Tothill Street.'
Ravenhill del.
Two views on one plate
March 1836

GATES
'Aldgate, Bishopsgate, Moorgate, Cripplegate, Aldersgate, Newgate, Ludgate, Temple Bar.'
160 × 225 mm
Dec. 1750

GEORGE'S FIELDS (ST)
'View of a Crescent as planned by Mr Dance and proposed to be built by Mr Hedger in St George's Fields in Honour of Mr Howard.'
Sept. 1786

GERRARD'S HALL
View of the crypt.
Fig. (1) on a miscellaneous plate.
T. Prattent del. & sculp.
Feb. 1794

GLOBE THEATRE
See Bankside

GREENWICH
'A View of London from One-Tree-Hill in Greenwich Park.'
179 × 242 mm
Jan. 1754

A reconstruction of the Old Palace of Placentia.
G. Hollis sc.
Jan. 1840

Greenwich Hospital. North view from the Isle of Dogs.
May 1748

GRUB STREET
General Monk's House.
April 1790

GUILDHALL
'A View of Guildhall in King Street, London.'
Buildings numbered with key below.
'1. Blackwell Hall. 2. Guild Hall Chappell. 3. Court of Requests Chamber. 4. Irish Society Chamber. 5. Comptrollers Office. 6. Comptrollers House. 7. Common Serjeants & City

Remembrancer's Offices. 8.
City Solicitor's Office. 9. St
Laurences Church.'
177 × 238 mm
Jan. 1751

'Cenotaph at Guildhall to the
Memory of the Earl of
Chatham.'
Cook del. & sculp.
May 1782

'A View of the inside of Guildhall
as it appear'd on Lord Mayor's
Day, 1761.' (Showing the
banquet held in honour of the
Royal Family in Coronation
year.)
180 × 230 mm
Dec. 1761

An ancient coffin found in
Guildhall Chapel.
Wood engraving in the text p. 3
July 1822

GUY'S HOSPITAL
Monument to the founder,
Thomas Guy.
Cook sculp.
June 1784

HACKNEY
'House at Clapton in which Mr
Howard was born, – lately
pulled down.'
T. Prattent del. & sculp.
Original drawing in Victoria &
Albert Museum.
June 1793

HAMPSTEAD
'House built and inhabited by Sir
Henry Vane.'
W. Davison del. J. Smith sc.
May 1828

HAREFIELD
View of Harefield Place.
L. Francia del. J. Greig sculp.
Jan. 1815

HART STREET
'Whittington's Palace, Hart
Street, Crutched Friars.'
Elevations of three sides.
T. Prattent del. & sculp.
July 1796

HAVERSTOCK HILL
Sir Richard Steele's Cottage.
(Original watercolour drawing in
the Nichols File, Folger Library.)
Jan. 1824

HAYMARKET THEATRE
Front elevations of the new and
the old Haymarket Theatres
Both elevations on one plate.
Hixon del. & sculp.
March 1822

HIGHGATE
Elevation of the Ladies' Charity
School at Highgate 'copied
from a scarce print'.
Figure (1) on a miscellaneous
plate.
July 1800

HIGHGATE HILL
See Maps & Plans

HOGARTH'S TOMB
Etching by John Thomas Smith
Nov. 1822

HOUSES OF PARLIAMENT
'A View of the House of Peers,
the King sitting on the
Throne, the Commons
attending him.'
J. Lodge delin & sculp.
178 × 240 mm
Jan. 1769

'Ground-plan of the two Houses of Parliament and adjoining Edifices, showing the extent of the Conflagration.'
Wood engraving in the text p. 481
Nov. 1834

HUNGERFORD MARKET
View of the old market looking towards the river with shops and stalls.
Aug. 1832

Bust of Sir Edward Hungerford on the market house.
Wood engraving in text p. 114
Aug. 1832

View of the River Front of the New Market with three sections showing Great Hall, Upper Court and Lower Court.
Sept. 1832

HYDE PARK
The New Barracks.
(1) View from Knightsbridge.
(2) View from Hyde Park.
Two views on one plate.
160 × 178 mm
July 1797

Elevation and plan of the Humane Society's Receiving-house in Hyde Park.
Figure (2) a, b on a miscellaneous plate.
April 1792

The Royal Humane Society.
(1) View of the Receiving house and a representation of the method of recovering persons from under the ice.
(2, 3) Copies of prints originally engraved and published by Pollard in 1787 from paintings by Robert Smirke representing (2) 'Case of suspended animation' and (3) 'Case of resuscitation'.
Three wood engravings on one plate.
April 1821

'Royal Humane Society's Receiving House in Hyde Park.'
Figure (4) on a miscellaneous plate.
F. Cary sc.
Feb. 1802

HYDE PARK TURNPIKE
B. Longmate Jnr. del. et sculpt.
1792 Supplement

IRONMONGER'S HALL
Front elevation
T. Holden Archt.
Nov. 1750

ISLINGTON
'North View of the Mansion of Sir Walter Rawlegh at Islington.'
(The building is the Pied Bull Inn.)
Jan. 1791

Coat of arms in stained glass in the Pied Bull supposed to have been the arms of one of Sir Walter Raleigh's naval companions.
Figure (5) in a miscellaneous plate
T. Prattent del. & sculp.
Oct. 1795

'The Three Hats Public House and other old houses at Islington.'
August 1823

'An old Public House at Islington.' (The Queen's Head.) With details of carvings on the building.
T. Prattent del. & sculp.
June 1794

JAMES'S PARK (ST)
See Maps & Plans

'Charles II and Nell Gwyn in St James's Park.'
Wood engraving. Illustration to *The Story of Nell Gwyn* by Peter Cunningham. June 1851

JEW'S HOSPITAL
See Mile End Road

JOHN'S (ST) GATE, CLERKENWELL
'The Wood Engraving in this page (by Mr Hughes) now used for the first time, represents accurately the present appearance of St John's Gate.'
Jan. 1820 Titlepage
(St John's Gate, being the publishing office of the *Gentleman's Magazine*, a representation of the gate had appeared on every monthly titlepage since its first number in January 1731 and ten different cuts were used over a period of 100 years. The early ones were very crude and it was not until January 1820 that an attempt was made to give an accurate picture of the gate.)

Arms and a date copied from the Gate.
Figure (c) on a miscellaneous plate
Dec. 1749

A date carved on the north side.
Figure (X) on a miscellaneous plate
March 1748

Various coats of arms on the Gate.
Figures (A–G) on a miscellaneous plate.
Schnebbelie del. Royce sc.
Oct. 1788

A door and various carved details.
Nov. 1813

Proposed restoration. South front.
Lithograph. W. P. Griffith del.
July 1845

JOHN'S (ST) SQUARE
See Clerkenwell

KATHARINE'S (ST) HOSPITAL
(1) View of all the Hospital buildings.
(2) The Master's House
(3, 4) Sculptured panels
C. J. Smith sc.
July 1828

KENSINGTON
'Water or bell tower, Kensington.'
Wood engraving from Faulkner's *History of Kensington*.
June 1821

KENSINGTON GARDENS
'Plan of part of Kensington Gardens from a survey made in 1729.'
Wood engraving. H. Dudley sc.
(Copied from Rocque's *Plan of the Royal Palace & Gardens of Kensington*.)
August 1838

KEW GARDENS
'The Great Pagoda.'
May 1763

'The Water Engine.'
July 1763

'A View of the Palace at Kew from the Lawn.'
154 × 284 mm
Aug. 1763

'A View of the Lake and Island, with the Orangerie, The Temple of Eolus & Bellona, and the House of Confucius in the Garden of Her Royal Highness the Princess Dowager of Wales at Kew.'
170 × 230 mm. Hulett Sculp.
1767 Supplement (The plate bears date Jan. 1768)

'The Mosque in Kew-Garden.'
Sept. 1772

'The House of Confucius in Kew Gardens.'
June 1773

LAMBETH PALACE
'Lambeth Palace as it appeared in the Autumn of 1829.'
May 1830

Gate House.
Wood engraving in text p. 394
May 1830

Interior of the Hall now the Library.
Drawn & Engraved by Robt. Wm. Billings.
Aug. 1834

LEADENHALL STREET
'An arch under a house in Leadenhall St. – part of the remains of the old church formerly on that spot.'
Figure (3) in a miscellaneous plate
T. Prattent del. & sc.
Oct. 1795

Old House on south side of Leadenhall St. Drawn in Feb. 1820 just before it was demolished.
Wood engraving. (Original pencil drawing by E. J. Carlos in Nichols File, Folger Library.
March 1826

Great fire in Leadenhall St. See Maps & Plans

LEWES' PRIORY INN, BERMONDSEY
Plan & view of the crypt with details of columns and windows.
C. Bigot
April 1830

View of the crypt, with plans and details.
Sept. 1832

Architectural details
185 × 223 mm
March 1833

LONDON BRIDGE
'The Outside of St Thomas's Chapel, on the great Pier of London Bridge fronting towards the Tower.'
Sept. 1753

'The Inside View of the Under Chappel of St Thomas within London Bridge.'
Oct. 1753

Three dated inscriptions from
London Bridge.
B. Cole sculp.
Detail from a miscellaneous plate
Oct. 1758

Two views of Nonesuch House.
Wood engravings p. 227
From *Chronicles of London Bridge*
by an Antiquary.
Sept. 1827

'Royal Pavilion erected for the
ceremony of opening London
Bridge.'
Wood engraving in text p. 127
August 1831

'View of the New & Old London
Bridges, from the Tower of St
Saviour's Church.'
Wood engraving p. 124
August 1831

Plan of Old & New London
Bridge.
Wood engraving in text p. 202
Mar. 1832

'Snuff-box, formed out of the old
foundation Piles of London
Bridge.
Designed by W. Knight and
drawn on wood by W. H.
Brooke, Esq. F.S.A.'
Mar. 1832

Old London Bridge during its
demolition Feb. 1832.
Lithograph
Mar. 1832

Old streets about London Bridge
– See Maps & Plans

LONDON HOSPITAL
'Ground Plot of the London
Hospital intended to be erected
in a Field near Whitechapel
Mount.'
Wood cut in text p. 103
Mar. 1752

LONDON WALL
Inscription on a stone in London
Wall.
Wood cut in text p. 369
July 1753

Specimen of the Masonry of
London Wall.
Wood engraving in text p. 607
June 1843

LOWTHER ARCADE
Wood engraving
Mar. 1831

LUDGATE
Roman sculpture found near the
site of Ludgate.
Fig. (1) Head of a female. (2)
Figure of Hercules. (3) An
Altar or Pedestal.
J.C. (John Carter) del. J.B. (J.
Basire) sculp.
Sept. 1806

MANSION HOUSE
(With St Stephen Walbrook.)
May 1748

Maps and Plans

'A New and Accurate Plan of the
Cities of London and
Westminster and Borough of
Southwark, with the New
Roads & New Buildings.'
Extent. Islington – Ratcliff –
Lambeth Palace – Hyde Park.
Scale: 6 ins.: 1 stat. mile.
This map was published in two

parts. Part (1) 404 × 339 mm.
Part (2) 404 × 331 mm.
Part (1) 1763 Supplement
Part (2) March 1764
(It does not appear in Darlington & Howgego *Printed Maps of London*, 1st edn, but is No. 123a in Howgego *Printed Maps of London*, 2nd edn)

'An Accurate Map of the Country Sixteen Miles round London Drawn and Engrav'd from an Actual Survey.'
Extent. Cheshunt – Purfleet – Hayes – Colsdon.
Published in four parts, each 212 × 202 mm
1764. January, April, May and June. To be bound together at the end of the volume.
(Not in Howgego)

BISHOPSGATE
'A Plan of ye Great Fire in Bishopsgate Street, Leadenhall Street and Cornhill etc. on Thursday Novr. 7th. 1765.'
Nov. 1765

CHARING CROSS
'Improvements in the Strand and Charing Cross.'
A plan showing the site of the future Trafalgar Square and the new buildings on the north side of the Strand.
Lithograph R. Martin.
March 1831

CORNHILL
'Plan of the late Fire in Cornhill.'
Wood cut on page 148.
April 1748

FINSBURY
Map of the Archery Grounds in Finsbury Fields showing Archery Marks.
Copy of an unidentified map 'temp. Eliz.' dedicated to Mr. R. Baker & Mr. R. Sharpe and bearing the arms of the Goldsmiths' Company.
215 × 127 mm
March 1832

LONDON BRIDGE
'Plan of ye Old Streets about London Bridge.'
98 × 83 mm
1760 Supplement

NEW ROADS
'A Map of the Surrey Side of the Thames from Westminster Bridge to the Borough. With a Plan for laying out the Roads to Black Fryars Bridge.'
200 × 260 mm
1766 Supplement

'A Plan of the intended New Road from Padington to Islington.'
B. Cole sc.
155 × 465 mm.
Scale: 900 ft. – 1 ml.
1755 Supplement

'A Map of the New Roads & etc. from Westminster Bridge.'
225 × 357 mm
May 1753

'Plan of a New Street, propos'd to be built from the Royal Exchange to London Bridge.'
98 × 96 mm
1760 Supplement

PARLIAMENTARY FORTIFICATIONS
'A Plan of London and Westminster Shewing the Forts

erected by Order of the
Parliament in 1643 & the
Desolation by the Fire in 1666.'
194 × 245 mm
June 1749

ROAD MAPS
London to Chester. 193 × 300 mm
Jan. 1765

London to Bristol & Bath.
180 × 206 mm
March 1765

London to Dover, Rye, Hythe,
Margate, Ramsgate & Deal.
185 × 290 mm
May 1765

London to Portsmouth,
Chichester, Southampton &
Pool. 125 × 290 mm
July 1765

London to Aberistwith via
Oxford & Worcester.
175 × 200 mm
Sept. 1765

London to Land's End via
Salisbury, Exeter, Plymouth
etc. 175 × 350 mm
1765 Supplement

London to Cambridge, Ely &
King's Lynn 180 × 145 mm
April 1766

London to Harwich. 180 × 120 mm
April 1766

London to York. 185 × 298 mm
June 1766

London to St David's.
170 × 285 mm
Feb. 1775

ROMAN LONDON
'Augusta Londinum, or first plan
(as presumed) of Roman
London.'

Wood engraving p. 516
June 1829

ST JAMES'S PARK
'Plan of the Improvements in St
James's Park.'
144 × 257 mm
Oct. 1827

WESTMINSTER HALL & ABBEY
'Plan of Westminster Hall &
Abbey. The Line of Procession
and Galleries etc. erected for
Spectators on the occasion of
the Coronation of His Majesty
George IV, July 19th. 1821.'
200 × 275 mm
Lithograph J. Wyld del.
July 1821

[End of Maps & Plans]

MARSHALSEA PRISON
'Inside View of the Palace Court
of the Marshalsea, Southwark.'
T. P(rattent) del. & sculp.
1803 Supplement

'North View of the Marshalsea
Southwark before the New
Buildings.'
T. P(rattent) 'from an Original
Drawing by Lieut. Orme in the
Possession of W. Jenkins Esq.'
May 1804

'Marshalsea before the New
Buildings were erected.'
T. P(rattent) del.
Sept. 1803

MARYLEBONE GARDENS
Plan of the Garden at Mary-le-
bone Park in the year 1659.
Figure (5) in a miscellaneous
plate.
June 1813

MIDDLESEX HOSPITAL
'South East View...'
J. Paine, Archt. Exemplar dedit.
　E. Rooker Sculp.
158 × 244 mm
Jan. 1757

MILE END ROAD
'Jew's Hospital, Mile End Road, Whitechapel.'
T. Prattent del. & sculp.
Dec. 1819

MONUMENT
The top of the Monument.
Figure (B) on a miscellaneous plate.
Dec. 1749.

'The Emblematical Figures on the monumental Pillar...'
Figure (A) on a miscellaneous plate.
Dec. 1749

NATIONAL GALLERY
Plan & Elevation of the 'Proposed New National Gallery'. (An impracticable suggestion which never materialized.)
Wood engraving in text p. 222
Aug. 1856

NEW RIVER
'Perspective View of the Arch under the New River at Bush Hill.'
157 × 237 mm
Oct. 1784

'View of the New River as conveyed through the Frame at Bush Hill.'
171 × 355 mm
Sept. 1784

NEW ROADS
Paddington to Islington/To Blackfriars Bridge/From Westminster Bridge/Royal Exchange to London Bridge.
See... Maps & Plans

NEWGATE PRISON
'The New Platform and Gallows in the Old Bailey.'
Dec. 1783

'Elevation of an intended new Building for the reception of Felons & Debtors, in the room of Newgate, Shewing the External part from North to South.'
'A Section, Shewing the Internal part...'
150 × 400 mm
Sept. 1762

'A Front View of the Goal as it is now erecting in the Old Bailey
135 × 302 mm
Aug. 1770

'The Windmill fixed on Newgate to work the Ventilators...'
A side elevation and detail of the sails.
April 1752

'Ventilators erected in Newgate...'
April 1752

Part of a board used by prisoners to effect their escape.
Wood cut on page 378
Aug. 1750

NEWINGTON BUTTS
'Rectorial House at Newington Butts.'
Rt. Lane fect.
1794 Supplement

NONESUCH PALACE
Copy of the engraving by Hoefnagel.
Geo. Hollis Sc.
Aug. 1837

NORTHUMBERLAND HOUSE
Elevation of the frontage.
Feb. 1752

NOTTING HILL
Stone coffin found by workmen digging foundations for the new buildings in Victoria Park, near the Hippodrome, on Notting Hill.
Wood engraving in text p. 499
Nov. 1841

OLAVE'S (ST) FREE GRAMMAR SCHOOL
View
G. Hollis Sc.
Jan. 1836

Seal of the School dated 1576.
Wood engraving in text p. 144
Jan. 1836

OLD FORD
'Gateway of King John's Palace at Old Ford.'
T. Prattent del. & sculp.
1793 Supplement

OLD JEWRY
'Mansion House, Old Jewry – formerly the residence of several Lord Mayors – at present it is occupied by The London Institution.'
(Original ink & wash drawing by T. Prattent in the Nichols File, Folger Library.)
1811 Supplement

OPERA HOUSE, HAYMARKET
'A View of the Dresses at the late Masquerade given by the King of Denmark.'
J. Lodge delin. et sculp.
Nov. 1768

PANTHEON, OXFORD STREET
'View of the great Saloon of the Pantheon, London, from the North Gallery.'
Drawn & Etched by Thos. Kearman
160 × 134 mm
Jan. 1835

PARSONS GREEN
'A picturesque old barn on the Fulham road near Parsons Green.'
Figure (3) on a miscellaneous plate
1801 Supplement

PAUL'S (ST) CROSS
'...as it appeared in 1621.'
Figure (3) on a miscellaneous plate.
Nov. 1784

PAUL'S (ST) SCHOOL
View from the churchyard of St Paul's Cathedral.
Half a double plate.
Sept. 1818

PITCHER'S COURT
'House in Pitcher's Court, Bell Alley. Once the residence of Robert Bloomfield.'
(Original ink & wash drawing in the Nichols File, Folger Library.)
Dec. 1823

POST OFFICE
Oct. 1829

PRYOR'S BANK, FULHAM
See – Fulham

PUTNEY
View of the Tete du Pont in
  Putney Fields.
Wood engraving p. 206
Sept. 1812

'East Prospect of Fulham Bridge,
  over the River Thames, from
  Fulham in Middlesex to Putney
  in Surrey.'
Fig. (IV) of a plate of 6 bridges.
30 × 436 mm
July 1751

REGENT'S CANAL
'Mouth of the Tunnel on the
  Regent's Canal, Islington.'
The western end of the tunnel
  with White Conduit House in
  the distance.
T. Bonnor del. & Sculp. (Original
  watercolour drawing in the
  Nichol's File, Folger Library.)
Aug. 1819.

REGENT'S PARK
Old Queen's Head and Artichoke.
Nov. 1819

ROAD MAPS
See... Maps & Plans

ROMAN LONDON
'Roman Remains found in
  Southwark. 1832.'
Plate illustrating 18 fragments
  and objects.
W. Taylor del.
July 1832 and May 1833

For 'Augusta Londinum'
  see... Maps & Plans

ROYAL EXCHANGE
Front as altered in 1820
Wood engraving in text p. 494
Nov. 1844

View of second Royal Exchange.
Wood engraving in text p. 493
Nov. 1844

SADLER'S WELLS
View of the old building
Wood engraving. Half a double
  plate
From Lewis's *History of Islington*.
July 1842

SANDFORD MANOR HOUSE,
FULHAM
'...formerly the Residence of Nell
  Gwyn.'
From Faulkner's *History of
  Fulham* and *History of
  Kensington*.
Wood engraving in text p. 243
Mar. 1813

SHOE LANE
See... Bangor House

SHOT TOWER
View of the Manufactory for
  Patent Shot established by
  Messrs Watts.
Wood engraving in text p. 77
From Allen's *History of Lambeth*
Jan. 1826

SHREWSBURY HOUSE, CHELSEA
Wood engraving. Half a double
  plate.
June 1829

SLOANE, SIR HANS. STATUE OF.
See... Apothecaries' Garden

SMITHFIELD
Obelisk proposed to be erected in Smithfield.
Wood engraving in text p. 23
Jan. 1752

SOMERSET HOUSE
'Front View of the Royal Academy, Royal & Antiquarian Societies etc. in the Strand.'
J. Carter del. J. Royce sc.
125 × 215 mm
Jan. 1779

'Back Front of the Royal Academy, Royal & Antiquarian Societies etc. in the Strand.'
Carter del. Royce sculp.
125 × 310 mm
Aug. 1779

'View of the Inner Part of Old Somerset House.'
Medland sculp. Ryley del.
The view shows the old building in the process of demolition.
Jan. 1798

SOUTHWARK
'Monteagle House near St. Saviour's Church, Southwark.'
Sept. 1808

'The Talbot Inn, Borough High Street.'
(Original ink & wash drawing by T. Prattent in the Nichols File, Folger Library.)
Sept. 1812

STAMFORD STREET, BLACKFRIARS ROAD
See...Benevolent Soc. of St Patrick

STANWELL
N. E. View of Stanwell Free School
Fig. (4) on a miscellaneous plate.
B. Longmate del. & scul.
Nov. 1793
See also...Mary (St) Stanwell

STATIONERS' HALL
'Painted Window presented by Mr Alderman Cadell to the Company of Stationers.'
Longmate del. & sculp.
Nov. 1814

STEPNEY
Dean Colet's house.
Half a double plate.
Sept. 1818

STOKE NEWINGTON
Rose and Crown Inn.
Wood engraving in text p. 389
From Robinson's *History of Stoke Newington*
Nov. 1820

STRAND
'Mr Alderman Pickett's Plans of Public Improvement.'
'View of the North Side of the Street from the East of Temple Bar.'
'View of the North and South Side from the West of St Clements Church.'
Two views on one plate
185 × 150 mm
W. Thomas sculp.
Dec. 1793

'The New Buildings in West Strand.'
The new block bounded by West Strand, William St. and Adelaide St. showing the facade

of the Lowther Arcade and the
'pepper pot' towers at the
corners.
Wood engraving p. 204
March 1831

SURGEON'S HALL
Front elevation
Nov. 1752

SYON HOUSE
G. Hollis del. et sc.
May 1847

TALBOT INN, BOROUGH HIGH
STREET
See... Southwark

TAVISTOCK SQUARE
'The New Gothic Chapel near
Tavistock Square.'
May 1802

TEMPLE
Middle Temple Hall
S. Rawle del. & sculp.
Oct. 1798
See also... Temple Church

THAMES
'A View of the Barges conducting
his Danish Majesty from
Whitehall to the Temple (Sept.
23rd. 1768) in his way to the
Mansion House.'
168 × 244 mm
Oct. 1768

THAMES STREET
View of Thames Street on fire on
the 13 Jan. 1714/15. (From a
drawing by Mr Gery Strong.)
Fig. (1) on a miscellaneous plate.
Dec. 1784

THEOBALDS PALACE
A reconstruction
Vinkenboone del. Hollis sc.
Feb. 1836

TOTTENHAM
'Tottenham High Cross, 1805.'
This plate had already appeared
in Robinson's *History of
Tottenham* for which it was
'engraved from an original
drawing' in 1818.
April 1820

Fountain 'near the Turnpike at
Kingsland'.
Wood engraving in the text p. 502
June 1822

TOWER
Representation of the White
Tower on the private seal of
William de Turri.
Wood engraving in text p. 246
Sept. 1825

Arms & inscriptions on the wall
of the Beauchamp Tower.
Mar. 1791

TOWER MENAGERIE
'Marco, a Lion in the Tower.'
160 × 170 mm
Feb. 1749

TWICKENHAM
'Mr Pope's Villa at Twickenham.'
S. Lewis del. F. Cary sculp.
1807 Supplement

UPPER MALL HAMMERSMITH
'Pavilion of Queen Catherine.'
Wood engraving. Half a double
plate.
(From Faulkner's *History of
Hammersmith*)
Sept. 1839

VAUXHALL GARDENS
'A Perspective View of the Grand
  Walk in Vauxhall Gardens and
  the Orchestre.'
164 × 263 mm
Aug. 1765

VINTNERS' HALL
Ancient painting at Vintners' Hall
  showing the History of St
  Martin.
J. Royce sculp.
April 1783

WALBROOK HOUSE
(Fig. (2) on a miscellaneous plate
T. Prattent del. & sc.
Oct. 1795

WANSTEAD HOUSE
Metz del. Heath sculp.
June 1830

WEST HAM
'West Ham Abbey Gate.'
The only remnant of the
  Cistercian Abbey of Stratford
  Langthorne at Bow.
T. Prattent del. & sculp.
Oct. 1793

WESTMINSTER BRIDGE
'The Elevation of the Centre
  Arch –'
190 × 180 mm
Feb. 1754

Elevation
T. Jefferys sculp.
One of 5 elevations of bridges on
  the same plate.
30 × 450 mm
1750 Supplement

'A South View of Westminster
  Bridge.'
T. Jefferys sculp.
Key below the view which
  includes Westminster Abbey
  and Palace.
1746 Supplement

'The engine for driving the piles
  of the new bridge at
  Westminster.'
Fig. (III) on a miscellaneous plate.
Sept. 1751

'Design'd for Westminster Bridge
  by Mr King.' Detail of the
  centre arch 'when the
  superstructure was proposed by
  the commissioners to be built
  with timber'.
Fig. (II) of a plate of 6 bridges.
July 1751

Sections and plans of piers and
  arches.
250 × 195 mm
1752 Supplement

WESTMINSTER HALL
South View – representing the
  manner of serving up the first
  course at the Coronation
  Banquet.
North View – with a
  representation of the Champion
  entering through the
  Triumphal Arch. (Coronation
  of George IV.)
Two wood engravings
212 × 173 mm
Aug. 1821

'A View of the Scaffolding in
  Westminster Hall.'
(The view, looking south, shows
  the Hall prepared for the trial
  of Lord Lovat.)
168 × 150 mm

Also a separate plate showing the
  seating arrangements for the
  trial.
March 1747

Architectural details
J. Basire sculp.
Nov. 1808

WESTMINSTER PALACE
Back view of the old buildings of
  the Exchequer which were
  removed in the spring of 1793
  taken from a window in the
  Ordnance Office in St Margaret
  Street.
1806 Supplement

'Three elevations of part of
  Whitehall Palace (now the
  Treasury) in the original and
  present states.'
J. Carter del. J. Basire sculp.
Dec. 1816

'Exterior & Interior views of the
  Princes' Chamber & Old House
  of Lords.'
Three engravings on one plate.
Dec. 1823

'Plan of the Powder Plot Cellar
  and other Buildings Adjoining
  the Old Palace, Westminster.
  (With a key on p. 209.)
Lithograph, 'Litho. 12 Fludyer
  St. Westm.'
Sept. 1825

'Ruins of St Stephen's Chapel,
  (House of Commons) Seen
  from the roof of Westminster
  Hall.'
B. W. Billings del. & sculp.
Sept. 1835

'Vestiges of Sculpture & Painting
  in St. Stephen's Chapel,
  Westminster, after the Fire of
  October 1834.'
R. W. Billings sc.
Jan. 1836.

WHITE CONDUIT HOUSE
View of the conduit which gave
  its name to White Conduit
  House tea garden and an
  inscription thereon.
Figs. (1, 2) on a miscellaneous
  plate
1801 Supplement

View of the old building
Wood engraving. Half a double
  plate.
From Lewis's *History of Islington*.
July 1842

See also – Regent's Canal

WINCHESTER HOUSE, BROAD
STREET
G. Hollis sc.
April 1839

WINCHESTER PALACE, SOUTHWARK
Plans, Elevations & Sections of
  the remains of Winchester
  Palace.
G. Gwilt del. Longmate sculp.
June 1815

North west view of ruins.
  Circular window. Details. Plan.
J. Carter del. J. Basire sculp.
Dec. 1814

'Remains of the Bishop of
  Winchester's Palace, Bank Side,
  Southwark.'
1791 Supplement

WINCHESTER PALACE, CHELSEA
Wood engraving in text p. 506
June 1822

WOODFORD
View of Hereford House,
 Woodford
Fig. (1) on a miscellaneous plate
T. Prattent del. & sculp.
Oct. 1795

WORMWOOD SCRUBS
'View of the Railroad bridges.'
The view shows where Wood
 Lane and the Paddington Canal
 cross the Birmingham, Bristol
 & Thames Junction Railway.
Wood engraving in text p. 283
 (From Faulkner's *History of
 Hammersmith*.)
Sept. 1839.

X. RICHARD HORWOOD'S PLAN OF LONDON: A GUIDE TO EDITIONS AND VARIANTS, 1792–1819

By PAUL LAXTON

THE story of the making of Horwood's plan of London, in so far as it can be reconstructed, has already been told in the companion introduction to the London Topographical Society's facsimile of the 1813 edition, *The A to Z of Regency London* (Publication no. 131, 1985). This article may be regarded as an appendix and an addendum to the *A to Z*. It has four sections.
    (a) Some further biographical material about Richard Horwood, which came to my attention too late to be included in the *A to Z*, is placed on record.
    (b) A checklist of changes made to the plates during the printing of the first edition of 1792–9.
    (c) A description of the main changes to the plates in the three editions published by William Faden in 1807, 1813 and 1819.
    (d) A list of the titles and imprints found in all the various editions of the map.
The main purpose is to enable the reader to place any copy of the plan – either the whole work or a single sheet – into its proper sequence. This in turn provides a guide to the nature and extent of Horwood's revisions: they add to our understanding of the map's publication history, of the process of production of large multi-sheet maps in general and of the topographical development of London between the 1790s and 1819. That last topic is ubiquitous in this paper, but only in so far as topographical changes are reflected in changes to the plates of the map; it is not discussed directly. However, the author hopes that by drawing attention to the continuous development of the

Horwood plates, he will stimulate *systematic and critical* study of this evidence for the building and rebuilding of London rather than the mere search for the origins of this or that individual house or street.

The introduction to the *A to Z* described the complex way in which copies of the first edition of the plan were assembled from 32 plates, which were in varying degrees altered by the addition or removal of material between their first printing and their last. Three plates appear to have remained unchanged, two of them have been found printed in six different states, and the rest lie somewhere between these two extremes. The number of recorded subscribers is 838, but it is obvious that Horwood did not have 838 prints pulled from each plate after it had been first engraved. So in theory we could imagine that no two copies of the plan were the same, a customer's order being made up from whatever limited stocks of each sheet were available at the time. In practice neither the revisions nor the printing were evenly spread throughout the period of production. Nevertheless, the author has found that only a few of nearly fifty copies he has examined were identical.

For those who have not spent many months comparing, square inch by square inch, 32 large plates against a master photocopy some explanation may be helpful. At several stages in the research the investigator discovers variations which he suspects, because of their frequent association with other changes to the plates, he may have missed on some earlier copy. Return visits and precise enquiries of curators clear up most of the ambiguities, and the sequence of changes gradually falls into place. However, the investigator is always left with the nagging suspicion that intervening states have been missed. And then, of course, there are copies to which he has not had access. It is expected, therefore, that this checklist will cause further states to be discovered. The author or the Editor of the London Topographical Record would be grateful to be told of any additions or errors.*

* Department of Geography, University of Liverpool, PO Box 147, Liverpool L69 3BX

(a) Richard Horwood in Staffordshire

In 1803, in one of his letters to Charles Taylor, Secretary to the Royal Society of Arts, Horwood referred to 'My Brother from Trentham'.† Trentham Hall in Staffordshire was one of the estates of the Marquis of Stafford, Granville Leveson-Gower: the estate's agent since 1774 had been Thomas Horwood. As part of his scheme to introduce new efficient management Thomas had Trentham re-surveyed by his brother Richard, who would have been aged 21.‡ Some maps associated with this survey, redrawn by Richard Horwood and dated 1782, survive – his only extant work apart from his plans of London and Liverpool.¶ This glimpse of a career in estate surveying, whilst establishing Horwood's credentials, leaves a question mark over the intervening nine years, between his work revising rural estate maps in 1782 and the start of his surveying in the West End of London.

(b) Variants of the plates in the first edition, 1792–9

All but 3 of the 32 plates have been found to have had at least two different states, so that the number of different copies of the whole plan, made up of combinations of plates in different states, is likely to be substantial. In order to make this checklist intelligible to those with only one copy of the plan before them, features added or (less frequently) deleted after the first state are usually described in the notes on the first state as absent or (less frequently) present. The descriptions of the states must therefore be read in sequence. Words or characters shown here in italics are transcribed directly from the plan, though superscript letters in contractions are not reproduced here in that form. N, S, E

† Royal Society of Arts, Manuscript Transactions 1802–03, section 17. Letter from 32 Norfolk Street, Liverpool, dated 21 April 1803
‡ A. D. M. Phillips, 'A note on farm size and efficiency on the North Staffordshire estate of the Leveson-Gowers, 1714–1809', *North Staffordshire Journal of Field Studies* 19 (1979), 31 and 38. R. Wordie, *Estate management in eighteenth-century England. The building of the Leveson-Gower fortune* (Royal Historical Society, 1982), 47–59.
¶ Staffordshire Record Office D593/H/3/32 and 344.

and W have been used throughout in preference to top, bottom, right and left.

Plate A1

1st state
> Has a small letter A below S border and lacks number 1 outside E border.
> Lacks house numbers in Nottingham Place, Nottingham Street, Northumberland Street (N of Nottingham Street), David Street, York Place, Upper Baker Street and East Street (Leading out of David Street).
> Lacks large building on N side of the New Road above the letters *RO*.
> Has *YORK PLACE* on S side of New Road.
> Lacks any houses in York Street E of Quebeck Street and in String Street.
> Lacks 2 houses S side of David Street below *ID*.
> Has *YORK PLACE UPPER* but lacks *NORTH* after *BAKER STREET*.
> Field boundaries and trackways in middle third of sheet as on Fig. 1.
> Lacks track running across Lisson Green WSW from Yorkshire Stingo.

2nd state
> Large letter A added below S border and figure 1 added outside E border.
> 2 houses added in David Street below *ID*.
> House numbers added in Nottingham Place, Nottingham Street, Northumberland Street (N of Nottingham Street), David Street, York Place (2 N blocks only), Upper Baker Street and East Street.
> Buildings added and shaded, on N side New Road above letters *RO*.
> Track across Lisson Green added.
> Field boundaries and trackways altered.

3rd state
> *NORTH* added after *BAKER STREET*.
> *UPPER* deleted after *YORK PLACE*.
> Premises added in York Street (7 on N, 5 on S) and in String Street (E side).
> *YORK BUILDINGS* replaces *YORK PLACE* in New Road.

Viewplate
> The viewplate was a separate copper plate and was replaced by another separate plate extending the fields in the middle of the plate further northwards. In theory the viewplate could have been printed on sheet A1 with the S part of the sheet in any of the three states described above. In practice it was printed only with the first two states.

Plate B1

1st state
  Lacks all parish boundaries.
  Lacks the circle in Fitzroy Square.
  Lacks house numbers in Fitzroy Square, Cleaveland Street, Warren Street, Fitzroy Street, Upper Titchfield Street, Chapel Street, Howland Street, Grafton Street, Tottenham Court Road, Charlton Street, Carburton Street and Upper Marylebone Street N side.
  Lacks *Longman & Broadreip* with pointing hand in Tottenham Court Road.
  Lacks *Fitzroy Market*, all its buildings and premises fronting Hertford Street.
  Lacks *Mortimer Market* and its buildings off E side Tottenham Court Road.
  Lacks *Jones, Taylor & Cos Manufactory* off E side Tottenham Court Road.
  Lacks *Grafton Mews* E of Fitzroy Square.
  Lacks *Holbrook C[ourt]* and *Holbrook B[uildings]s* E of Fitzroy Market.
  Lacks 3 houses (numbers 41–43) on W side Upper Titchfield Street.
  Lacks 2 houses on N side Buckingham Street (E end) and 11 houses adjacent on W side Cleaveland Street N of Buckingham Street.
  Lacks houses on S side Warren Street W of Fitzroy Street (except for a group of 5 on E corner of Conway Street).
  Houses in N block E side Upper Titchfield Street are not joined in a continuous terrace.
  Marginal sheet numbers and letters smaller than in later states with the figure aligned E-W not N-S.
  Lacks premises in Pancras Street standing alone beyond number 6 on N side.
  Has house numbers in Chapel Street (SE corner).

2nd state
  Parish boundary (Marylebone/St Pancras) added S centre to NW.

3rd state
  Longman & Broadreip with pointing hand added in Tottenham Court Road.

4th state
  House numbers added in Tottenham Court Road N of Frederick Street to the boundary of the viewplate.
  Marginal plate numbers and letters re-engraved at larger size and the numbers aligned N-S.

5th state
  Circle added in Fitzroy Square.

6th state
Boundary of Street Giles's Parish added in SE.

House numbers added in Tottenham Court Road (S of New Road from Paddington), Fitzroy Square, Cleaveland Street, Warren Street, Fitzroy Street, Upper Titchfield Street, Chapel Street, Howland Street and Grafton Street.

Premises in Pancras Street added.

10 houses added on S side Warren Street with 3 adjacent in Conway Street (W side) and adjacent in Fitzroy Street (W side).

Viewplate
The view is found with states 1 and 2. See note under Plate A1.

Plate C1

1st state
Lacks *Dukes Row* just S of Somers Town.

2nd state
Dukes Row added.

Plate D1

1st state
Lacks *Eagle Court* off St John's Lane.

2nd state
Eagle Court added.

Plate E1

1st state
Lacks *STREET* close to S border W of White Cross Street.

Lacks *Lewingtons B*[*uildings*]. in a court NE of St Luke's Hospital.

Lacks building S of *OLD* in Old Street with the names *Morman Smith & Sawmaker* and *Moreland & Co. Bricklayers*.

Lacks *Cox's Cooperage* and *Mr. Chippingdales Cooperage* with associated buildings near centre of S border.

2nd state
STREET added.

3rd state
Lewingtons B. added.

4th state
The four names of firms lacking in the 1st state now added with the buildings associated with them.

Red Lion Market replaces *White Cross Square* off S end White Cross Street.

Plate F1

1st state
Lacks *Webbs Build[ing]s* off S side Hare Street.
Lacks *Mr. Allports Nursery* above *E* in *HACKNEY ROAD*.
Lacks *New Turville St.* off Cock Lane and *Half Nicols St* off New Nicols Street.
Lacks *Plow Yard* off Shoreditch in SW corner.
Lacks the following near Great Pearl Street by S border: *White Horse Court*, *Pig Alley*, *Vine Place*, *Crown Co[urt]* and *Popes Head Co[urt]*.

2nd state
All 11 names missing in the first state now added.

Plate G1

No variants observed

Plate H1

1st state
Lacks parish boundary between the title and SE corner of the plate.

2nd state
Parish boundary on E side of plate now added.

Plate A2

1st state
Lacks *HIGH* in NE corner.
Lacks numbers and key to Earl of Leicester's house in Grosvenor Square.
Lacks the mews in blocks E and W of Baker St near N border.
Lacks gardens and mews between Gloucester Street and Life Guards Stables.
Lacks all houses on W side of String Street and 4 adjoining houses in Durweston Street.
Lacks 5 houses on E side String Street immediately S of Durweston Street.
Lacks 4 houses on N side Durweston Street above *EET*.
Lacks 2 most northerly houses on W side of Great Cumberland Street and the wall or fence behind them.
Lacks chapel [Brunswick Chapel], 4 adjoining houses, with mews and 3 buildings behind, on N side of Upper Berkeley Street.

Lacks 13 houses in Gloucester St: 7 on E side immediately N of King Street, 6 on W side immediately S of Dorset Street.
Lacks lodge and short path into Hyde Park from Oxford Street.
Lacks path across Hyde Park from S border to opposite Kings Street mews.
Lacks avenue of 8 trees in Park Lane near Upper Grosvenor Street.
Lacks figure *2* outside E border.
Lacks *Greens Repertorium* in Edwards Street S of Manchester Square.

2nd state
HIGH added in NE corner.
2 added outside E border.
Earl of Leicester's house now has numbered rooms and a key to the numbers.
Avenue of 8 trees added in Park Lane.
Path added across Hyde Park.

3rd state
Greens Repertorium added in Edwards Street S of Manchester Square.

4th state
Houses added in String Street, Gloucester Street and S side Durweston Street.
Lodge and short path into Hyde Park from Oxford Street added.
A chapel [Brunswick Chapel] and adjacent buildings added.

5th state
Mews added in blocks E and W of Baker Street and between Gloucester Street and Life Guard Stables, the latter mews with gardens.
Four houses added on N side of Durweston Street above *EET*.

Plate B2

1st state
Has small marginal plate numbers and letters, the numbers on their side aligned E–W.
Lacks all parish boundaries.
SHUG LANE printed upside down at the E end of Piccadilly.
Lacks *THAYRE STREET* (and the street itself) running N out of Hinde Street.
St James's Market is depicted as 9 regular shaded blocks of stalls.
All the names of firms (listed below under the 3rd state) and the pointing hands indicating their location are lacking.

2nd state
Parish boundaries added except for boundary of St Giles's parish (NE corner).

3rd state

The following firms are now named and their locations indicated by a small pointing hand:

Abrm. Daykin (Duke Street).
Bickley & Lardinors Ironmongers (Berners Street)
Wren Coach Maker (Wigmore Street)
Richardsons, Sadlers (Oxford Street)
Irwin Hatter (Oxford Street)
Watson Linnendraper (Oxford Street)
Shrader Grd. Piano Forte Maker to her Majesty (Princes Street)
Jenden Linnendraper (Conduit Street)
Ganers Piano Forte & Harp Maker (Broad Street)
Houston & Co. Grand Piano Forte Manufacturers (Wardour Street)
Patrick & Co. Linn. Draprs (Air Street)
Stewarts Tin Manufactory (Piccadilly)

Second sequence of house numbers added in Berkeley Square and Bruton Street.

Stalls in St James's Market extended northwards and some of the rows filled in by shading.

St Giles's parish boundary added in NE corner.

4th state

Marginal plate numbers and letters re-engraved at a larger size with the numbers realigned N–S.

5th state

SHUG LANE (upside down) at E end of Piccadilly replaced by *TITCHBORNE STREET*.

Watson Linnendraper (between the two Es of *OXFORD STREET*) and *Wren Coach Maker* (over *RE* in *WIGMORE STREET*) deleted.

Stippling in the gardens in the centre of Hanover Square, Cavendish Square and Berkeley Square replaced by a symbol representing grass.

Many mews and other back buildings highlighted by re-engraving or over engraving (e.g. Adams Mews, Blenheim Mews).

Plate C2

1st state

Lacks colonade on N side Drury Lane Theatre.
Lacks *Princes C[ourt]* between Great Suffolk Street and Whitcomb Street.
Lacks *Falconbridge C[ourt]* N end Crown Street.
Lacks boundary wall behind 11 Bedford Square running to N border.

2nd state

Falconbridge C and *Princes C* added.
Colonade added to Drury Lane Theatre.

HORWOOD'S LONDON PLAN 223

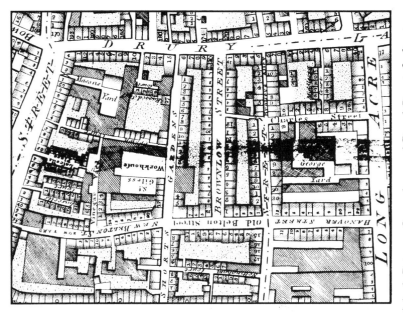

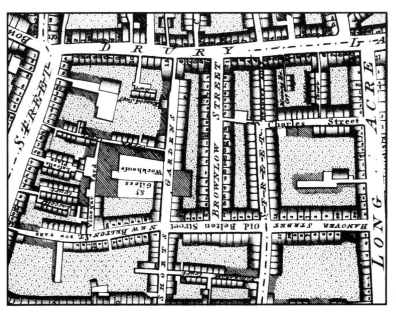

Fig. 1. Part of Horwood's plan: plate C2. Scale 0.85 × original. Between Long Acre and Broad Street. Left: 1799 edition. Right: 1807 edition. These details show characteristic 'tidying up' and re-numbering by Faden's engraver.

3rd state
Boundary wall added behind 11 Bedford Square.

Plate D2

1st state
Lacks stippling in NE St Paul's Cathedral churchyard.
Lacks parish boundary from Lincolns Inn New Square to W border.
Lacks *North Hoare & Co.* at 35 New Bridge Street.
Lacks *Crane Co[ur]t.*, *Green Abor Court* and courts containing 16 houses and one other building, all off Upper Thames Street at E border.

2nd state
Stippling added NE of St Paul's to E border.
Parish boundary added near W border.
North Hoare & Co. added.

3rd state
Crane Cot. and *Green Arbor Court* added, with associated buildings in courts.

Plate E2

1st state
Lacks *Little Moorfields* and *Broker Row* along E side of Moor Fields and Bethlem Hospital.
Has *WHITES C.* in a court off Little Brittain near W border.
MR. WHITBREADS BREWERY in small stamped roman letters in the courtyard of a shaded building near N border.

2nd state
Mr. Whitebreads Brewery prominently engraved (over a building which is now unshaded) in place of small stamped letters.
WHITES C. changed to *Cross Keys Sq.*

3rd state
Little Moorfields and *Broker Row* added.

Plate F2

1st state
Lacks parish boundary N of the Tower of London between Coopers Row and E side of Little Tower Hill.
Lacks *Mr. Coope* in a yard N of Whitechapel off Osborne Street.
Lacks *Montague Co[ur]t* and *Masons Co[ur]t* off E side of Bishopsgate Street.

HORWOOD'S LONDON PLAN 225

Lacks *King Street, Hand Co[urt]., Creechurch Lane* and *New Co[urt]* N of Leadenhall Street.
Lacks *Union Row* in Sion Square SE of Whitechapel churchyard.
Lacks *Vine Street* of W side of Minories.
Lacks *Hog Yard, Watts Squ[are]* and *Well Street Co[urt]* W of Well Close Square.
Lacks *Jones's Sail Cloth Manufactory* with a group of several large industrial buildings and two boundary walls or fences, all in Gowers Walk (SE centre of the plate).
Lacks *Royalty Theatre and Maud's Sugar House* W of Well Close Square.

2nd state
Parish boundary added N of Tower of London.

3rd state
All names and buildings missing in the 1st state now added **except** *Royalty Theatre* and *Maud's Sugar House*.

4th state
Royalty Theatre and *Maud's Sugar House* added.

Plate G2

1st state
Lacks *Fox's* in *Fox's Lane* by Shadwell churchyard (S border).

2nd state
Fox's added.

Plate H2

1st state
Lacks *Whetstones Yard* by the Thames at W border.
Lacks *MR WHITINGS GARDEN* with several boundary walls or fences W of London Street in SW of the plate.
Has *Mr. Woodwells* by canal S of Rose Lane.

2nd state
Whetstones Yard added by the Thames at W border.
MR WHITINGS GARDEN with several boundary walls or fences added.
Mr. Woodwells, by canal S of Rose Lane, deleted.

Plate A3

1st state
Has a smaller letter *A* above N border than in later states.

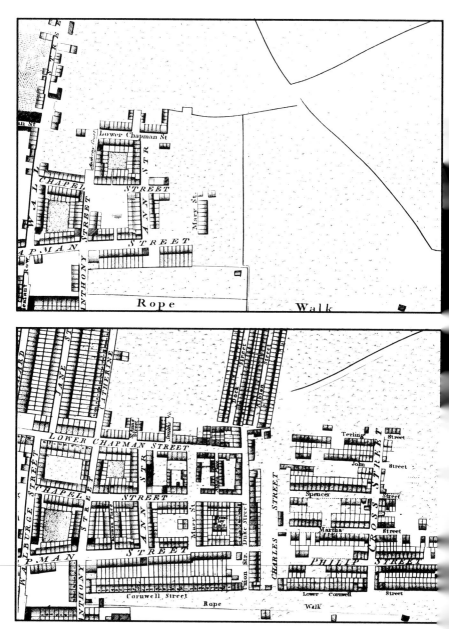

Fig. 2. Part of Horwood's plan: plate G2. Scale 0.62 × original. Parish of S[t] George in the East, north of Cable Street. Top: 1799 edition. Bottom: 180[7] edition.

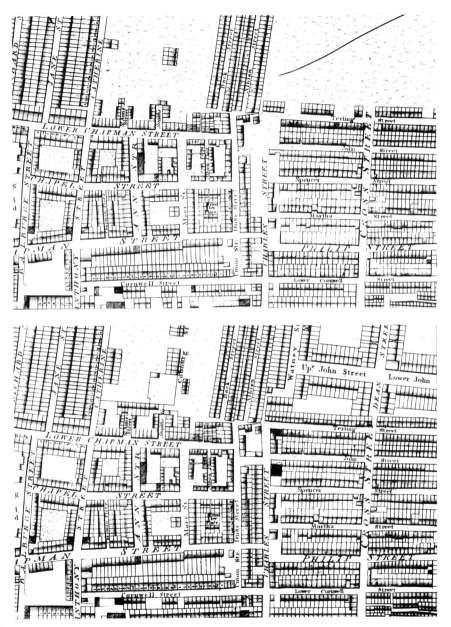

Fig. 2(cont). Top: 1813 edition. Bottom: 1819 edition. Comparison with Ordnance Survey 1:2500 London streets 63 and 64 (revised 1894) reveals the general topographical reliability of Horwood's/Faden's representation of houses.

Lacks two buildings S of the word *Turnpike* near E border.
Lacks garden walls of houses NE of turnpike bar by E border.
Lacks *Morleys Floor Cloth Manuf[act]ory* S of *T* in KNIGHTSBRIDGE

2nd state
Marginal letter *A* enlarged.

3rd state
Building and garden walls added S and NE of turnpike bar.
Morleys Floor Cloth Manufory added S of *T* in KNIGHTSBRIDGE.

Plate B3

1st state
Lacks parish boundary along E side of Buckingham House.
Street running E from St James's Square named *CHARLLS*.
Buildings between reservoir in Green Park and Piccadilly lie S of letters *AD* in *PICCADILLY*.
Lacks *Delap Co[urt]* off York Street by E border.
Has small marginal number *3* on R (and possibly in other margins but uncropped copies have not been observed).
Lacks house numbers in Upper Eaton Street.
Lacks building between Buckingham Gate and St James's Park [the Guard House].
Path across Green Park S of *MARTINS* stops short of The Mall under letter *S*.
Parish boundary in St James's Park lies N of canal at the point where it intersects the E border.

2nd state
Marginal plate number(s?) enlarged.
Parish boundary in St James's Park lies in the canal at the point where it intersects the E border.
House numbers added in Upper Eaton Street.

3rd state
Parish boundary completed across E front of Buckingham House.

4th state
CHARLLS corrected to *CHARLES*.

5th state
Delap Co added.

6th state
Path in Green Park now reaches The Mall.

Buildings N of Green Park reservoir now below letters *CA* in *PICCADILLY*.
The Guard Houses added near Buckingham House but not named.

Plate C3

1st state
    Lacks parish boundary between St Margaret's and St John's Westminster.
    Parish boundary in St James's Park lies N of canal at the point where it intersects the W border.
    Lacks *Villers* in *Villiers Street* near centre of N border.
    Lacks *Smithwaites Timber Yard* E side Narrow Walk.
    Lacks the names of several riverside yards N of College Street in Lambeth. See list under 2nd and 3rd states.
    Lacks two docks and the names of several riverside yards between College Street and Bridge Road. See list under 4th state.

2nd state
    Smithwaites Timber Yard added on E side Narrow Walk.
    The following names added in yards N of *COADES...Manufactory* in Narrow Lane:
        Messrs. Peach & Larkin Barge Builders
        Weble & Ciff Iron Foundry
        Mr Hearne Barge Builder
        Mr Harris
        Bazings English Timber Yard
        Mr. Hodgsons Timber Yard
        Mr. Tomlings Timber Yard
    Parish boundary in St James's Park lies in the canal at the point where it intersects the E border.

3rd state
    The following added over several yards N and S of College St: *THE PROPERTY OF GEO. BIGGIN ESQR.*
    Parish boundary between St Margaret's and St John's added.

4th state
    The following names added in yards S of College St:
        Tan Yard
        Mr Roche's Timber Yard
        Mr. Adams's Timber Yard
        Mr. Smith's [N of the existing words *TIMBER YARD*]
        Mr. Harris's Timber Yard
        Mr. Sander's Timber Yard
        Mr. Burnham's Coal Wharf
        Mr. Adams's Timber Yd

The words *TIMBER YARD* S of *Mr Roche's* deleted and replaced by a dock.
A dock inserted between *Mr. Adams's* and *Mr. Smith's*.

5th state
NARROW (in *NARROW WALK*) moved northwards and replaced by *Pedlars Acre*.

Plate D3

1st state
Lacks *Ebenezar Place* off St George's Road in S centre of the plate.
Lacks the following along the fronts of buildings in Great Surrey St: *Phenix Row*, *Warwich Row*, *Valentine Place*, *Burrows Builds.*, *St. George's Place* and *Commerce Row*.

2nd state
Ebenezar Place added.
The six names lacking in 1st state added along Great Surrey Street.

Plate E3

1st state
Lacks *Broad Yard* on N side of Blackman Street.
Lacks *Haglins Gatew[a]y* above *TR* of *TOOLEY STREET* S of Hays Wharf.
ALMS HOUSES, of Park Street, Southwark, punched in small letters upside down.

2nd state
Broad Yard added.
Haglins Gatewy added.
Alms Houses correctly engraved to replace incorrectly punched letters.

Plate F3

1st state
Lacks *Turnpike* at S end of Parker Row in S centre of the plate.
Lacks *Five Garden Row* SE of St John's Church, Southwark.
Allen & Cos. Brewery in courtyard of a large shaded building in Nightingale Lane, Wapping.

2nd state
Turnpike added.
Five Garden Row added.
Allen & Cos. BREWERY engraved over the building itself, which is now unshaded, rather than the courtyard.

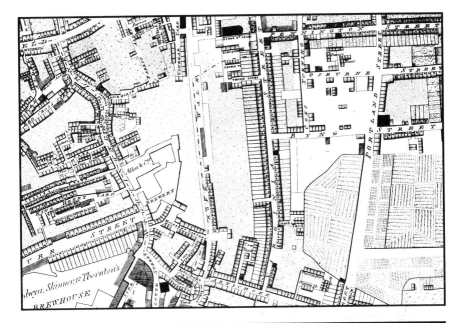

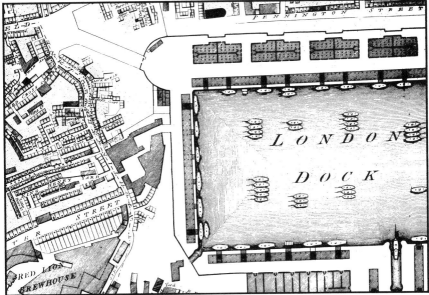

Fig. 3. Part of Horwood's plan: plate F3. Scale 0.51 × original. Nightingale Lane and London Dock. Top: 1799 edition. Bottom: 1807 edition.

Plate G3

1st state
> Lacks the following names on riverside premises on the N bank of the Thames:
> *Pichard & Co. Brewery*
> *Gun Dock*
> *Phoenix Wharf*
> *New Crane Dock*
> *Shadwell Dock*
> *Cole Stairs*
> *Cookes Wharf*
> *Morgans Wharf*
>
> Lacks the following names on riverside premises on the S bank of the Thames:
> *Mr. Hodges Cooperage*
> *Mast Makers*
> *Timber Wharf*
> *Lead Pipe Manufactory*
> *Rogers Timber Yard*
> *Boat Yard*
> *Wilkinsons Wharf*

2nd state
> All the names lacking in the 1st state now added except *Wilkinsons Wharf* near the *I of ROTHERHITHE*

3rd state
> *Wilkinsons Wharf* added.

Plate H3

1st state
> Lacks *Co[urt] after Limekiln* near Limekiln Dock.

2nd state
> *Limekiln Co[urt]* now correctly named.

3rd state
> *MR. BATSONS YARD* replaced by *COX, CUSTING & Co.* over two oval shapes.

4th state
> *CUSTING* altered to *CURLING*.

Plate A4

1st state
   Court W of Sloane Square is unnamed.
   Lacks house numbers in Lower Sloane Street.
   Lacks *SMITH STREET*, 21 houses in Smith Street, 13 houses in a street off Smith Street, and *Morley's Floor Cloth Manufactory* with its buildings.

2nd state
   House numbers added in Lower Sloane Street.
   Smith Street added with all associated additions to the plate.

3rd state
   Court W of Sloane Square named *H. Woods Stone Yard*.

Plate B4

1st state
   Has a small marginal *4* outside E border.
   Wharf on Chelsea Water Works canal is marked simply *Stone Wharf*.
   Lacks row of 5 houses with gardens NW of Upper Belgrave Place.
   Lacks *Roberts B[uilding]s* and *Poyds W[al]k* W of Chelsea Water Works canal off Avery Farm Row.

2nd state
   Larger marginal *4* engraved outside E border.
   Row of 5 houses and gardens added.
   Gardens and parish boundary altered near E border just S of marginal number *4*.
   Roberts Bs and *Poyd Wk* added.

3rd state
   A. Wilson's added above *Stone Wharf*.

Plate C4

1st state
   Lacks *Mr Vidlers Yard* and *Wake Mason* off Millbank Row.
   Lacks parish boundary in SW corner of the plate.

2nd state
   Wake Mason added.
   Mr Vidlers Yard added, but upside down.

3rd state
   Parish boundary added in SW corner of the plate.

4th state
: *Mr Vidlers Yard* now right way up.

Plate D4

1st state
: Lacks *Mr Farmers Manufactory* at S border.
: Lacks *Weymouth Str.* by E border N from Prospect Row.
: Lacks boundary wall or fence on N side at N end of Lambeth Walk.

2nd state
: *Mr. Farmers Manufactory* added.
: *Weymouth Str.* added.

3rd state
: Wall or fence added off Lambeth Walk.

Plate E4

No variants observed.

Plate F4

1st state
: Lacks *Augustus Row* and *Parkers Build[in]gs* in Grange Road near N border.

2nd state
: *Augustus Row* and *Parkers Build[in]gs* added.

Plate G4

1st state
: Text of the *EXPLANATION* end with the words *at a trifling expense*.

2nd state
: Three lines of text in smaller italics added below the text of the *EXPLANATION* as follows: *The Proprietor thinks it his Duty to state to the Public, that he never pledged himself to show the interior or extent of the back parts of Premises or in any | way to distinguish property unless specifically required, nor in an undertaking of such Magnitude and difficulty wou'd it have been possible in any length of | time – But if any Gentleman wishes to have his Property or Premises more particularly shewn. The Proprietor will make any addition required at the least possible expense.*

Plate H4

No variants observed.

Notes on variants of the first edition

A In some cases states have been distinguished by the marginal plate numbers and letters. Information on these is necessarily incomplete since many extant copies have been cropped to the border in order that the sheets could be pasted together for mounting or binding.

B This list of variants will almost certainly be incomplete. Several variants were found only after many copies had already been examined and there has not been time to check all copies again. The author would be pleased to receive corrections or information which reveals further variants so that the list may be revised.

C The author would also be pleased to receive (in confidence if necessary) lists of the states of plates in any copies that he has not seen. A complete list of copies examined is given below.

D Many of these variants may be observed by examining the reproduction of Horwood's plan published in *The Mask* in 1926 and obviously taken from a copy printed when all the plates were in early states. The reproduction by the London Topographical Society (Publication no. 106, 1966) shows the first edition in its final state with the single exception of plate H3 which is reproduced in its 3rd (penultimate) state.

E A tabular summary of the number of variants is shown below. The vast number of possible combinations (nearly 7×10^{13}) is, of course, meaningless: even if 1,000 copies were printed it is unlikely that corrections were being made to the plates so frequently that all or most copies were unique. The calculation does, however, served to remind the bibliographer that there is a high chance that two copies may be different in at least some detail.

	A	B	C	D	E	F	G	H
1	3	6	2	2	4	2	1	2
2	5	5	3	3	3	4	2	3
3	3	6	5	2	2	2	3	4
4	3	3	4	3	1	2	2	1

$A1 \times B1 \times C1 \ldots F4 \times G4 \times H4 = 69{,}657{,}034{,}752{,}000$

(c) *Main changes to the plates in the Faden editions of* 1807, 1813 *and* 1819

These changes were so extensive that it would be fruitless to attempt to describe them comprehensively in words. This list is intended to give an idea of the general extent of Faden's revisions. It also enables a reader with only one copy of the plan to hand to compare that copy in general terms with other editions or, in the case of the 1813 and 1819 editions, to attribute the contents of the sheets to earlier revisions.

To date the only variants I have observed in the Faden editions are in plates C2 and G1 in the 1813 edition, though with so few extant copies (and no one library holds copies of the 1807 and 1813 editions) this is not to say that Faden did not revise the plates from time to time.

Because these are descriptions of general changes and not observed in the same detail as the changes in the First Edition it is not claimed that they are comprehensive. In a few cases the comment 'No other changes noted' appears in brackets; this does not give an absolute guarantee that there were no changes at all. Words in italics denote literal transcriptions from the map. The figures in parentheses in the following notes are approximate numbers of houses added to the plate. Due to the inevitable imperfections of photocopies, the author's fallible powers of concentration, and to uncertainty as to whether some buildings are intended to be taken as houses or, for example, stables or workshops, some figures are more approximate than others. The persistent rounding to the nearest 5 in the 1819 notes is an indication of the margin of error.

In the necessarily abbreviated notes which follow, differences observed between one state of a plate and its predecessor are described in a form which may give the impression that they are a systematic record of new building and demolition. Though this may be generally true, it is important to appreciate that some of the changes represent corrections of errors on the map, for example, where developments had not been properly recorded on the previous edition, or where Faden had inserted new developments from proposed plans which were modified when executed. The word 'houses' is used where 'premises' might be more correct: some buildings may be business premises or public houses, though most would certainly be occupied as dwellings. Again, the addition of a 'house' to the plate does not *necessarily* mean that it was constructed since the date in the imprint of the previous state of the plate. Thus on the 1819 edition some 'new' houses may have been built before 1813 and some give the appearance of being in the course of construction when surveyed at whatever intermediate date Faden's surveyor recorded them. No attempt has been made to comment on these developments by reference to local sources or, for example, the *Survey of London*.

General revisions made by Faden to the plates for the 1807 edition

This was the most thorough revision the plates even had: in addition to the subtraction or addition of buildings and public works (approximately 9750 individual houses and countless other buildings were added to the 30 plates surviving from the first edition) the style of the map was significantly changed. Nine innovations were systematically applied to all plates:

1. All plates were given new imprints.
2. The letters and numbers giving sheet references were placed in circles.
3. Business premises, such as warehouses, lying behind the street frontage were delineated in much greater detail. In particular Horwood's vague and fuzzy fading out of shading was replaced by sharp and seemingly accurate building lines.
4. Names of business proprietors were removed from yards and other premises, especially in Lambeth, Bermondsey and Rotherhithe.
5. Architectural details were added to many churches and other public buildings; steps and rows of columns, for example, to represent porticos and main entrances, and apses or other protruding structures to avoid the appearance of stark rectangularity.
6. Public buildings, and some private ones, were systematically named. In the case of churches *St John's Churchyard* would become *St John's Church and Churchyard*, with *Church and* over the building itself. It is noticeable that several meeting houses (some anonymous, some perhaps newly established in existing buildings) were named. The names of buildings were often engraved over heavy cross hachuring giving very poor legibility, especially in all forms of photographic reproductions.
7. Names were added to many fields, gardens, burial grounds, and areas of water such as ponds.

8   Many unnamed yards and alleys were labelled.
9   Numbers were added to many hundreds of houses unnumbered on the first edition.

Plate A1

1807

Regent's Park is largely unchanged. Several field names have been added and the name *ST. MARY LE BONE PARK*. Several buildings added in the area of Willens Farm. About 425 houses have been added to the plate: Alpha cottages and adjacent developments (80 houses); field named Lisson Green is replaced by houses (120); in and around Circuit and John Streets (62); Nottingham Place and Northumberland St (20); in and around Baker Street N (129). Nottingham, Northumberland and York Mews added. *St. Marylebone* and *Chapel* added to workhouse in Northumberland Street.

1813

This was either **an entirely new plate** or it was totally re-engraved.

1819

Canal resited. Cumberland Square removed and replaced by villas. *REGENTS* [sic] engraved across the park. The lake in Regent's Park re-engraved; the water now stops short of the E border.

Plate B1

1807

Field names added. *ST. JAMES'S* added above *BURYING GROUND*. Circular reservoir on E side Tottenham Court Road added. About 210 houses added to plate: around Henry and Eden Streets which are new (102); Mary Street and Place (19); Southampton Place (12); Southampton Court (13); Norton and Carburton Streets (19); at the E end of London Street and around Mortimore Market (25); corner of Carburton Street and Cleaveland Street (28). Warren Mews added.

1813

This is either **a completely new plate** or has been totally re-engraved.

1819

End of the park lake (W border) removed. Clarendon and Bridgewater Streets added in NE corner with associated development.

Plate C1

1807
Addition of Russell Square, Tavistock Square and adjacent proposed street blocks. Bedford Nurseries are now on what will become Euston Square. Over 1550 houses added: N of Upper Guilford Street and S of *ISLINGTON* [Euston Road] (1060); Sommers Town and around Clarendon Square (370); around Russell Square and Keppel Street (120). There are some new field boundaries. Extensive new labelling and house numbering.

1813
Euston Square and area of houses N of it to the Polygon added. About 520 houses added in open area N and E of Foundling Hospital, including Lucas and Sidmouth Streets and Mecklenburgh Square; around Euston Square and Seymore Place (320); elsewhere (25).

1819
Streets and about 385 houses added: NW corner N of Clarendon Square (60); N of Russell Square and Keppel Street (25 plus mews); Brewer, Smith, Weston, Denton and adjacent streets (150); Edmund, Norfolk and contiguous streets (170); others in smaller scattered blocks (40). St Pancras New Church [foundation stone 1 July 1819] replaces 25 houses including Seymour Place South. *City Lt. Horse Stables* added E side Grays Inn Lane. *Bedford Nursery Grounds* named. Several street names added including *MONTAGUE STREET NORTH, TORRINGTON STREET, Drapers Place and Claremont Place.*

Plate D1

1807
Field names prominently added. Smith Street/Northampton Square scheme laid out. Lodge and Governor's House added to House of Correction. St John's Church named. Reservoirs added at New River Head Water Works, including the *NEW RESERVOIR*. Reservoir opposite Winchester Place renamed *Upper Pond*. About 530 houses added: Northampton Square, King, Wynyatt and adjoining streets (275); around City Road in NE corner (175); small developments elsewhere (80).

1813
Smith Street/Northampton Square area completed (about 170 houses added) and houses numbered. Houses added in Grays Inn Road (9), Charles Street (16) and City Road around York Place (42). Providence Chapel replaces a pond in Grays Inn Lane.

1819

BRAYNES ROW replaced by *EXMOUTH STREET*. Wellington Square and associated streets (with about 225 houses) N of Exmouth Street replace *NORTHAMPTON FIELD*. Middleton Street and Guildford Street East added. New reservoir in Hanging Field. Regents Canal added NE corner. Word *PENTONVILLE* removed at N of plate. Several small housing developments (115) added in far N of sheet. Clerkenwell New prison added in Corporation Row. Coach Manufactory added W of House of Correction. Upper John Street and Wilson Street and about 50 houses added off Grays Inn Lane.

Plate E1

1807

Fields and gardens named in N of sheet (e.g. *Gardens Belonging to Wenlock's Barn Prebendary*). *MR CHAMPION'S* removed before *Vinegar Manufactory*. *GENUINE BEER BREWERY* replaces properties on E side of Golden Lane. Finsbury Market and adjacent streets laid out in SE corner. Streets and about 1150 houses added to the plate: Haberdashers Street and Place and Aske Terrace (64); off Ivy Lane (13); around Coffee House Walk (39); Myrtle and Gloucester Streets (37); Walbrook Place (32); Mount Pleasant (11); Plumbers, Trafalgar, Ebenezer and Providence Streets (88); Moffat Street and Westmoreland Place (193); City Terrace, Union Row and Nelson Place (78); Windsor Terrace and Windsor Place (73); Nelson Street, houses N of Johns Row, and Back of Pines Row (85); Georges Square (18); Luke, Mark, Matthew, James and Cross Streets (154); Paul Square (19); [Northampton Terrace] and adjoining houses (26); Brunswick Place (32); Agnes Street and Crescent, St Leonards Square and Lower Tabernacle Walk (74); several blocks replace a brewery in Old Street S side W end (65); in other places (47). There are also houses removed around Finsbury Market and the Quaker burial ground near Banner Square.

1813

Cross, Nile, Union and Allerton Streets added (120). Additional houses in Plumbers Street (10) and Nelson Street (29). Haberdashers Street numbered. Ebenezer Street now Collingwood Street. Northampton Terrace named. Word *Gardens* removed near George Square. [No other changes noted.]

1819

Regents Canal Bason now dominates NW quarter. About 830 houses added, notably: Provost, Moneyer, Union and Cross Streets with Warners Buildings, York Place and Walbrook Row (248); Ironmonger, Radnor and Galway Streets with Ratcliffe Row S side [replacing gardens of *Ironmongers Company* in 1813] (246); Baldwin Street, Bath Buildings and Pool Terrace (121). Union and Allerton Streets, wrongly aligned on 1813 plate, now repositioned. Old Street Square now Batholomew Square. Collingwood Street now Collingwood Street **and** Ebenezer Street.

Plate F1

1807
Many gardens labelled and names added as described in the list of general revisions. About 645 houses added in scattered developments, notably: Mount, Nelson, Collingwood and Duke Streets (211); around and behind Old and New Nichols Streets (114); along Hackney Road (105); Essex and Huntingdon Streets (46); Union Buildings (37); Shacklewell Street (22).

1813
Many houses added off Hackney Road: Brunswick and Cambridge Streets and Weymouth Terrace (109); Cole Harbour, Catherine, Henrietta, Nelson and Jane Streets (267). New Houses off Church Street N and E of Thorold Square: Tyrell, Satchwell, New Tyssen, Charlotte and George Streets (about 220 houses added), between Church Row and James Street near St Matthew's church about 50 scattered houses added.

1819
About 330 houses added in scattered blocks, including: in NE corner N of Hackney Road (69); Winchester and Carlisle Streets E side of Bethnal Green Workhouse (32); in and around Half Nichols Street (72); Bird Cage Walk area (34); Brunswick Street (24); Busby and Fuller Street area (45).

Plate G1

1807
Jews added to *Burial Grounds* SW and SE corners. Colour manufactory added on W edge of plate. Ponds, footpaths and gardens labelled. *Mad House*, *Cornhill Workhouse* and *Soap Manufactory* named around BETHNALL GREEN. About 185 houses added in scattered developments, mostly around Green Street (70) and North Street (30).

1813
Building added in field NE of Patriot Square [Jewish School]. Much housing added to SE corner of plate: Ann, Charles, Devonshire, James, West, Essex and York Streets and houses in Globe Lane (about 345 in all). About 65 houses, including John Street, added to E centre of plate around Green Street [Globetown].

1819
> *Jews Chapel*, previously isolated and unlabelled, now named, and adjoining school added on N side of it. *Jews Alms Ho* added in Devonshire Street. Canal added in NE corner of plate. Addition of approximately 600 houses scattered individually or in small blocks in four main areas: Globe Lane with Charles, Ann, Devonshire and Essex Streets; Coventry and Pitt Streets; N of Old Bethnal Green Road; E centre of the plate.

Plate H1

1807
> Title, oval plaque, scale bar and compass rose deleted. Grove Road and all associated developments added, including Bing Street. *Jews Hospital and Burial Ground* added SW corner. *Foot Path from Mile End Old Town to Bow &c.* and footpath itself added. Twig Folly and its houses [Globetown] added (about 30 houses). Nursery E centre of plate labelled. Stratford/Mile End parish boundary named in SE. Some field boundaries added, paths named and isolated houses added.

1813
> Ploughed field added in SE corner. About 50 houses added in Norton Street/East Street development [Globetown]. 42 houses added S of Ewings Buildings in Mile End Old Town and a further 25 near Bencrofts Place. Regents Place added (36). [No other changes noted.]

1819
> *REGENTS CANAL* runs down W half of sheet. Several field boundaries added just E of canal N of Mile End Old Town. About 40 houses added in six scattered blocks.

Plate I1

1807
New plate.

1813
> East London Water Works, with two ponds and a reservoir, added. New houses S of the two orchards in SE. Round pond N of St Leonard's Church and a rectangular water area N of dye house removed. Charles Street added (23). Houses added in field W of *Bull Shed Buildings* (98). Dyers Row added (24).

1819
> Virtually unchanged. A farm building W centre (one of a pair) removed. School added and labelled N centre. Minor changes to buildings at SW end of Dyers Row in SE of plate.

Plate K1

1807
New plate

1813
See list of titles and imprints. *The Three Mills* replaced by *Brewery* in small mapped area in SW otherwise unchanged.

1819
See list of titles and imprints. [No other changes noted.]

Plate A2

1807
In Hyde Park double avenue of trees replaced by less regular arrangement, outer circle added to reservoir and footpaths altered and/or named. Many houses have now been numbered. About 100 houses added in Paddington Parish. John, Newnham, Cato, Molyneaux, Shouldham, Homer and Harcourt Streets added (about 190 houses) and about 85 houses elsewhere. Coach manufactory added in Adams Street West. Montague Square added. The numbering and list of rooms in the Earl of Leicester's House in Grosvenor Square have been removed.

1813
Connaught Place added. Bryanston Square added and the surrounding blocks (including Montague Square) built up (about 300 houses plus blocks where individual houses are not distinguished, suggesting that they may be merely proposed developments). Very few changes S of Oxford Street.

1819
Two main areas of change. Connaught Place reconfigured, Upper Bryanstone and Upper Seymour Streets extended W of Edgware Road, and a street off Edgware Road now runs into Tomlins New Town. Seymour Place, Crawford Street and Montague Place re-engraved and some new houses added. Otherwise little change.

Plate B2

1807
Decorative gardens have been engraved in the middle of Cavendish, Golden, Berkley and Hanover Squares. Queen Ann Street is now Foley Place. *Foley House* is now *Gloucester House*. *Town Hatter* (N side Oxford Street) deleted. *Workhouse* now *St George's Workhouse*. Houses added on Union Street at the back of the Middlesex Hospital. *Stewarts Tin Manufactory* (E end Piccadilly) and names of other business premises deleted.

1813
Chapel added at corner of Thayr and Hinde Streets. Number 25–27 Albermarle Street and the chapel (?) to the N of them added. Pantheon in Oxford Street now hachured. [No other changes noted.]

1819
Langham Place and Regent's Street added, with colonnade and associated new frontages and street works. Letter *R* [in Tottenham Court Road] re-engraved in NE corner.

Plate C2

1807
Very considerable re-engraving and labelling especially of back buildings which Horwood had left incomplete and vague. Many streets numbered for the first time. *THE QUEENS MEWS* replaces *THE UPPER MEWS*. Gardens added in Lincoln's Inn Fields. Houses added around Great Store St (65) and Montague Street and Bedford Place added (93 houses added, 13 removed). Parish boundary formerly on W and S of Leicester Square now across the diagonal and labelled. Additional building on NW corner of British Museum. Additions and changes to buildings in Covent Garden.

1813
STRAND BRIDGE, the bridge (eventually Waterloo Bridge) in faint outline, and fine lines out into the river beside the bridge all added. Changes to plans of Drury Lane and Covent Garden Theatres. Area across Princes Street from St Ann's Burial Ground re-engraved – *Wellington S.* has replaced *Edmonds Cot.*

1813 *b*
A second state of C2 has an addition between Leicester Square and Castle Street – *St Martins National School* with access to Castle Street.

1819
Approach road to Waterloo Bridge cut through from Strand and named *WELLINGTON STREET*. *LYCEUM* replaced by *ENGLISH OPERA HOUSE* and the building enlarged. *Theatre* N side Wych Street replaced by *Olympic Theatre*. Some new houses in Bloomsbury. *BOW ST. PETER ST.* replaced by *Museum Street*.

Plate D2

1807
Fleet Prison re-engraved in detail (chapel shown). St Paul's named *CATHEDRAL OF ST PAUL* and the *Statue* named outside W front. Circular enclosure round St Clement Danes indicated, as is a

pecked line showing the proposed widening of the road to form a circus round the church. Livestock pens shown in West Smithfield Market. *Chapel* added at Bridewell Hospital and named. Old Ludgate Prison in Giltspur Street now shown as law courts. Buildings numbered in Grays Inn Square and Holborn Court. Skinner Street cut through leaving an island of properties at Snow Hill. Fleet Prison buildings re-engraved. *St Andrews Workhouse & Burial Ground* added E side of Shoe Lane. *LINCOLNS INN OLD BUILDINGS* named. Chapel added E side of Fetter Lane. Many other minor changes.

1813

The pecked circle round St Clement's church in the Strand has been removed, and twelve additional unnumbered premises form an arc on the NW from the Strand to St Clement's Lane. Houses S of the church, previously in a straight row, are now set back to form a crescent.

1819

Minor changes only. Island of properties at Snow Hill now numbered. New houses E side Fleet Market, corner of Turnagain Lane. New buildings added in Furnivals Inn. New building S side Session House in Old Bailey. *National Central School* named in Baldwyns Place N edge of plate.

Plate E2

1807

Moorfields gardens and avenues [site of Finsbury Circus] cleared and E half of Bethlem Hospital removed. Bank of England detailed, hachured and renamed simply *THE BANK*. St Paul's Cathedral and many churches named. E and W sides of Mansion House given bays. Extensive minor changes in back streets including addition of St Giles Cripplegate Workhouse in Moor Lane. Post Office in Lombard Street hachured. St Helens Place (20) added on E side Bishopsgate Street. Far SW river front *Timber Yard* changed to *Steel Yard*.

1813

City boundary added across large building (*Mr Whitbread's Brewery* on first edition) between White Cross Street and Grub Street. Houses added at N end of Princes Street opposite Bank of England. [No other changes noted.]

1819

Bethlem Hospital removed from N side London Wall. Finsbury Place South, the London Institution, an RC chapel, and an oval garden, all erected in the open space [Moorfields] N of London Wall. *NEW CUSTOMS HOUSE* replaces several blocks between the River

Thames and Lower Thames Street. Wide street N of new Customs House replaces Water Lane, Area N of Crown Street in NE corner of plate reconfigured. Earl and Clifton Streets added as part of new developments [S of Finsbury Market on plate E1].

Plate F2

1807

Tower of London now shown in detail (see plate F3). Tobacco warehouse E of the Tower now replaced by the Royal Mint. Back buildings added and square laid out between Mansell and Leman Streets [Goodman's Fields]. Gardens labelled off Church Lane. Quakers Burial Ground named in NE corner. East India Warehouses in Petticoat Lane, Fenchurch Street and Coopers Row extended and/or named. About 510 houses added, mostly on E quarter of the plate, notably: between Whitehorse Lane and Charlotte Street (200); S side of White Horse Lane E of Church Lane (185); Thomas, Christian, Somerwood, William and Langdale Streets (80).

1813

Two large housing developments: Chicksand, Finch, John and George Streets (180); Sever and Mary Ann Streets N of Cable Street (90). Infilling housing elsewhere including between Bell Lane and Rose Lane (22 additional houses but Kings Head Court removed).

1819

About 250 houses added in scattered small blocks, most extensively in what were in 1813 gardens off E side Church Lane. Constitution Brewery removed from W side of Bell Lane. *ROPE WALK*, running S out of White Horse Lane removed (E side of plate).

Plate G2

1807

Commercial Road added right across the plate. About 1110 houses added, mainly in the following streets: Richard, Jane, Catherine, Dock, Albion, Lower Chapman, Mary, Cornwell, Union, Duke, Charles, Terling, John, Spencer, Martha, Philip, Lower Cornwell, Cross, Lucas, John (a second John Street), Mercers Row (with adjacent courts and rows), John (a third John Street, in NW corner of the plate), Queen Anne, St Thomas, Mount and Mount Row. New buildings, and some 20–25 houses removed, in the glass house off Brook Street. Some houses removed or converted in Denmark St in SW corner.

1813

Glass House off Brook Street now a *Foundery* and a *Cooperage*. Houses added in the following streets: Terling, John, Spencer and Martha.

HORWOOD'S LONDON PLAN 247

The following streets (all W centre) newly laid out with houses: Oxford, New, Rutland, Suffolk, Norfolk, Nelson. Church Road and E ends of Terling, John, Spencer and Martha Streets laid out with new houses. Jubilee Street and its houses added in centre of plate. Lisbon Street added N centre. George and Silver Streets added S of Stepney Green. (About 800 houses added in total.)

1819

Stepney Chapel added [St Philip, built 1818–19]. Windmill removed N of Commercial Road in centre of plate. Many new houses added especially in 5 blocks: around Jamaica Street and Arbour Square N side of Commercial Road (230 on a previously greenfield site); further W on N side of Commercial Road, including Sidney Street (85); around Sutton and Dean Streets S side of Commercial Road (100); Lucas Street and Vinegar Lane (80); around George Street (115). About 160 houses added in other scattered blocks. (Total houses added to the plate is about 800.)

Plate H 2

1807

Commercial Road and East India Dock Road added. Some new houses in SE of Salmons Lane. Soundings added in River Thames. Three rope-walks added: two E of Stepney churchyard, one in NE corner. Patent Cable Manufactory added. About 410 houses added including three main developments: between Gun Lane, Three Colt Street and Commercial Road in Limehouse (150, besides 10 removed); around Cow Lane, Church Row and Stepney Square in the NW of the plate (114); NW of canal bridge at Limehouse (63). Maritime Alms Houses added in NE corner.

1813

Extensive area of housing development added between Salmons Lane and the two new rope-walks in the centre of the plate: Catherine, John, James, Edward, George, Rosetta and Vine Streets (220). Stepney Square now named Trafalgar Square. Houses added in: Mill Place and Island Row (27); NW of canal bridge at Limehouse (64); around Trafalgar Square and John and Edward Streets [the ones N of Cow Lane] (225).

1819

The Regent's Canal and its *BASON* near the Thames now runs N to S down the plate. About 190 houses added in scattered blocks in NW corner and in centre of the plate around Catherine and Samuel Streets.

Plate 12

1807
New plate.

1813
Ashton and Union Streets and adjacent developments nearly all added (about 160 houses). Three small groups of houses added (15 houses). Building removed below word *Road* S centre. [Otherwise plate unchanged.]

1819
Three windmills removed from NE corner of plate. Works removed at junction of North Street and East India Dock Road. About 50 houses in small groups added in SW and SE corners and scattered in NW quarter.

Plate K2

1807
New plate.

1813
Barking Road added E of River Lea. Iron bridge over Lea named and detailed. Four houses and another building added, and other buildings repositioned, on peninsula SE corner. Edna Place and other houses added in SW corner (about 80 houses and cottages). Two new and one reshaped building in fields N of dock entrance. [Otherwise plate unchanged.]

1819
New industrial buildings at *The Four Mills* in NW corner. River banks re-engraved in SE. Several changes around East India Dock especially *East India Cos. Premises* at E end of the dock.

Plate A3

1807
Deer Pound added and named N of Serpentine. Names added to roads and paths in Hyde Park: *THE KINGS PRIVATE ROAD, THE PUBLIC ROAD, Foot Path to Grosvenor* and *Foot Path to Chesterfield Gate.* Several buildings added to W of barracks on N side of Knightsbridge and *LIFE GUARDS* added above *BARRACKS. Foot Guards* added before *Knightsbridge Barrks. The Ring* (E centre) named, and *Grove Ho.* (centre) replaces *The Rural Cattle.* Street name *KNIGHTSBRIDGE* moved further E. Cadogan Square and 10 houses added. Houses added in SW: Brompton Crescent (12), Exeter Place (20), S side New Street (8) and a group around Charlotte

Street, Cross Street and Wonder Wat Place (54 and mews). Chapel [Halkin Street] added but not labelled. Wire manufactory added in SE corner. Cannon Brewery added N side Knightsbridge.

1813
Chapel [Halkin Street] labelled. In Sloane Street numbers 172–[17]7 have replaced two houses on 1807 edition: houses to S still numbered 172–174. Cadogan Place and Upper Cadogan Place added with about 50 houses. In New Road last three houses added at N end W side.

1819
Belgrave Square and surrounding streets shown faintly in outline and unnamed. Some new houses on E side of Sloane Street N of Cadogan Square. Name *Halkin Street* added and chapel in it redrawn in new shape. Chesterfield Gate added to E side of Hyde Park.

Plate B3

1807
CONSTITUTION HILL and *RESERVOIR* added in Green Park. Octagon in St James's Square now a circle. *Chapel Royal* named in St James's Palace. Fences and *CANAL* added in St James's Park. Belt of trees and short fence added along N edge of The Queen's Gardens. Armoury added S side St James's Park. *Almshouses* in James Street now *Emanuel Hospital*. Houses numbered in the York Street area. Area round Hamilton Street W of Park Lane re-engraved (10 large houses replace 29 smaller ones). 101 houses added, notably off Grosvenor Place, Grosvenor Street and Ranelagh Street (39). 25 houses removed in George and Catherine Streets.

1813
Green Coat Pl. added. S and W of The Queen's Gardens there are small groups of houses added in a few places including *Ranelagh Place* (total 33 houses). [No other changes noted.]

1819
Circle in St James's Square now surrounded by a square garden. New path and bridge across St James's Park. New road added S centre N of Pimlico Wharf. Streets added in outline, including King's Road (unnamed), with some new houses around Chester Street. Numbers added to some houses in Duke and South Audley Streets. Houses added in Palace Street. An enclosure added in Green Park S of the first letter *I* in *PICCADILLY*.

Plate C3

1807
Trees added in middle of St James's Park and in Great Deans Yard. Guild Hall added N of Westminster Abbey and named. Among names added are: *Treasury Yard, Blue Board Yard, Phoenix Fire Engine, Coddicks Row, Duke St Chapel, Little Deans Yard, Records* [Abbey Chapter House], *Cock Yard, Swan Yard* and *Spread Eagle Yard* and many others. *Cartwright Street* changed to *Cartaret Street*. Ordnance Office and 39 other premises N and E of St Margaret's church removed. The *Gt Gun* added on N side of Horse Guard Parade.

1813
Westminster Square laid out (but not named) with gardens in the centre N of St Margaret's church. *NEW* added before *PALACE YARD*. *Wooden Bridge Stairs* now *Palace Yard Stairs*. Outline of the future Waterloo Bridge added. *Coddicks Row* replaced by *Whitehall Place*. [No other changes noted.]

1819
Buildings around Horse Guards Parade numbered and listed. Waterloo Place laid out N side of Pall Mall (but Carlton House still standing). Waterloo Bridge named (*IDGE* on this plate). Name of opera house moved from the street to the building itself. Many buildings cleared from E side of Princes Street, now shown as parkland. Whitehall Place now numbered and reconfigured. *Transport Office* replaced by *Board of Control* near Cannon Row. Westminster Guildhall realigned and *Westminster National School* added N of it. Houses added in Park Place in SE corner.

Plate D3

1807
Extensive labelling of institutions (especially chapels in institutions) and manufactories. *New Bridewell* in Great Suffolk Street now a *Soap Manufactory*. *The New Cut* named. Open ground S of Charlotte Street now has *NELSON SQUARE so to be called when built upon*. School for the Blind added on S side of St George's Road. At least 940 houses added scattered across the plate but notably in: Charles, Oakley, Gloucester, Jurston and Collate Streets (187); Duke and Webber Streets and Anns Place (94); Dover, Flint, Gun, St Martin, Dean and Upper Dean Streets (247). *Judsons Street* changed to *Jetsom Street*. Chapel added to Female Reform in St George's Road. Kings Bench Prison shown in detail: *State House* and other buildings named.

1813

New *BETHLEM HOSPITAL* and its large building replaces School for the Blind and Nursery Row in St George's Road. *Garden* deleted SE corner. [No other changes noted.]

1819

Very extensive revisions. Waterloo Bridge Road added from the Thames to London Road. Industrial buildings near the Thames removed (timber yards, vinegar and woollen cloth manufactories). Commercial Road and streets to the S laid out. Many new streets and houses in S of the sheet; North and South Streets; Frances Street; around Nelson Square; N and S of Borough Road. Two new buildings added to New Bethlem Hospital. Great Union Street laid out and many new houses added in SE corner. Some houses removed in Kings Bench Walk, Providence Row, St Georges Crescent and Ebenezer Place.

Plate E3

1807

Buildings of brewery detailed and the name *ANCHOR BREWERY* added. Borough Water Works on London Bridge named. *Mr Edwards's* deleted from tenter ground in SW and many other proprietor's names removed, especially in the tan yards in the SE of the plate. Many industrial and commercial premises detailed and tidied up. Extensive labelling of courts and yards, and of institutions such as *GUYS MAD HOUSE*. Only 8 additional houses noted. Paths added in Bermondsey churchyard.

1813

Swan Street, Great Dover Street and an unnamed thoroughfare have been cut into existing built up areas, rope-walks and gardens E of Blackman Street. About 350 houses added, mostly in following streets: Bath Terrace (21); William, John, Rodney (all new), New Lant and Great Suffolk Streets W centre (126); Richardson Street and Wellington Place (72); Baal, Elim, Zephon and Wilderness Streets (55); Castle Street and houses to NW (57). Path through letter *E* in Bermondsey churchyard removed.

1819

Bridge Street and Southwark Bridge added. *GREAT SUFFOLK STREET EAST* and *SWAN STREET* named but part of the latter extending S of the former (on 1813 edition) has now been removed. About 250 houses added: Devonshire Street in SW corner, around Nelson Street, NW of Castle Street, around Elim Street, and N side of Union Street.

252 HORWOOD'S LONDON PLAN

Plate F3

1807
London Dock, its *BASON* and associated warehouses and sheds added in place of several streets and about 695 houses, notably Osburne, Byng, Virginia and Portland Streets, Cox's Turning and Artichoke Lane. Pennington Street survives but all houses on the S side have gone. Soundings and mudbanks shown in the River Thames. *Goodwyn, Skinner, & Thornton's BREWHOUSE* changed to *RED LION BREWHOUSE* and the buildings of the brewery altered and shaded in. Only about 20 houses added to the plate. Tower of London shown in detail. [The first edition carried the comment *THE TOWER The Internal Parts not distinguished being refused permission to take the Survey.*]

1813
Large area of warehouses of irregular shape added on E side of Nightingale Lane and at W end of Pennington Street, removing about 100 houses. Two rose symbols added in the Thames near entrance to London Dock Basin; presumed to be buoys. Small outer basin and lock placed between Thames and SW corner of London Dock. E end of Redmaid Lane replaced by dock buildings. George and Fendall Streets added in SW of plate (22).

1819
Minor changes only. N of the Thames: 2 courts demolished on N side East Smithfield; houses demolished S side New Street; army stores in Lower East Smithfield now smaller. S of the Thames: Engine Manufactory added S of plate; 7 houses removed E edge; about 20 houses added S side New Walk; 8 houses added S side George Street.

Plate G3

1807
Soundings and mudbanks added in River Thames. London Dock and associated large buildings added: about 265 houses removed mostly in Shard, Bird, Torrington, Byng, Osborn and Pennington Streets. Strip of about 50 houses cleared from N side of Milk Yard. Rotherhithe Basin and four proposed dry docks added. Albion Street added with about 40 houses including adjacent developments. 15 houses added in Deptford Lower Road. *Calanders Gardens* added.

1813
London Dock warehouses extended eastwards: about 50 houses removed. Outline of further developments between Gravel Lane and New Gravel Lane. *ROTHERHITHE BASON* now *GRAND*

S[URREY] OUTER DOCK[S]. Proposed dry docks between Thames and Rotherhithe Basin removed. Outlines of Kings Road, Upper Clarence Street, Kings Row, Maling Place, Hawley Place and Morrison Street added. Outline of developments between Keppel Street and New Gravel Lane added. Watercourse in SE corner re-engraved (widened) and named *Mill Pond*. About 40 houses added in S Rotherhithe.

1819

Few changes. N of the Thames about 25 houses demolished in scattered blocks, especially just E of Wapping Church. S of the Thames about 17 houses added off S side Albion Street and a new boat builders premises W of Rotherhithe Church.

Plate H3

1807

Soundings and mudbanks added in River Thames. Shore of the Isle of Dogs S of Limehouse Causeway wholly re-engraved. Canal from Rotherhithe basin added. Entrances to India Docks added. *Rope & Sailcloth Manufactory* added in Limehouse. Only about 20 houses added to the plate.

1813

Commercial Docks (with their acreages) added. *GRAND SURRY INNER DOCK* replaces the canal from the Rotherhithe Basin which was narrower. Deptford and Greenwich Road added on the Isle of Dogs with associated developments. Isle of Dogs shoreline largely re-engraved again. New Buildings added off Cow Lane (29).

1819

Very little change. Block of commercial buildings added SW of Commercial Dock. A few riverside buildings added or removed. Windmill removed from SE corner. Some houses added around Limehouse Causeway.

Plate I3

1807
New plate.

1813

About 280 houses added all but 35 (James Street in SW) in small blocks across the N of the plate. Two dry docks added in SE. Some additional dock buildings at each end of the inwards dock. Outline of sheds or warehouses on S side outwards dock. [No other changes noted.]

1819

About 100 houses added in scattered blocks especially in NE corner (about 50) and in Kings Road (12). Two isolated houses close to N wall of West India Dock have been removed.

Plate K 3

1807
New plate.

1813
[Plate unchanged except for new imprint.]

1819

About 110 houses replace two rope-walks in Brunswick Street. Mast Pond added N of Old Dock. Warehouses and sheds added S and W sides of East India Dock (N) and other buildings between East India Dock (S) and the Thames. Bow Creek banks re-engraved. Entrance basin to East India Dock now divided by a pier. Several buildings inserted around Bow Creek.

Plate A 4

1807

Footpath added across Chelsea Common. Ranelagh Gardens removed. Terrace and names of the Gardens added in Chelsea Hospital. *CHELSEA* added in River Thames. *ROYAL HOSPITAL ROW* now *JEWS ROW*. Royal Military Asylum and York Hospital added. About 87 houses added mostly in The Kings Road and around Sloane Square.

1813

Some buildings added on Chelsea Common. Cadogan Terrace and Ellis Street added (57). Jubilee Place (19) and South Street (35) added. Houses added in Lower George Street (16) and elsewhere (50).

1819

Several new streets and about 600 houses added in 3 main groups: Manor, Wellington and Collingwood Street (140); replacing Chelsea Common and Chelsea Common Field are College Street and Place, Marlborough Square, Frances, Keppel, Wellesley and Leander Streets and John, James, Jubilee and Cumberland Places (395); around Cadogan and Sloane Terraces and in Simon Street (50); streets laid out in NE.

Plate B4

1807
  Most prominent change is the addition of names: *REACH* added in River Thames. Mill Bank Distillery added. *Manufactory* deleted after White Lead. Many names added including: *NEAT HOUSE GARDENS, Osier Beds, Garden Grounds, Cuts to supply Chelsea Water Works and Pimlico Works*. Gardens round Military Hospital (newly named) replace pond on first edition. Only 12 houses added.

1813
  Only significant changes are in Tothill Fields which is re-engraved: pond removed and streets laid out with *PLAY GROUND FOR THE WESTMINSTER SCHOLARS*; everything S of Rochester Row (including *New Rochester Row*) and E of gardens along the parish boundary is new. About 50 houses added in this area of the plate. Addition of small dock off the Thames and two small buildings just above the new name *New Ranelagh. Eccleston Street* and numbers 5–13 added. 18 houses added on S side of Belgrave Street and W centre of plate. [No other changes noted.]

1819
  NEW VAUXHALL BRIDGE Road replaces *PROPOSED NEW ROAD*. Some other new streets no longer marked in broken lines. Four groups of houses added: Eccleston Street and Ebury Place (30); Garden and Chapter Streets and Hide Place (80); Rochester and Bell Streets (40); Willow Street and Wellington Place (30).

Plate C4

1807
  Added between Kennington Lane and Workhouse Lane are *VAUXHALL PARK* and broken lines indicating a proposed circus (though not on the site of Vauxhall Circus) and streets leading to it. *HAMLET OF NINE ELMS* and riverside buildings added in SE. Other changes few and of the general kind.

1813
  THE NEW PENITENTIARY and its octagonal perimeter wall added but no buildings shown. Regent Street and all the associated streets laid out to W of New Penitentiary. Medway Street and two new Meeting Houses added. *Mr Copland's Yard* (Ferry Road) added. Pond and fence outside W side Vauxhall Gardens removed. Vauxhall Circus, Tyers Street, Little Park Street, and associated houses and streets added. Some new buildings between Vauxhall Walk and the river. *Thomas* [sic] *Bank Wharf* and *New Bridge Street* added. Projected *VAUXHALL BRIDGE* and broken lines indicating the

proposed bridge added. [Note that the word *Projected* has been scored out in the Guildhall Library copy.]

1819

The buildings of Milbank Penitentiary now shown, with six wings and a chapel at the centre. Vauxhall Bridge (previously in outline) shown with turnpike at each end. Approaches to the bridge, Vauxhall Bridge Road, *NEW ROAD* along embankment to Mill Bank Row, and a road from Vauxhall to Kennington Oval inserted. N of the Thames: a button manufactory and about 140 houses added in Regent Street and adjoining streets; Page and Earl Streets added. S of the Thames: about 50 houses added; about 12 demolished next to a burial ground now named as Lambeth Burial Ground; George and Henry Streets inserted; Vauxhall Circus (E centre) removed; *Intended Coach Field* removed N of Vauxhall Gardens; Kennington Oval reshaped; *Coal Wharf* and *Brewhouse* labelled S of Vauxhall Bridge.

Plate D4

1807

Characteristic new labelling but all other changes associated with housing development. About 680 houses added, notably in the following: on White Hart Fields in Market Street area (90); around Hanover, Hampton, Penton and Francis Streets (227); Clayton Street (60); Weymouth Street (98); other scattered developments (207).

1813

About 370 houses added in: Devonshire Street, Park Street and development just S of Elizabeth Place (108); Regent Street and unnamed street crossing it, both laid out just N of Elizabeth Place (18); Thomas Street and associated building added in NE corner (92); Bird Street (56, including 48 houses in courts); Brooks Street (28); William Street area in SW corner (30); Nile Place off Weymouth Street (19); Union and New Streets (17). [No other changes noted.]

1819

Houses inserted in seven places: John and Sir Streets and Arnold Place – all new streets (50); Regent and Doris Streets (50); Devonshire Street (50); Mansion House Place and Wilsons Buildings – new streets (50); off William Street in NE corner (10); N side of Prospect Place N edge (15). Houses in Newton terrace in SW corner re-engraved and numbered.

Plate E4

1807
About 870 houses added, mostly in small blocks: around Kent Road and E of The Paragon and Flint Street (343); on Locks Fields (N of Richmond Place) especially Lion Street and Hawksworth Place (327); Walworth Fields S of Richmond Place (135); on Walworth Common (67). *ASYLUM FOR THE DEAF AND DUMB* added with its buildings. *LOCKS FIELDS* and *WALWORTH FIELDS* and other open spaces named. *MEADOW LAND* added in SE corner. Newington/Camberwell parish boundary named.

1813
A further 800 houses added, some in new streets but much infilling: Albany New Road (43); Trafalgar and George Streets (93); Kings Row, Mount and Horsley Streets (70); Nelson Place, Northampton and North Streets (107); William, Charles, Adam and County Terrace Streets, County Terrace and Ayliffe Buildings (173); scattered across Locks Fields (99), Lion, Union and New Streets (40); and several smaller developments. Canal added in SE corner along the Camberwell/Newington boundary, which has been realigned.

1819
About 380 houses added, scattered all over the plate; three largest developments in and around John Street (66) in S, Townsend Street (30) in NE centre, and Hunter and Potier Streets (65) on N edge. School and adjacent buildings added S of Bermondsey New Road.

Plate F4

1807
Many fields named, including *BERMONDSEY SPA*. House in Fort Place now named *Philanthropic Society*. A mansion and its landscape gardens laid out in The Kent Road. *TANNERIES* added in Grange Road. Only about six houses added to plate.

1813
The mansion in The Kent Road, but not the gardens, removed. About 70 new houses, including those in Albany New Road in SW corner, but little major change to the plate.

1819
A few minor changes only. About 60 houses added: Crimscott Street and along S side The Kent Road and Eastman Row on N side. Two rows of houses re-engraved. Isolated building added in gardens just E of *Engine Manufactory*.

Plate G4

1807
All material in the *EXPLANATION* deleted after the words *erased and corrected*. Many fields and gardens named. *ROGUES LANE* changed to *CORBETS LANE*. About 25 houses added to the plate. Footpaths or bridleways added and named. Glue manufactory (SW) and windmill (centre) added.

1813
Two streams added in NE corner in place of continuous body of water. Canal and area of water [Evelyn's Pond] added E of Trundleys Lane in SE corner. *Halfpenny Hatch to Greenland Dock* re-engraved for no apparent reason. Scattered buildings added in Blue Anchor Road. About 40 houses added to the plate in scattered blocks.

1819
Augusta Place (with 7 houses) added N edge. Below the Explanation is added *NB. The new roads and Bridges are coloured Yellow. TRUNDLEYS LANE* and the area of water in SE corner both removed.

Plate H4

1807
Soundings and mudbanks added in River Thames. Isle of Dogs wholly re-engraved. *DEPTF* [part of *DEPTFORD*] resited. New Dock, West Dock and Canal now surround Greenland Dock. Grand Surrey Canal runs right down the plate. Details of Royal Dockyard added. About 40 houses added to the plate. Kent and Surrey county boundary named boldly.

1813
Evelyn, West and East Ponds added around Surrey Canal. Greenland Dock, West Dock and New Dock now named Commercial Dock Numbers 1, 2 and 3 respectively: the Canal is the East Country Dock. West Dock reshaped and several changes are made to dockside buildings. About 110 houses added such as those W centre of plate and in S around Surrey Canal. Footbridge in Windmill Lane altered and name placed on the bridge itself.

1819
Area of water and the words *EVELYN POND 22A 0R 10P* removed. Five houses removed N of Commercial Dock No 1. 5 houses added in Windmill Lane and words *Foot Bridge* moved onto the bridge itself. 22 houses (including Greenfield Street) and a chapel (?) added near elbow of Surrey Canal in S.

Plate I4

1807
New plate.

1813
All houses along the River Thames (about 20) are newly added. [No other changes.]

1819
Chain Cable Manufactory with several buildings added SW beside the Thames. A few other small buildings added along the lane to the Ferry House. [No other changes.]

Plate K4

1807
New plate.

1813
[Plate unchanged except for new imprint.]

1819
Two unidentified buildings removed and about 75 houses added: all changes in S of the plate near Greenwich. *Iron Stores* labelled near Norfolk College. [Otherwise plate unchanged.]

(d) *List of titles and imprints in all editions of Horwood's map*, 1792–1819

First edition title

[In oval panel on plate H1. Scale and scale bar below oval]

PLAN | of the Cities of | LONDON and WESTMINSTER | the Borough of | SOUTHWARK, | and PARTS adjoining | Shewing every HOUSE. | By R Horwood | Scale of 20 Chains or a quarter of a Mile. | [scale bar: six inches]

Second edition title

[On new plate K1]

A plan | of the CITIES of | LONDON & WESTMINSTER, | with the | Borough of Southwark | including their adjacent | SUBURBS, | In which every Dwelling House is described & numbered, | Surveyed and first published by | RICHARD HORWOOD, | MDCCXCIX. | ——— | Second Edition | For the purpose of rendering this NEW EDITION of Mr Horwoods Plan of London &c complete | and correct as far as the nature of the work would admit; the whole extent of the | Plan has been carefully re-surveyed throughout; from which examination all the New Buildings, | Alterations & other Improvements are now inserted to the present time. | THIS EDITION has likewise been augmented with eight new copper plates, extending the Plan | eastwards to the River Lea; thereby comprehending those important commercial objects the | LONDON, WEST INDIA and EAST INDIA DOCKS, | by the proprietor WILLIAM FADEN, | Geographer to His Majesty & to His Royal Highness the Prince of Wales, | Charing Cross. | and Published by him, June 4th 1807.

Third edition title

[Plate K 1]

'Third Edition' replaces 'Second Edition' on line 13, and the following is added after the date 'June 4th 1807' which remains unaltered:
———— | Scale of 20 Chains or a quarter of a Mile | [scale bar] | THIRD EDITION 1813.|

Fourth edition title

[Plate K 1]

'1819' replaces '1807'. 'FOURTH EDITION' replaces 'THIRD EDITION 1813.' but 'Third Edition' on line 13 remains unchanged.

First edition imprints

Publisher's imprint

On most plates the sheet numbers were engraved first, leaving little space in which to insert the publisher's imprint.

A1	Published as the Act directs Febry. 17. 1794
A2	Published as the Act directs January 23d, 1794
A3	Published as the Act directs Jany. 28th. 1794
A4	Published as the Act directs April 10th. 1794
B1	Published as the Act directs Octr: 25th 1793
B2	Published as the Act directs June 22d: 1792
B3	Published as the Act directs January 2d. 1795
B4	Published as the Act directs January 2d. 1795
D4	Published as the Act directs January 2d. 1799
Others	PUBLISHED AS THE ACT DIRECTS BY R. HORWOOD MAY 24 1799 (C2 and C3 lack the full point in R HORWOOD: H1 has a full point after the 24)

Draughtsman's imprint

A1	Horwood Delint.
A2	Horwood Delint.
A3	Horwood Delint.
A4	Horwood Delint.
B1	Horwood Delint.
B2	*Horwood Delint.*
B3	Horwood Delint.
B4	Horwood Delint.
D4	Horwood Delint.

Engraver's imprint

A1	Spear. Sculpt. Star Alley Fenchurch-Street
A2	Spear. Sculpt. Star Alley Fenchurch-Street
A3	*J. T.* Sculpt.
A4	J. T. Sculpt.
B1	Spear. Sculpt. Star Alley Fenchurch-Street.
B2	*Spear. Sculpt. Star Alley, Fenchurch-Street.*
B3	J. T. Sculpt.
B4	J. T. Sculpt.
D4	J. T. Sculpt.
H1	Ash. Sculp. Adams Place, Borough. [Ash's workshop is marked on plate E3]

Second edition imprints

All plates have completely new publisher's imprints and carry no draughtsman's or engraver's imprints. They are in the centre below the border except for plates in the bottom row where they appear in the centre above the border.

On the 32 original plates:

Second edition. Published by W. Faden [sheet number in a circle] Charing Cross June 4th 1807.

On the 8 sheets added in the east:

Published by W. Faden [sheet number in a circle] Charing Cross June 4th 1807.

Third edition imprints

The same as the second edition but 'Third' replaces 'Second' and '1813' replaces '1807'.

Fourth edition imprints

The same as the third edition but 'Fourth' replaced 'Third' and '1819' replaces '1813'.

My thanks are due to Francis Herbert, Ralph Hyde and John Phillips for their assistance on many points, to the many map librarians who have made copies of Horwood's map available to me, to Christopher Cromarty for taking the photographic illustrations, and to the Corporation of London for allowing the reproduction of maps held in the Guildhall Library and the Greater London Record Office.

This paper is based upon examination of all the copies of Horwood held in the following locations. The number of copies, some of them incomplete and many cropped or damaged, is given in parentheses.

First edition 1792–9
British Library (5), Guildhall Library (6), Greater London Record Office (7), Westminster Public Library (6), Marylebone Public Library (2), University of London Library (2), Royal Geographical Society (2), Royal Institute of Chartered Surveyors (2), Royal Society of Arts (1), India Office Library (1), Phoenix Assurance Company (4), Bodleian Library, Oxford (1), Brasenose College, Oxford (1), Cambridge University Library (2), Liverpool Public Library (1), Private copies (3).

Second edition 1807
British Library (1), Guildhall Library (1, two sheets missing), University of London Library (1).

Third edition 1813
Guildhall Library (1), Greater London Record Office (1), Marylebone Public Library (8 damaged sheets only).

Fourth edition 1819
British Library (1), Guildhall Library (4 sheets only), Greater London Record Office (3, one has only 14 sheets), Westminster Public Library (1).

Further copies are known in the following locations but have not been examined:

First edition
London Library (1), Royal Library, Windsor Castle (1), Library of Congress, Washington (1), University of Texas, Austin (1), University of Indiana, Bloomington (1), Duke University, North

Carolina (1), Church of St Katherine Cree, London (1), Private copies (several known from auction records and dealers catalogues).

Second edition
Lambeth, Minet Library (copy listed in James Howgego, *Printed Maps of London circa 1553–1850*, second edition 1978, p. 157, but apparently missing in 1984).

Fourth edition
Private copy (incomplete: auctioned 27 July 1987), University of Indiana, Bloomington (sheet F3 bound with their first edition).

XI. THE MUSEUM IN DOCKLANDS

By CHRIS ELMERS

SINCE 1981 the Museum of London has been actively pressing for the setting up of a major new museum – the Museum in Docklands – to deal with the exciting story of the capital as the greatest maritime, industrial and commercial centre that the world has every known. Docklands, then largely undeveloped, was selected as the future home for a Museum of London's 'working history' because it was seen to be the key to much of London's past, present and future commercial success. Over the following years, the arguments for establishing such a Museum – the important, but all too little known, story of working London; the riches of the Museum of London's collections; the creation of a major visitor and tourist attraction; employment potential and future financial viability – have been well rehearsed in an extensive series of reports and consultants' studies. Following the Museum's failure to secure accommodation in the Grade I listed sugar warehouses, of 1802–3, on the North Quay of the West India Dock, and an alternative site at Blackwall, the London Docklands Development Corporation (LDDC) in 1988 committed two historic tobacco warehouses, at the Royal Victoria Dock, and £6 million to the Museum project, subject to formal approval from the Department of the Environment.

The impetus to build up the Museum of London's collections of port associated material was the direct result of the dramatic changes in cargo-handling technology which took place in the 1960s and 1970s. These changes – containerization in particular – had seen the main focus of port activity move progressively downstream to Tilbury, leaving in its wake empty docks and warehouses, unemployment, and vast areas of derelict land. The impact of this decline spread outwards to other industries, such as flour milling, ship repairing and barge building, which all but collapsed. In 1979 it was decided that the Museum would have to act very quickly if it were to save a representative collection of material relating to London's important maritime and port activities. This was a timely decision, for within two years the two remaining docks within the Greater London area – the once mighty West India and Millwall Docks and the Royal Group of Docks – had closed.

Often fighting against overwhelming odds, but with valuable help from the Port of London Authority and the LDDC, museum staff have worked hard to ensure that important 'workaday' aspects of London's culture have been saved for future generations to appreciate and enjoy. The collections include thousands of objects, ranging from cranes, winches, trucks, trolleys, weighing equipment, gauging tools and signs, to the humble docker's hooks and comprehensive groups of ship repairing equipment, salvage gear, coopers' tools, rigging and sailmaking equipment, ships chandlers' gear and milling machinery. Significant Thames craft were identified for acquisition and a 1924 ex-steam tug (the *Knocker White*), a Wey Navigation barge (the *Perseverance IV*), a 1937 diesel launch tug (the *Varlet*) and a 1920s watermans skiff were all added to the collection. From upriver, the Museum acquired Hammerton's Twickenham Ferry skiff, a hiring skiff and a pleasure punt, to help complete a typology of Thames craft. Other important maritime items include an early ship's deck-house, a large number of anchors, mooring and navigation buoys, navigational instruments, ship models and the name-plate from the ill-fated pleasure steamer, the *Princess Alice*, sunk in 1878 with the loss of over 600 lives.

Beyond its collection relating to the industries and commerce of the port and river, the Museum has the contents of some 65 workshops in store. This fine collection ranges, alphabetically, from artificial flower making to wheelwrights, and spans the scale of industrial production, from the manufacture of fine watches to the production of heavy forgings for ships. Although most of London's important trades, and manufacturing districts, are now represented in the Museum's collections, it is sad to record that they have grown as a direct result of London's continued de-industrialization. Besides the contents of workshops and factories, the Museum also has an impressive collection of stationary engines, road vehicles and equipment from gasworks and electricity generating stations. Taken together, the Museum's range of working history material is of national importance and probably forms the largest collection in the world.

Pending the opening of the new Museum – which is planned for 1992 – some items from the collections are on display at the Visitor Centre, at the Royal Victoria Docks, which is the focus of the highly successful Museum of London coach tours of Docklands. The Museum is also responsible for the management of the Port of London Authority Archive and Library Collection, at Poplar, with its wealth of printed and manuscript material, pictures, photographs, and films. To this has been added the Museum of London's field records of many workplaces, associated buildings, and Docklands' sites, as well as an extensive collection of material relating to recent developments in the area. In addition, the Library also houses over 200 tape-recorded interviews relating to all aspects of life and work in Docklands. Museum staff also operate a very popular mobile exhibition trailer, 'Museum on the Move', which services local schools and events in Docklands.

The full potential of the collections of course, will only be realized when the new museum is open. Then an estimated 500,000 visitors a year will be able to enjoy its displays and participate in lively demonstrations of working vessels, cargo handling and craft workshops. Much work, however, still needs to be done to realize the full potential of the project and bring to fruition the valuable work already undertaken by the Museum of London, with much financial support from the LDDC and the Government's Community Programme Scheme. Significant levels of capital funding still need to be raised from both the public and private sectors.

London's working history is unique and a permanent home to tell its story is long overdue. With the continued help of past and present supporters, and the recruitment of new ones, the Museum in Docklands is set to become a reality as one of the most exciting heritage and tourist projects of the 1990s.

XII. CENTRE FOR METROPOLITAN HISTORY

By HEATHER CREATON

ESTABLISHED in 1987 by the Institute of Historical Research (University of London), the Centre for Metropolitan History has now settled into its own premises at 34 Tavistock Square and is home to several flourishing new projects on London history.

The Centre, under the direction of Dr Derek Keene, provides a forum for the exchange of ideas about metropolitan history at seminars, conferences and other meetings. It undertakes original research into the society, economy, culture and fabric of London, with regard to its role both within the British Isles and in the world at large. Scholars from other parts of the world participate in its activities, and studies are undertaken comparing London with other cities.

Although the Centre is new, its range of interests and coverage already covers a wide spectrum of London history. Current projects, supervised jointly by the Director of the Centre and a scholar expert in the field, range in date from 1250 to 1986 and in subject from agriculture to finance, and from disease to office life. Among them is *The Social and Economic Study of Medieval London*. Documentary sources are used to reconstruct the history and topography of houses and other properties in the City of London, revealing trends in its development up to the Great Fire of 1666. In addition to findings already published, this study has also generated computerized biographical indexes of about 12,000 London inhabitants from the twelfth to the seventeenth centuries, and other listings. At present the study is working on the parishes surrounding the Bank of England. The project has been funded mainly by the Economic and Social Research Council, by an anonymous donor, and by an appeal sponsored by the Bank of England.

The project *Feeding the City: London's impact on the agrarian economy of southern England, 1250–1350* measures the impact on its hinterland of the capital's demand for food and other supplies by analysing manorial account rolls and inquisitions post mortem to build up a picture of monastic and lay estates. The results will be based primarily on a representative series of estates within the study area. The principal findings will be published as a monograph describing and analysing the patterns of agriculture, marketing and communications in a large area of southern England at this period, with particular reference to the impact of the capital city. A by-product of the research will be a detailed inventory of manorial accounts for the sample counties up to 1350. The project is funded by the Leverhulme Trust, and employs two research assistants who are jointly supervised by Dr Bruce Campbell of Queen's University, Belfast. Other research deals with later historical periods, notably *Epidemics and Mortality in the Pre-industrial City: Florence and London compared*. This examines the effect of epidemic disease on two of the major cities of seventeenth century Europe, comparing 'normal' patterns of mortality with those for the crisis plague years and using computer analysis for record linkage. It is supervised by Dr Keene and Dr John Henderson, of Wolfson College, Cambridge.

Also progressing well is *From Counting-House to Office: the evolution of London's central financial district, 1690–1870*, a study of the changing nature of the City as mixed residential and commercial areas gave place to purpose-built office blocks and exchanges and as workers moved out to the suburbs to live. The work is jointly supervised by Dr Martin Daunton of University College, London, employs one research assistant, and is funded by the Economic and Social Research Council. Two further projects are now in the planning and fund raising stage: *Office Life and Environment, 1870–1980*, to be jointly supervised by Mr Robert Thorne of English Heritage, and *The Jobbing System of the London Stock Exchange: an oral history*, to be jointly directed by Dr David Kynaston. The former will examine employment patterns and business practice as expressed in office design. The latter is concerned with the culture of the stock market in the City of London before the 'Big Bang' of October 1986. This event brought to an end the distinctive way of life of the jobbers on the Stock Exchange, a group which has left few written records of its activities.

In addition to organizing research projects such as these, the Centre provides a practical service for those interested in the history of London by carrying out bibliographical work and publishing news of work in progress. Two projects currently reflect this side of the Centre's activities:

The Bibliography of Printed Works on London History to 1939. There is at present no general bibliography for students of London and this publication, due in 1992, aims to fill the gap. It provides a selective listing of relevant works published to date, and is undertaken in co-operation with Guildhall Library.

Register of Research in Progress. The Centre circulates questionnaires to those working on any aspect of the history of London and maintains a list of current research topics. The first list appeared in the *London Journal* in the autumn of 1988, and will be updated at intervals.

The Centre's first series of *seminars* on metropolitan history had a successful start in the academic session 1987–8, and met again on alternate Wednesdays in the autumn and spring terms of 1988–9. The themes this year were 'Metropolitan politics' and 'Out-of-doors in the metropolis'.

Conferences are also held from time to time. One-day meetings during the autumn of 1988 included *The research potential of human skeletal remains in London* (with the Museum of London) and *Archivists and Historians in London* (with the London Archive Users' Forum).

Further information about the Centre and its work is available from: Miss Heather Creaton, Deputy Director, Centre for Metropolitan History, 34 Tavistock Square, London WC1H 9EZ.

XIII. WILLIAM FRANCIS GRIMES
1905–1988

OBITUARY

By RALPH MERRIFIELD

PROFESSOR William Francis Grimes, CBE, DLitt, FSA, known to his many friends as 'Peter', died on Christmas Day, 1988, at the age of 83. His nickname, bestowed by Audrey, his second wife, reflects his appreciation of music rather than any similarity to the hero of Britten's opera. Few of his archaeological friends knew that in his younger days he had been a promising violinist. He took a First in Classics at the University of Wales, and obtained an MA with distinction for archaeological work on the legionary tile and pottery factory at Holt in Denbighshire. He was appointed Assistant-Keeper in Archaeology at the National Museum of Wales in 1926, and wrote the *Guide to the Collections Illustrating the Pre-history of Wales*, published in 1939 and reissued as *The Pre-History of Wales* in 1951. He left the Museum in 1938 to join the Ordnance Survey as Assistant Archaeology Officer under O. G. S. Crawford, but with the coming of war was seconded to the Ministry of Works and given a new task. This was to excavate and record archaeological sites threatened by destruction through the requirements of national defence. They included the still puzzling Late Bronze Age and Iron Age site at Heathrow ('Caesar's Camp'), excavated prior to runway construction.

By the end of the war, Grimes was one of the most experienced excavators in the country, and could fairly claim to have been its first full-time 'rescue' archaeologist – a profession that has since expanded to offer so many young archaeologists somewhat precarious employment. His appointment in 1945 as Director of the London Museum, in succession to Mortimer Wheeler, brought him to the capital,

William Francis Grimes.

where rescue archaeology was now confronted with its greatest challenge. More than a seventh of the area of the walled Roman and medieval city had been destroyed by enemy action and now awaited rebuilding. There was a unique opportunity for archaeological investigation, and the Society of Antiquaries of London took the initiative, first by sponsoring a season of trial excavation in 1946, and in the next year by establishing the Roman and Medieval London Excavation Council to organize a more extended programme. Grimes was its inevitable choice as Director of Excavations, and the work continued from July 1947 until December 1962. The difficulties were enormous, since no site could be excavated before the bombed ruins were cleared, and lack of funds precluded the use of machinery. The Council did its best with fund-raising from City and other London institutions and from individuals, but in those years of war-weariness failed to raise the necessary enthusiasm, and in the published list of donations only twenty-six amount to more than £100. The enterprise in fact was kept going only by an annual grant from what was then called the Ministry of Public Buildings and Works (later part of the Department of the Environment). This was sufficient to keep at work a small team of labourers, such as Grimes had been accustomed to use during the war years. He, however, was no longer a full-time excavator. He had the London Museum to bring to life again in its new premises at Kensington Palace, and his status in the archaeological world inevitably led to other demands on his time that could not be refused – membership of the Royal Commission for Ancient and Historical Monuments for Wales from 1948 and a close involvement with the Council for British Archaeology, as Secretary from 1949 to 1954 and as President from 1954 to 1959. Then in 1956 he was appointed Director of the Institute of Archaeology and Professor of Archaeology in the University of London.[1]

It is, of course, his contribution to our knowledge of the early topography of London that is of greatest interest to the readers of this volume, but it has to be recognized that only a small proportion of his time could be devoted to excavation, and the circumstances dictated that only very small areas could be excavated. The strategy determining where these small areas were to be was therefore of the greatest importance. The city defences, as a linear feature, gave good opportunities for this sort of investigation, and dating evidence that has not been superseded was recovered for the Roman wall of *c.* 200, and the thirteenth century (or later) medieval hollow bastions near Cripplegate. Grimes's policy of strategic placing of his cuttings was dramatically justified by his discovery of the Roman fort that preceded the city wall in this area. An initial cutting where the city wall changed direction revealed the characteristic corner of a fort, and its topographical relationship with the medieval (and modern) Silver Street, Addle Street and Cripplegate showed at once where the west and south gates would be found. Perhaps because it was revealed in such an orderly and logical way, this discovery seems to have given Grimes particular satisfaction.[2] Yet, ironically, the find with which his name will always be most closely associated, and which was the occasion (though obviously only one of many reasons) for his award of the CBE, was the result of pure chance. A cutting made across the stream-bed of the Walbrook, a central feature of Roman London, to investigate its size, depth and the nature of its banks, at the only place where its east bank could be conveniently sectioned, revealed a substantial Roman building with apsidal western end adjacent to the stream. It was reported in the Press as a probable basilican temple, but lay visible for a considerable time without attracting any further attention, while Grimes proceeded with the deep and difficult excavation through the stream-bed that was his primary purpose. It was not until the discovery of the head of Mithras (the first of a number of important marble sculptures to be found) revealed the cult practised in the temple, that the public imagination was captured; suddenly thousands of people were clamouring to visit the site. Popular demand for preservation of the building led to a promise by the owners that it should be reconstructed nearby at their own expense; for this Grimes's advice was invited but subsequently ignored. In his own words:

The result is virtually meaningless as a reconstruction of a mithraeum. It is exasperating that so much money should have been spent to such poor purpose.[3]

The somewhat negative results of the search of the Excavation Council for evidence of Anglo-Saxon London foreshadow precisely the results of the much more extensive investigations of the last fifteen years. Like his successors Grimes noted that there was a gap in the archaeological evidence for occupation within the walled city until the *later* Saxon period, though he accepted of course that there must have been some occupation from about 600 in the neighbourhood of St Paul's. Yet even the large bombed area immediately north of the Cathedral failed to produce any evidence. Residual pre-Norman pottery that could be dated everywhere in the City was consistently of *late* Saxon date, as were the only datable domestic Saxon structures found – sunken hut-sites in Cannon Street south-east of St Paul's. Traces of similar huts were found in Bucklersbury and Addle Street.[4] In addition the plan of the first (Saxon) church of St Alban Wood Street, traditionally the chapel of King Offa in the eighth century, was traced. Grimes pointed out that this chapel and the traditional site of the palace of Ethelbert were near the centre of the Roman fort, which may have attracted Saxon kings to this area, though characteristically he warned against the dangers of dabbling in such speculation.[5] Subsequent investigation has confirmed that Anglo-Saxon occupation within the city walls was limited to an ecclesiastical and royal presence until the ninth century, and that the busy port described by Bede in the eighth century must be sought further west.

Other City churches investigated by Grimes included St Swithun London Stone and (notably) St Bride's in Fleet Street, where the plans of five successive churches were recovered, ranging from the first stone church in the eleventh century to Wren's post-Fire church.[6] Outside the City excavation on the site of the great Carthusian monastery of Charterhouse led to a drastic revision of the plan as previously reconstructed, with the church a larger structure extending further to the south than had been thought.[7] On the site of the Cluniac Abbey of St Saviour in Bermondsey a substantial part of the plan of the east end of the Abbey Church was recovered at the junction of Abbey Street and Tower Bridge Street.[8] Medieval secular buildings investigated in the City included Neville's Inn at Windsor Court, an undercroft west of the tower of St Mary-le-Bow in Cheapside and Brewers' Hall at the junction of Aldermanbury and Addle Street.[9] It must be remembered also that Grimes made a major contribution to the study of medieval London by his investigations of the later history of the city wall, in which he demonstrated that the masonry of one rebuild could be closely paralleled by a mid-fourteenth century wall in Westminster Abbey, while a hitherto unknown bastion which he discovered in St Giles Cripplegate churchyard (11A) was stratigraphically thirteenth century or later.[10]

It is an impressive catalogue of achievement by a single archaeologist who was without trained archaeological assistance until he was joined in 1953 by Audrey Williams, formerly of the Verulamium Museum, who subsequently became his wife. Grimes was always acutely conscious of the fact that the complex stratigraphy of central London made close supervision of unskilled or semi-skilled excavators essential if grave errors were to be avoided. An attempt to use week-end amateur volunteers in 1949 proved unsatisfactory, but a group that formed at the time of the Mithras temple excavation in 1954, under the leadership of George Rybot of Shell was more cohesive and acquired the self-discipline that Grimes demanded. For many week-ends, that in Grimes's words were 'at once hilarious and purposeful', it did useful work in filling in the details of the Roman fort defences in Noble Street. It owed much to George's leadership and engaging personality. After 1956 groups from the University of London Archaeological Society also did similar work on the city wall near Cripplegate.

Yet it is doubtful if Grimes could ever have permitted any excavation, amateur or professional, on which more depended than the dotting of 'i's and the crossing of 't's in a story that he could already read, without the constant supervision that he often could not provide. For he was a perfectionist who found it difficult to delegate; he was fully aware that few archaeologists were his equal in delicate work with the trowel, in the photography of the features that were revealed, and in drawing of sections, which he was one of the first to insist should be drawn as they were seen, not interpreted prematurely by drawing in diagrammatic layers. Under his directorship large-scale excavation of the kind that might have been achieved by a Mortimer Wheeler – or, to cite only one archaeologist of a more recent generation, a Barry Cunliffe – probably would have been impossible for him, even if the necessary funds had been available, unless he had also been able to secure the services of more archaeologists whom he could trust as completely as he did Audrey. Nevertheless more would have been done if a much more generous allocation of funds had permitted the proper use of machinery in site-clearance and back-filling, all of which had to be done by hand. Yet possibly a Wheeler would have obtained that financial support by skilful exploitation of the media, from which the more scrupulous and conscientious Grimes always shrank. To him the great upsurge of public interest that followed the discovery of the Mithraeum had 'a curiously nightmarish quality', though he faced the situation with courage and apparent equanimity; one suspects that to Wheeler it would have been a golden opportunity.

There was a certain antipathy between these two great archaeologists that was very properly concealed from those who witnessed their occasional encounters. It was due in part perhaps to the fact that one had followed closely in the footsteps of the other – at best a difficult relationship that invites invidious comparisons – but much more to the clash of their characters, temperaments and philosophies, which were diametrically opposed. Wheeler was flamboyant and self-assertive with brilliant flashes of insight, and these qualities had served archaeology well; Grimes was single-minded and conscientious in his pursuit of truth, with a preference for solving problems by orderly, logical steps, and a distrust of hypotheses that had not been thoroughly tested. Wheeler was capable of great kindness, particularly in response to favours received or in redress of unacknowledged wrongs, but was by nature exploitative; Grimes could be stern when he considered principles had not been upheld, but was invariably kindly and considerate, and reluctant to make demands on others. Wheeler had made himself very much an Establishment figure, by distinguished military service in two wars as much as by archaeological achievement; Grimes, like so many of his fellow-countrymen, had sympathies with the Left in politics and no admiration for military glory. It was the old clash between Cavalier and Puritan.

Yet Grimes was by no means puritanical except in his intellectual approach and in the integrity of his principles. He was great fun at a party, and in his own way a dandy, with the invariable red carnation in his button-hole, and a beautifully embroidered silk waistcoat on any moderately festive occasion. His book *The Excavation of Roman and Medieval London* is full of touches of humour, and he does not disdain the little human anecdotes that reveal the light side of those difficult years. He disparagingly nicknamed the book 'Comic Cuts', but it is in fact a mine of accurate information on his archaeological work, in which, with characteristic conscientiousness every sentence has been considered in order to convey the exact truth or shade of probability as he saw it. For this reason any student of London archaeology will be obliged to study it with the attention that a theological student gives to Holy Writ.[11] For, sadly, it is all we have. It constitutes an admirable interim report, but no definitive report has been published on any of Grimes's London excavations.

For this he has of course been criticized, and it is a topic that cannot be evaded by his friends and admirers in any assessment of his life and work. How was it that one so conscientious, and one moreover who had been responsible for training so many young archaeologists, should have failed to set an example in what most would regard as an archaeologist's ultimate duty? It would not have been surprising if he had been unable to make substantial inroads into the vast backlog that awaited him, for only one of the projected three volumes of reports on his wartime work had been published.[12] The total task was impossible in one lifetime; yet it is sad that not a single definitive report on the important work of the Excavation Council in the City of London has been brought to publication by Grimes himself, if only to set a standard for subsequent reports by others who will have to take over the task at some future date. All his notebooks, drawings and other records were deposited before his death in the Museum of London, which already held the finds from the excavations. The data he recovered are therefore secure and have gone into what archaeologists like to call 'the Archive' – a term that implies greater order and accessibility than is likely to be achieved until more appropriate accommodation and a staff to deal with it can be acquired. The problem is not of course confined to the records and finds of the Roman and Medieval London Excavation Council, which are very small in compass when compared with those engendered by the much more extensive excavations carried out by the Museum's archaeological departments in the last fifteen years. They have a creditable record of publication but definitive reports that include all specialist reports have so far appeared only for a minority of the sites excavated. This is a country-wide problem arising from the great increase in rescue excavation that unfortunately coincided with a similar increase in the cost of publication. Grimes's failure to meet his full obligations has to be judged in the light of the continuing difficulties of his successors, who have much greater resources in staff. He did for a considerable time, however, have assistance with the finds from a succession of excellent research assistants; and when Joanna Bird completed her pottery report on the Bucklersbury House site, there seemed every prospect that this excavation report at least, with its important investigation of the Walbrook and Mithraeum, would soon be ready for publication, particularly as Professor Jocelyn Toynbee's report on the works of art from the temple had long been to hand.[13] But alas, it never materialized; if it had, the continuing interest in the temple would have ensured good sales, and it would not only have deflected criticism but would also have won great kudos for the author. Was this perhaps the very reason that Grimes, following the dictates of his puritanical conscience, hesitated to go ahead with it when so much of his earlier, less spectacular, work remained unpublished? Was there perhaps also an element of defiance in his reluctance to make a single gesture to placate his critics? For Wheeler had once tactlessly upbraided him on this account, and coming from a man who had no claim to authority over him it had never ceased to rankle. Human motivation is often mixed, but the principal reason for Grimes's unwillingness to take an opportunity that seemed to be well within his grasp was probably his perfectionism. If he was to do one definitive report on his London work, it had to be a model of its kind, and the adverse circumstances of the excavation itself may have precluded this, as he hints in his book.[14]

After his retirement in 1971 he returned to Wales to live in Swansea, but continued for several years as adviser to the Corporation of London on the conservation of the surviving remains of the city wall and their presentation to the public. He also helped to establish the City of London Archaeological Trust, with the purpose of raising funds from voluntary sources to assist his successors, and remained its President until 1986. Nevertheless, the archaeology of South Wales now became his main interest. Throughout the years in London he had never cut himself off from his Welsh roots. He had been a member of the Royal Commission on Historical Monuments for Wales from 1948 to 1978, and its Chairman from 1967. It was on the occasion of a presentation to him from his colleagues and the Commission's staff, when he retired from this office in 1978, that his wife died with tragic suddenness from a heart attack in the hotel where they were staying. Nevertheless he courageously went through this social occasion without a word to its organisers of his loss and the shock he had just received. Always considerate of others, he was characteristically determined not to spoil their enjoyment of an event that had been carefully planned. He made the necessary speeches and those present were only aware that he did not seem quite his normal self, and they suspected that he was unwell.

Grimes was also actively concerned with the work of the Field Studies Council, of which he was Chairman from 1966 to 1975 and subsequently its President. For many years he left London during the summer vacation to carry out field surveys in Pembrokeshire.

The high regard in which he was held by local archaeologists became apparent at his funeral. By his express wish the service was completely secular. It was accompanied by music from Beethoven's Sixth Symphony, and there were two addresses, the first by Beatrice de Cardi as a friend of many years associated with his London activities, and the second by Dillwyn Miles, an old friend from Pembrokeshire, who also read Dylan Thomas's *And Death Shall Have No Dominion*. It was a charming tribute that the members of the local archaeological society all wore red carnations. South Wales incidentally produced the greatest possible variety in weather for the occasion – brilliant sunshine, pouring rain, sleet, thunder and lightning, and a spectacular rainbow.

Apart from the health problems that seem to be inseparable from survival into the eighties, Peter's last years were evidently happy. He was married a third time – to Molly, an old family friend, who survives him, as also do a son and daughter of his first marriage. Everything indicated that he had settled down contentedly in the warm-hearted Welsh community that he had left for so many years of strenuous work and heavy responsibility elsewhere.

If I may end on a personal note, I shall not forget the sympathy and kindness Peter Grimes went out of his way to express in a dark hour of my own professional life, when every way ahead seemed closed, and I would gladly have taken early retirement or any other means of escape. Fortunately it proved to be the proverbial dark hour before a dawn in which the decision to go ahead with the Museum of London opened up new horizons, and it soon became clear that we were also on the threshold of a quite unexpected renaissance of London archaeology. But it is in such dark hours that we need friends.

Professor Grimes served on the Council of the London Topographical Society from 1947 until his death in 1988. He was chairman from 1961 till 1974, and thereafter vice president from 1974 till 1988.

1. This is by no means a complete account of the demands on Grimes's time and attention during his time in London. He also served as President of the Royal Archaeological Institute and of the London and Middlesex Archaeological Society, which owes to him the inauguration of its Archaeological Research Committee. He was a member of the Royal Commission for Historical Monuments for England as well as Wales, and served on the Ancient Monuments Boards of both countries.
2. W. F. Grimes, *The Excavation of Roman and Mediaeval London* (1968), 17–39, 238–9.
3. *Ibid.*, 92–117, 230–7.
4. *Ibid.*, 153–60.
5. *Ibid.*, 203–7. The later development of the church was also traced, and three phases recognised, 207–9.
6. *Ibid.*, 199–203, 182–97.
7. *Ibid.*, 175–80. See also M. D. Knowles and W. F. Grimes, *Charterhouse* (1954).
8. W. F. Grimes, *loc. cit.*, 210–17.
9. *Ibid.*, 163–72.
10. *Ibid.*, 78–90; 71–8.
11. Here I speak from experience. In writing a recent book on Roman London (*London, City of the Romans*, 1982) I found myself constantly asking 'What *exactly* does Grimes say about this?', and invariably found enlightenment in its pages.
12. *Excavations on Defence Sites, 1939–45* (1960).
13. The latter was eventually published separately in 1986 as Special Paper No. 7 of London and Middlesex Archaeological Society, *The Roman Art Treasures from the Temple of Mithras*, by J. M. C. Toynbee.
14. W. F. Grimes, *loc. cit.*, 237–8.

BIBLIOGRAPHY

By ORTRUN PEYN*

1928

A Beaker burial from Ludchurch, Pembrokeshire. *AC* 83 (1928), 338-43.
Corston Beacon: an Early Bronze Age cairn in south Pembrokeshire [with Cyril Fox]. *AC* 83 (1928), 137-74.
A 'dug-out' canoe from Llyn Llydaw, Snowdon, Caernarvonshire. *BBCS* 5 (3) (1928), 283.
An Early Bronze Age burial from Stormy Down, Pyle, Glamorgan. *AC* 83 (1928), 330-7.
An Early Bronze Age burial from Stormy Down, Pyle, Glamorgan. *BBCS* 4 (2) (1928), 173-4.
Polished celts from Kingsbridge, Pembroke. *BBCS* 5 (3) (1928), 272-3.
A Romano-British brooch from Penmaenmawr. *AC* 83 (1928), 194-5.
A stone axe from Trefor Quarry, Caernarvonshire. *AC* 83 (1928), 195-7.

1929

Beaker burial at Llanharry, Glamorgan. *BBCS* 5 (1) (1929), 80-1.
Burial mounds in the parish of Llanboidy, Carmarthenshire. *AC* 84 (1929), 325-32.
Dug-out canoe from Llandrindod Wells, Radnorshire. *BBCS* 5 (1) (1929), 85.
A fragmentary stone axe from Sker, Glamorgan. *AC* 84 (1929), 147-51.
A holed axe-hammer from Dolgelly, Merioneth. *AC* 84 (1929), 150-1.
La Tène I brooch from Merthyr Mawr, Glamorganshire. *BBCS* 5 (4) (1929), 370.
Scandinavian object from Castlemartin, Pembrokeshire. *BBCS* 5 (4) (1929), 370-1.
A stone axe from Kenfig Burrows, Glamorgan. *AC* 84 (1929) 149-50.

1930

Caerleon. *BBCS* 5 (3) (1930), 278.
Caerwent. *BBCS* 5 (3) (1930), 278-9.
Holt, Denbighshire: the works-depot of the Twentieth Legion at Castle

* Abbreviations are given at the end of this list.

Lyons (London: Society of Cymmrodorion, 1930) (Y Cymmrodor: the magazine of the Honourable Society of Cymmrodorion; vol. 41).
A leaden tablet of Scandinavian origin from south Pembrokeshire. *AC* 85 (1930), 416–17.
Polished stone axes from Cowbridge and Welsh St Donats, Glamorgan. *AC* 85 (1930), 210–12.
Recent finds of prehistoric implements from Wales. *AC* 85 (1930), 414–16.
Stone axe from Chepstow, Monmouthshire. *BBCS* 5 (3) (1930), 274.

1931

A bronze casting in the Guilsfield, Montgomeryshire, Hoard. *AC* 86 (1931), 358–61.
A bronze casting in the Guilsfield, Montgomeryshire, Hoard. *BBCS* 5 (4) (1931), 393–4.
Bwrdd Arthur: Llanfihangel-din-sylwy, Anglesey. *BBCS* 5 (4) (1931), 392.
Caerleon. *JRS* 21 (1931), 215–16.
The early Bronze Age flint dagger in England and Wales. *PPSEA* 6 (4) (1931), 340–55.
Excavations at Caerwent, Monmouthshire, 1930. *AC* 86 (1931), 210–15.
The Llandrindod Wells dug-out boat. *TRS* 1 (1931), 10–16.
A new Beaker from Brecknockshire. *BBCS* 5 (4) (1931), 392.
Notes on recent work in prehistoric archaeology in Wales. *PPSEA* 6 (4) (1931), 383–4.
A polished celt from Blaina, Monmouthshire. *BBCS* 5 (4) (1931), 391–2.
Polished stone axe from Meline, Pembrokeshire. *AC* 86 (1931), 361.
Recent finds of stone axes in Wales. *BBCS* 6 (1) (1931), 94.
Romano-British pottery from Crocksydam Camp, Warren, Pemb. *BBCS* 5 (4) (1931), 394–5.
South Wales megaliths – survey. *BBCS* 6 (1) (1931), 88–9.
The T.C. Cantrill Collection. *BBCS* 6 (1) (1931), 93–4.
Two dug-out boats from Wales. *AnJ* 11 (1931), 136–44.

1932

Bronze Age incense cups probably from Cardiganshire. *BBCS* 6 (2) (1932), 195–6.
Bronze Age pottery from Dolgelly, Merioneth. *BBCS* 6 (2) (1932), 195.
The coins found during the 1923 excavations at Caerwent (Venta Silurium), Monmouthshire [with V. E. Nash-Williams]. *BBCS* 6 (2) (1932), 179–94.
Incense cups from Cardiganshire. *AC* 87 (1932), 409–11.
A new Beaker from Brecknockshire: correction. *BBCS* 6 (2) (1932), 196.
Prehistoric archaeology in Wales since 1925. *PPSEA* 7 (1) (1932), 82–106.

Recent finds of stone implements. *AC* 87 (1932), 406–11.
Report on material from the excavation of the earthwork of Castell Odo, Aberdaron, Caernarvonshire [appendix to 'Castell Odo', by C. E. Breese]. *AC* 87 (1932), 383–5.
Surface flint industries around Solva, Pembrokeshire. *AC* 87 (1932), 179–92.

1933

Bronze Age pottery from near Dolgelly, Merioneth. *BBCS* 6 (4) (1933), 287.
Human burials, Llangan, and St Mary Hill parishes, Glamorgan. *BBCS* 6 (4) (1933), 291.
A new Upper Palaeolithic site in southwest Wales. *BBCS* 6 (4) (1933), 286–7.
Notes on excavations in Wales in 1933. *PPSEA* 7 (2) (1933), 270–1.
Priory Farm Cave, Monkton, Pembrokeshire. *AC* 88 (1933), 88–100.
Recent finds of stone implements. *BBCS* 6 (4) (1933), 287–8.

1934

Bronze implements. *BBCS* 7 (2) (1934), 332–3.
The Cottrell Park standing stone, St Nicholas, Glamorgan. *TCNS* 67 (1934), 104–8.
Pembrokeshire survey. *BBCS* 7 (2) (1934), 334–5.
South Wales megalithic survey. *BBCS* 7 (2) (1934), 335–6.
A third-century hoard of Roman coins found with a burial at Ilston, Gower. *BBCS* 7 (2) (1934), 209–19.

1935

A Bronze Age burial at Knighton, Radnorshire. *BBCS* 8 (1) (1935), 94–5.
Bronze Age burial from Llandegla, Denbighshire. *BBCS* 8 (1) (1935), 95.
Carmarthenshire in the Old and Middle Stone Age [with L. F. Cowley]. In: *A History of Carmarthenshire* (ed. by John E. Lloyd). Cardiff: Printed for the Carmarthenshire Society, 1935. Vol. 1, Pp. 24–39.
Coygan Cave, Llansadyrnin, Carmarthenshire [with L. F. Cowley]. *AC* 90 (1935), 95–111.
Finds from a peat-bog near Cwrt Mawr, Llangeitho, Cardiganshire. *BBCS* 8 (1) (1935), 94.
Notes on excavations in Wales in 1935. *PPS* 2 (1935), 144–6.
Notes on recent finds of perforated axe-hammers in Wales. *AC* 90 (1935), 267–78.
A prehistoric hearth at Radyr, Glamorgan, and its bearing on the nativity of beech (*fagus sylvatica* L.) in Britain [with H. A. Hyde]. *TCNS* 68 (1935), 46–54.

Preliminary note on a Bronze Age burial mound near Knighton, Radnorshire. *TRS* 5 (1935), 78.
Recent books on British archaeology. *Ant* 9 (36) (1935), 424–34.
Recent finds of prehistoric implements. *BBCS* 8 (1) (1935), 95–6.
Report on several small fragments of pottery from Dyserth Castle site not examined by Mr Stuart Piggott. *AC* 90 (1935), 213–14.
Report on the pottery from Rhos Ddigre. *AC* 90 (1935), 213.
The Roman legionary fortress at Caerleon in Monmouthshire: report on the excavations carried out in the Town Hall field in 1930. *AC* 90 (1935), 112–22.
The Tanglannau (Mallwyd) bronze hoard. *BBCS* 8 (1) (1935), 96–7.

1936

A barrow on Breach Farm, Llanbleddian, Glamorgan. *TCNS* 69 (1936), 49–68.
Bronze Age burials near Crug-coy, Llanarth, Cardiganshire. *BBCS* 8 (3) (1936), 271–2.
Finds of stone and bronze implements. *BBCS* 8 (3) (1936), 272–3.
The long cairns of the Brecknockshire Black Mountains. *AC* 91 (1936), 259–82.
Map of south Wales showing the distribution of long barrows and megaliths. Southampton: Ordnance Survey [1936?], pp. 1–56.
The megalithic monuments of Wales. *PPS* 2 (1936), 106–39.
Notes and drawings of beakers. In: *A Local Survey of the Early Bronze Age* [by J. B. Calkin], *PBNS* (1935–6), 46–56.
Notes on excavations in Wales in 1936. *PPS* 2 (1936), 226–8.
Pentre Ifan burial chamber, Nevern, Pembrokeshire. *BBCS* 8 (3) (1936), 270–1.
Reports on a Roman pottery making site at Foxledge Common, Wattisfield, Suffolk. Pt. 4: Remarks on the kiln. *PSIANH* 22 (1936), 187–9.
Two Bronze Age burials from Wales. *AC* 91 (1936), 293–304.

1937

A Bronze Age dagger from near Penmachno, Caernarvonshire. *AC* 92 (1937), 174–5.
A deer antler implement from Llanddona, Anglesey. *AC* 92 (1937), 172–4.
Early man in the Cardiff district. In: *The Book of Cardiff*. Oxford: O.U.P. [for the National Association of Head Teachers], 1937, pp. 46–56.
[Note by WFG quoted in]: Parc le Breos chambered cairn, Gower by H.H.H. *AC* 92 (1937), 172–6.
Pembrokeshire survey. *BBCS* 8 (4) (1937), 386–7.
Pentre Ifan burial chamber, Pembrokeshire. *BBCS* 9 (1) (1937), 82–4.

A round barrow of the Bronze Age near Jacket's Well, Knighton, Radnorshire. *TRS* 7 (1937), 23–9.

1938

A barrow on Breach Farm, Llanbleddian, Glamorgan. *PPS* 4 (1938), 107–21.
The excavation of Gorsey Bigbury. Section III: The pottery. *PUBSS* 5 (1) (1938), 25–45.
Ty-isaf long cairn, Breconshire. *BBCS* 9 (3) (1938), 283–5.

1939

Bronze Age urn from Pen-y-lan, Llandyssul, Carmarthenshire. *CA* 78 (1939), 114–16.
The excavation of Ty-isaf long cairn, Brecknockshire. *PPS* 5 (1939), 119–42.
Guide to the collection illustrating the prehistory of Wales. Cardiff: National Museum of Wales and the Press Board of the University of Wales, 1939.
Meini Gwyr, Carmarthenshire. *BBCS* 9 (4) (1939), 373–4.

1940

A newly discovered Carmarthenshire megalith. *AC* 95 (1940), 80–3.
The Sutton Hoo ship burial. 7: The salvaging of the finds. *Ant* 14 (1940), 69–75.

1943

Excavations at Ffridd Faldwyn Camp, Montgomery, 1937–39 [by B. H. St. J. O'Neil and W. J. Kemp. In this article Section 'F. Finds' by W.F.G.]. *AC* 97 (1943), 51–5.
A grooved hammer from Rhandirmwyn, Carmarthenshire. *AC* 97 (1943), 235–6.

1944

Excavations at Stanton Harcourt, 1940. *Oxon* 8–9 (1943–4), 19–63.
Museums and the future. *Ant* 18 (69) (1944), 42–9.

1945

Early man and the soils of Anglesey. *Ant* 19 (1945), 169–74.
The earthwork at Vervil near Merthyr Mawr [with H. J. Randall]. *AC* 98 (1944–5), 241–7.

Maiden Castle [review of *Maiden Castle, Dorset*, by R. E. M. Wheeler]. *Ant* 19 (73) (1945), 6–10.

1946

A find of the Early Iron Age from Llyn Cerrig Bach, Anglesey. *Ant* 20 (77) (1946), 13–15.
Linear earthworks: methods of field survey [with C. Fox and B. H. St. J. O'Neil]. *AnJ* 26 (1946), 175–9.
Prehistoric period. In: *A Hundred Years of Welsh Archaeology* (ed. by V. E. Nash-Williams). Gloucester: Cambrian Archaeological Association [1946?], pp. 24–79.
A stone axe from Iver, Buckinghamshire. *RB* 14 (6) (1946), 361–4.

1947

Roman London. *CJ* 42 (1947), 379–82.
Roman London [summary of lecture]. *CAP* 44 (1947), 20–2.

1948

Pentre-Ifan burial chamber, Pembrokeshire. *AC* 100 (1948), 3–23.

1950

Contributions to a field archaeology of Pembrokeshire. 1: The archaeology of Skomer Island. *AC* 101 (1950), 1–20.
The Roman and Medieval London Excavation Council: Roman London, 1949–50: the Cripplegate Fort. *AN* 3 (1) (1950), 60–1.

1951

Aspects of Archaeology in Britain and Beyond: Essays Presented to O. G. S. Crawford, ed. by W.F.G. London: Edwards, 1951. Pp. 144–71: The Jurassic way across England.
The coins found during the 1925 (south wall) excavations at Caerwent (Venta Silurum), Monmouthshire [with V. E. Nash-Williams]. *BBCS* 14 (1951), 242–9.
The London Museum: a Short Guide. London: Trustees of the London Museum, [1951].
The Prehistory of Wales. Cardiff: National Museum of Wales, 1951.

1952

Air photography and archaeology. *TLMAS* 11 (1952), 1–9.
An early wall at the hall of the Merchant Taylors' Company. *TLMAS* 11 (1952), 85–6.
The La Tène art style in British Early Iron Age pottery. *PPS* 18 (1952), 160–75.

1954

Charterhouse: the Medieval Foundation in the Light of Recent Discoveries [with David Knowles]. London: Longmans, 1954.
London's unique Mithras temple: the amazing Walbrook site described – its importance and its implications. *ILN* 225 (9 Oct. 1954), 594–6.
The Roman and Medieval London Excavation Council, 1953–54 (St Bride's Church site, Friday Street, Walbrook). *AN* 5 (2) (1954), 48.
The scientific bias of archaeology. In: *The Advancement of Science* (British Association) 40 (1954), 343–6.
Where the Mithraeum will stand, and some notable finds. *ILN* 225 (16 Oct. 1954), 636–7.

1955

The Council for British Archaeology: the first decade. *AN* 5 (8) (1955), 1–7.

1956

Excavations in the city of London. In: *Recent Archaeological Excavations in Britain* (ed. by R. L. S. Bruce-Mitford). London: Routledge & Kegan Paul, 1956, pp. 111–43.
The prehistory of Caldey [with A. D. Lacaille]. *AC* 104 (1956), 85–165.

1957

A catalogue of the prehistoric finds from Worcestershire [by C. N. S. Smith; revised and edited by W.F.G.]. *TWAS* 34 (1957), 1–27.

1958

Archaeology and the university. *University of London, Institute of Archaeology Annual Report and Bulletin* 13 (for 1955–6, publ. 1958), 37–48.
Capel Garmon Chambered Long Cairn, Denbighshire. London: H.M.S.O., 1958.

Some smaller settlements: a symposium. In: *Problems of the Iron Age in Southern Britain : Papers given at a CBA conference held at the Institute of Archaeology, December 12 to 14, 1958* (ed. by S. S. Frere). London: University of London, Institute of Archaeology, [n.d.], Occasional paper 11] pp. 17–28.

1959

Roman Britain in 1958. [Section on] London. *JRS* 49 (1959), 125–6.

1960

Excavations on Defence Sites, 1939–1945. 1: *Mainly Neolithic – Bronze Age.* London: H.M.S.O., 1960.

1961

The prehistory of Caldey, pt. 2 [with A. D. Lacaille]. *AC* 110 (1961), 30–70.

1962

Prehistoric and Roman Pembrokeshire [In: Report of the Summer Meeting of the Royal Archaeological Institute at Tenby in 1962.] *AJ* 119 (1962), 308–50 (includes Nanna's Ave, p. 322; Clegyr Boia, p. 336; St David's Head promontory fort and Carn Llidi, pp. 336–7; Pentre Ifan, p. 341; Moel Trigarn, p. 341; Meini-Gwyr, p. 342; Ffynnan-Brodyr, p. 342; Castlemartin rath, pp. 347–8; Fishponds Camp, Bosherston, p. 348; Dry Burrows barrow group, p. 348.

1963

The stone circles and related monuments of Wales. In: *Culture and Environment : Essays in Honour of Sir Cyril Fox* (ed. by I. Ll. Foster and L. Alcock). London: Routledge & Kegan Paul, 1963, pp. 93–152.

1964

Excavations in the Lake Group of barrows, Wilsford, Wiltshire, in 1959. *BIA* 4 (1964), 89–121.
From the first inhabitants to the first metalworkers. In: *The Land of Dyfed in Early Times* (ed. by Donald Moore). Cardiff: Cambrian Archaeological Association, 1964, pp. 10–11.
The problem of the Raths of Pembrokeshire. *Ibid.*

Wales and Ireland in prehistoric times: some meditations. *AC* 113 (1964), 1–15.

1965

The archaeology of the Stamford region. In: *The Making of Stamford* (ed. by Alan Rogers). Leicester: Leicester University Press, 1965, pp. 15–33.

Neolithic Wales. In: *Prehistoric and Early Wales* (ed. by I. Ll. Foster and G. E. Daniel). London: Routledge & Kegan Paul, 1965, pp. 35–69.

1967

Sir Cyril Fox (1882–1967) [Obituary]. *AC* 116 (1967), 208–10.

1968

The Excavation of Roman and Mediaeval London. London: Routledge & Kegan Paul, 1968. Pp. xxi + 261; 32 pages of plates and illustrations.

1971

The Temple of Mithras in Walbrook London. In: *Britannia Romana: conferenze organizzate dall'Accademia Nazionale dei Lincei in collaborazione con la British Academy.* (Accademia Nazionale dei Lincei; anno 368, 1971, quaderno N. 150: Problemi attuali di scienza e di cultura). Pp. 29–36, plates 1–11.

1972

Lily Frances Chitty. In: *Prehistoric Man in Wales and the West: Essays in Honour of Lily F. Chitty* (ed. by Frances Lynch and Colin Burgess). Bath: Adams & Dart, 1972, pp. 1–3.

1976

Time on our side?: a survey of archaeological needs in Greater London: report of a joint working party of the Department of the Environment, the Greater London Council and the Museum of London [W.F.G. et al.]. London: DoE, GLC, MoL, 1976.

1979

Archaeology in the City of London. Summary of address to meeting of University of London Convocation, 13 May 1980 Minutes, 21 October 1980, pp. 20–3.

Heath Row. *CArch* 67 (1979), 238.

A history of implement petrology in Britain. In: *Stone Axe Studies: Archaeological, Petrological, Experimental and Ethnographic* (ed. by T. H. McK Clough and W. A. Cummins). London: C.B.A., 1979 (C.B.A. Research report, 23), pp. 1–4.

1980

Foreword to *Archaeology in the Ordnance Survey 1791–1965* by C. W. Phillips. London: C.B.A., 1980.

1986

Preface; and 'Introduction: the archaeological background'. In: *The Roman Art Treasures from the Temple of Mithras* by J. M. C. Toynbee. London: London & Middlesex Archaeological Society, 1986 (London & Middlesex Archaeol. Soc. Special Paper, 7), pp. iv, 1–4.

Sources

(1) CBA's bibliographical publications
(2) Annual reports of the National Museum of Wales (from J. Kenyon)
(3) Cards from the library of the Institute of Archaeology
(4) Soc of Antiquaries catalogue
(5) A Romano-British bibliography (55 B.C.–A.D. 449)/by Wilfrid Bonser. – Oxford: Blackwell, 1964. – 2 vols.
(6) Cumulated indexes of the journals in which especially Grimes' early articles appeared

List of Periodical Abbreviations

AC	Archaeologica Cambrensis
AJ	Archaeological Journal
AN	Archaeological Newsletter
AnJ	Antiquaries Journal
Ant	Antiquity
BBCS	Bulletin of the Board of Celtic Studies
BIA	Bulletin of the Institute of Archaeology [of the University of London].

LIST OF PUBLICATIONS

CA	Carmarthenshire Antiquary
CAP	Classical Association Proceedings
CArch	Current Archaeology
CJ	Classical Journal
ILN	Illustrated London News
Oxon	Oxoniensia
PBNS	Proceedings of the Bournemouth Natural Science Society
PPS	Proceedings of the Prehistoric Society
PPSEA	Proceedings of the Prehistoric Society of East Anglia
PSIANH	Proceedings of the Suffolk Institute of Archaeology and Natural History
PUBSS	Proceedings of the University of Bristol Spelaeological Society
RBEA	Records of the Buckingham Society of East Anglia
TCNS	Transactions of the Cardiff Naturalists' Society
TLMAS	Transactions of the London and Middlesex Archaeological Society
TRS	Transactions of the Radnorshire Society
TWAS	Transactions of the Worcestershire Archaeological Society

XIV. PHILIP DAVID WHITTING, GM
1903–1988

By CHRISTOPHER THORNE

PHILIP WHITTING, who died on December 14 1988 aged 85, was a distinguished scholar, a legendary teacher, and a man whose integrity and selflessness made him an inspiration to many who were fortunate enough to know him. Educated at Bradfield and at St John's College, Oxford, he subsequently made himself into a leading student of Byzantine history, and of Byzantine numismatics in particular. His achievements in this field are reflected in part in his substantial volume on *Byzantine Coins* (1973), in part in the remarkable collection of over 10,000 of those coins that he built up from the late 1920s onwards and that he made available in 1970 to the Barber Institute of Fine Arts at the University of Birmingham. (A commentary on a selection from this collection, written by Dr John Kent of the British Museum, was published by the University in 1985.)

Whitting's co-authored *Introduction to Sasanian Coins* (1985) illustrates the continuance of his work into old age. Moreover, his interests and scholarship were by no means confined to matters numismatic. Topography and local history were other areas towards which he directed his zeal, as exemplified by a volume which he instigated, edited, and wrote in part, *A History of Hammersmith* (1965) – a book which also serves as a reminder of the degree to which he identified with and worked for the community in which he lived.

Above all, however, it was in his everyday teaching of the History VIIIth at St Paul's School from 1929 onwards that his massive erudition and powers of enquiry were most regularly on display. To generations of able young Londoners he opened up new historical vistas, while at the

AN OBITUARY

Philip David Whitting, G.M.

same time imparting new standards of analysis and exposition. (It was his custom to tear up his own notes at the conclusion of a course, preparatory to thinking through the subject anew before he spoke on it again.) And when his protégés moved on to university – many of them bearing an open scholarship or exhibition to Oxford or Cambridge; not a few destined to become scholars in their own right – all too often they were to encounter there teaching that fell far short of the rigour, the depth, and the imagination to which they had become accustomed in West Kensington.

Whitting's achievements at St Paul's brought him extensive tributes when he retired in 1963. He had left a lasting mark, not only on many of his pupils, but also on those like myself who had been privileged to work alongside him as a colleague. But his impact as a teacher had reached, and continued to reach, wider still, encompassing those innumerable audiences of amateur historians around the country to whose meetings, however distant their location or sparse their attendance, he brought the same degree of care and commitment that he contributed to gatherings of his academic peers. The work that he carried out on behalf of the Historical Association deserves special mention in this context.

During the Second World War, Whitting became a Squadron Leader in the R.A.F., working on intelligence matters in the Mediterranean sphere of operations, where he could bring to bear that first-hand knowledge of Byzantine regions that he had accumulated during school vacations in the 1920s and '30s. In the opening period of the conflict, however, his service had taken the form of Civil Defence in the Hammersmith area. And it was there, during the Blitz of October 1940, that he repeatedly put his life at risk in order to examine and report on unexploded bombs, actions for which he was awarded the George Medal.

Whitting was a man of great earnestness, as well as courage. He could be bluntly scornful of the pretentious and the slovenly, as of those who sought short cuts to what could be paraded as achievement. His atheism could surge vigorously to the fore when confronted by what he took to be

Philip Whitting by Edward Halliday
(by courtesy of Birmingham University)

a mere religiosity on the part of others. Yet at the same time he possessed a ready sense of humour, and a discriminating palate to accompany an appetite that in its turn matched his solid frame. Above all, his life as a batchelor (complete with cherished cats) was enriched by many strong and lasting friendships – though those involved would say that the enrichment came above all from him to them. Philip Whitting's concern for others and his unostentatious efforts on their behalf were as remarkable as his learning, his ability to awaken enthusiasm in those around him, and a bravery that brought him through not only the Blitz but also a battle with cancer in the 1970s. In the final phase of his life his remarkable physical stamina at last deserted him, and the power of his mind, too, eventually became dimmed. But the essential nature of this rare and noble man shone clear to the very end.

[This obituary appeared in *The Guardian* on 20 December 1988; the society is grateful for permission to reprint it here.]

Editor's note

Philip Whitting was a man who seemed, both physically and mentally, to be caste in a mould larger than that used for other human beings. He served the London Topographical Society for thirty years, joining the Council as Hon. Secretary in December 1958 and filling that post until the October of 1966, when he became a Vice-President. He remained in that office until his death in December 1988. The depth of his scholarship, the strength of his mind, and the firmness of his judgement were relied upon by the whole Council; with Philip's massive frame seated at the Committee table, we knew that we had a sheet anchor. The Council, and the whole Society, are the poorer by his absence; we shall miss him greatly.

A. L. S.

PUBLICATIONS

By R. H. THOMPSON

The arrangement is modelled on Philip Whitting's bibliography for Marjorie Honeybourne in Volume 24 of the *London Topographical Record*.

*, items which have not been examined.

AMATEUR HISTORIAN, vol. 1, 1954. 'The marking of battlefields', pp. 371–4.
Vol. 4, 1958/9. 'A battlefield memorial at Stamford Bridge', pp. 84–5.
Vol. 5, 1962/3. 'Book reviews: from the churches', pp. 155–6.
 Book reviews: *The church in Anglo-Saxon England*, by C. J. Godfrey, p. 270.
Vol. 6, 1963/5. 'Numismatics and local history', pp. 2–7.
 Book reviews: around London, pp. 97–9.
 Book reviews: *An introduction to the study of history*, by V. H. Galbraith, p. 104.
 Book reviews: *A history of Helmsley Rievaulx and district*, edited by J. McDonnell, pp. 137–8.
 Book reviews: *Men of Kent in the Dark Ages*, by Frank Jenkins, p. 176.
 Book reviews: *Monks and civilisation*, by Jean Décarreaux, p. 210.
Vol. 7, 1967. 'Book reviews: *The cost of living in 1300*, by Daphne Harper...[and two other Farnham Papers]', pp. 166–7.
Subsequently LOCAL HISTORIAN (q.v.).

ANGLO-SAXON COINS: *studies presented to F. M. Stenton on the occasion of his 80th birthday, 17 May 1960*, edited by R. H. M. Dolley (London: Methuen, 1961). 'The Byzantine Empire and the coinage of the Anglo-Saxons', by P. D. Whitting, pp. 23–38, pl. iii.

BARBER INSTITUTE OF FINE ARTS. *Catalogue of an exhibition of coins from the 'Mardin' hoard of Byzantine folles, many with Islamic countermarks, recently acquired by the Barber Institute*, [exhibited] in the Barber Institute of Fine Arts, University of Birmingham, 1977 [Birmingham: the Trustees, 1976 (printed)]. [14] pp. 'Addenda and corrigenda', [1] p., as insert.
A selection of Byzantine coins in the Barber Institute of Fine Arts, principally from the collection of Dr P. D. Whitting, [by] John Kent (Birmingham: The Trustees, 1985). 80 pp.

BRITISH ASSOCIATION OF NUMISMATIC SOCIETIES. *The Yearbook of the*

B.A.N.S., no. 8, 1962. 'In memoriam Ronald G. Bartlett, 1921–1961', pp. 8–9.
'Black Gate Museum, Newcastle upon Tyne: (a) Byzantine coins', pp. 24–5 (Notes on numismatic collections in museums, no. 2).
Cunobelin: the yearbook of the B.A.N.S., no. 9, 1963. 'Week-end course, 1962', pp. 12–13.
'Obituaries: Dr F. Parkes Weber', p. 14.
No. 10, 1964. 'Week-end course, 1963', pp. 11–12.
'The annual congress of the British Association of Numismatic Societies, Cheltenham, 1963', pp. 13–15.
No. 12, 1966. 'Dr John Walker, C.B.E., F.B.A.', pp. 10–11.
'Short Cross coins in the Ashmolean, Birmingham, Fitzwilliam and Yorkshire museums', p. 50 (Notes on numismatic collections in museums, no. 6).
No. 14, 1968. 'Beginning on the Byzantines', pp. 16–26, pls. ii–iii.
No. 15, 1969. 'Obituaries: Albert Baldwin', p. 4.

BYZANTINE COINS (London: Barrie & Jenkins; New York: Putnam, 1973). 311 pp. (The world of numismatics).
* *Monnaies byzantines* (Fribourg: Office du Livre, 1973). 311 pp. (L'Univers des Monnaies, vol. 2).
* *Die Münzen von Byzanz* (München: Ernst Battenberg, 1973). 319 pp. (Die Welt der Münzen, 2).
* *Byzantine coins*. 2nd edn., edited by J. P. C. Kent. Forthcoming.

BYZANTIUM: [*a BBC Radio guide to Byzantine history and art*], producers and editors Adrian Johnson and George Walton Scott; consultant for the series Philip Whitting (London: British Broadcasting Corporation, 1968). 52 pp.
'The Byzantine world', pp. 2–11.

BYZANTIUM: *an introduction*, edited by Philip Whitting (Oxford: Blackwell by arrangement with the British Broadcasting Corporation, 1971). xiv, 178 pp., 6 maps, 8 pls.
'Introduction', pp. ix–xiv.
'Byzantine art and architecture', pp. 135–63, pls. 1–8.
Byzantium: an introduction, edited by Philip Whitting. New edn. (Oxford: Blackwell, 1981). xiv, 178 pp., 8 pls.
'Introduction to the new edition', pp. vii–xiii.
'Byzantine art and architecture', pp. 135–63, pls. 1–8.

THE CAMBRIDGE MEDIEVAL HISTORY. VOL. IV: *The Byzantine Empire*, Part II: *Government, church and civilisation*, edited by J. M. Hussey with the editorial assistance of D. M. Nicol and G. Cowan (Cambridge: University Press, 1967). 'Description of the plates' – '42: Coins', selection and notes by P. D. Whitting, pp. xxxix–xl.

[CATS]. 'Cats in Cyprus', [2] pp. [Trained by Basilian monks to kill snakes; extract among the Whitting Room offprints in the Barber

Institute; possibly from the Cats Protection League journal *The Cat*, c. 1936].

CYPRUS NUMISMATIC SOCIETY. *Numismatic Report*, 1972. 'A VIIth-century hoard from Cyprus', by P. J. Donald and P. D. Whitting, pp. 44–6. [Originally published in *Numismatic Circular*, 1967.]

EAST YORKSHIRE LOCAL HISTORY SOCIETY. *Coins, tokens and medals of the East Riding of Yorkshire* (York: the Society, 1969). [4], 80 pp., xvi pls. (East Yorkshire local history series, no. 25).

EDINBURGH FESTIVAL SOCIETY. *Masterpieces of Byzantine art*: [*catalogue of an exhibition*], sponsored by the Edinburgh Festival Society in association with the Royal Scottish Museum and the Victoria & Albert Museum; director, David Talbot Rice. 2nd edn. (Edinburgh: University Press, 1958). Exhibition held in Edinburgh 23 Aug. to 13 Sept., London 1 Oct. to 9 Nov., 1958. 'Coins', pp. 77–92. 'Mr Whitting... undertook to prepare the section on coins and to lend the exhibits' (Foreword, p. 6).

FULHAM AND HAMMERSMITH HISTORICAL SOCIETY. *John Salter's map of Hammersmith: some notes and a gazetteer* (London: the Society, 1971). 15 pp. 'The map' by Philip Whitting and Pamela Taylor, pp. 6–9.
'Gazetteer' by Philip Whitting, pp. 9–15.

FULHAM HISTORY SOCIETY. *A history of Fulham to 1965*, by members of the Society; edited by P. D. Whitting (London: the Society, 1970). 330 pp., map, xvi pls.
'Introduction', pp. 9–20.

HAMMERSMITH HISTORICAL RECORD, no. 4, Oct. 1959. 'Hammersmith tokens of the 17th century', pp. 1–3.
'The beginnings of public health in Hammersmith', pp. 3–7.

HAMMERSMITH LOCAL HISTORY GROUP. *Western roads out of London, extracted from John Ogilby's 'Britannia', 1675*, [by P. D. Whitting] ([London]: the Group, 1955). [4] pp., plate.
Map of the parish of Hammersmith, 1853, by A. J. Roberts..., reproduced in four sections: 1..., 2...(issued in 1956), 3..., 4...(issued in 1957) ([London]: the Group, 1956[–7]). [2] pp., 4 pls. 'Some notes on a map of the parish of Hammersmith in the county of Middlesex, 1853', by P. D. W[hitting], [1957], [2] pp.
An exact survey of the city's of London, Westminster, ye Borough of Southwark and the country near ten miles round, by John Rocque, London, 1746: Hammersmith section ([London]: the Group, 1959). [2] pp., map. 'Notes on place-names and sites, [by] E. J. M[iller], P. D. W[hitting], pp. [1–2].
Panorama of the Thames, circa 1830 [*by John Clark, drawing master*]:

Hammersmith section ([London]: the Group, 1960). [4] pp., iv pls.
'Descriptive notes', [by] E. J. M[iller], P. D. W[hitting], [4] pp.
A history of Hammersmith, based upon that of Thomas Faulkner in 1839; edited... by Philip D. Whitting (London: the Group, 1965). vii, 273 pp., viii pls.
'Preface', pp. 1–5.
'Backcloth to Hammersmith', pp. 9–20.
'The parish church of St Paul', pp. 57–67.
'A riverside walk', pp. 101–11.
'Health and hospitals', pp. 197–202.
News letter, no. 8, winter 1963. 'A new history of Hammersmith', p. [1].
No. 13, spring 1965. 'What the new History hopes to achieve and what it cannot', p. [2]. [Report of a talk.]
No. 22, summer 1967. 'Fulham and Hammersmith: an interpretation of local loyalty', pp. [3–4]. [Report of a talk.]

A HANDBOOK OF THE COINAGE OF THE BYZANTINE EMPIRE, by Hugh Goodacre. [1st edns.] reprinted as a complete volume (London: Spink, 1957). 'Addenda' [compiled by R. A. G. Carson, J. P. C. Kent, A. Veglery, P. D. Whitting, and G. Zacos], 8 pp.

HISTORICAL ASSOCIATION. *History*, new series, vol. 36, 1951. 'Reviews: *Three Byzantine saints*, by Elizabeth Dawes and N. H. Baynes', p. 113.
Vol. 41, 1956. 'Reviews: *Civil Defence* (*History of the Second World War, U.K. civil series*), by T. H. O'Brien', pp. 278–9.
Vol. 42, 1957. 'History books for schools, I', pp. 81–5.
'Reviews and short notices: *Inventory of British coin hoards, A.D. 600–1500*, by J. D. A. Thompson', pp. 211–12.
'Reviews...: *Bedouin command*, by Peter Young', pp. 270–1.
Vol. 43, 1958. 'Reviews...: *The Byzantine world*, by J. M. Hussey', pp. 126–7.
Vol. 44, 1959. 'Reviews...: *Roman history from coins*, by Michael Grant', p. 35.
Vol. 45, 1960. 'Reviews...: *Guide to numismatics*, by C. C. Chamberlain', p. 301.
Vol. 46, 1961. 'Reviews...: *Constantinople in the age of Justinian*, by Glanville Downey', p. 122.
Vol. 49, 1964. 'Reviews...: *Istanbul and the civilization of the Ottoman Empire*, by Bernard Lewis', p. 230.
Vol. 52, 1967. 'Reviews...: *The Cambridge medieval history, Vol. IV: The Byzantine Empire, Part I*, edited by J. M. Hussey', pp. 179–81.
'Reviews...: *Byzantium: the imperial centuries*, by Romilly Jenkins, [and], *Byzantine East and Latin West*, by D. J. Geanakoplos', pp. 306–7.
Vol. 53, 1968. 'Reviews...: *The Cambridge medieval history. Vol. IV: The Byzantine Empire, Part II*, edited by J. M. Hussey, [and] *Byzantine*

studies: the proceedings of the XIIIth International Congress, edited by J. M. Hussey...[and others]', pp. 67–8.
Vol. 54, 1969. 'Reviews...: *The city of Constantinople*, by Michael Maclagan', pp. 83–4.
Teaching of history leaflet, no. 4. *Russia: notes on a course for Modern schools*, compiled by P. D. Whitting with the help of members of the Teaching of History Committee of the Association ([London]: the Association, [1947]). 20 pp. [There was a 4th edition in 1966.]
No. 22. *Coins in the classroom: an introduction to numismatics for teachers* (London: the Association, 1966). 46 pp., front., [4] pls.

An INTRODUCTION TO SASANIAN COINS, by David Sellwood, Philip Whitting & Richard Williams (London: Spink, 1985). [2], 178 pp., [16] pp. of pls.

JOURNAL OF HELLENIC STUDIES, vol. 79, 1959. 'Notices of books: *A handbook of the coinage of the Byzantine Empire*, by Hugh Goodacre', pp. 223–4. [Reprinted in the *Numismatic Circular*, 1960.]

LOCAL HISTORIAN [previously AMATEUR HISTORIAN, q.v.], vol. 8, 1968/9. 'Book reviews: *Archives and local history*, by F. G. Emmison', p. 109.
'Book review: business and industrial histories', pp. 187–9.
Vol. 10, 1972–3. 'Book reviews: *Borough of Twickenham Local History Society Papers 1 to 22*', pp. 88–9.
'Book reviews: *Local History Records*..., [published by] The Bourne Society', p. 305.
Vol. 12, 1976/7. 'Book reviews: *Lincolnshire History and Archaeology*, vol. x, 1975', pp. 115–16.

LONDON AND MIDDLESEX ARCHAEOLOGICAL SOCIETY. *Transactions*, vol. 22 (no. 3), [1970]. 'Local, metropolitan and national history; adapted from an address delivered to the London and Middlesex Local History Committee's Annual Conference held at Guildhall on 23rd November 1968', pp. 41–6.

LONDON NUMISMATIC CLUB. *News Letter*, vol. 4 (nos. 4 & 5 i.e. 5 & 6), Dec. 1964. 'The reverse types of Byzantine solidi', pp. 82–3.
Vol. 4 (no. 20)/Vol. 5 (nos. 1–2), June/Dec. 1968. 'Mint practice at Byzantium', pp. 223–5.
Vol. 5 (no. 3), March 1969. 'Byzantium in crisis', pp. 25–6.
Vol. 6 (no. 1), Jan. 1974. 'The fantastic coinage of the mediaeval Turks', pp. 12–17.

LONDON TOPOGRAPHICAL RECORD, vol. 24, 1980. 'Marjorie Blanche Honeybourne: an obituary', compiled from information supplied by Philip Whitting...[and others], pp. 203–8.
'List of publications by Marjorie Blanche Honeybourne, with a note on her sketch maps and plans', compiled by Philip Whitting, pp. 208–10.

THE 'MARDIN' HOARD: *Islamic countermarks on Byzantine folles*, [by] N. M. Lowick, S. Bendall [and] P. D. Whitting ([London]: A. H. Baldwin & Sons, 1977). 2–79 pp.

MINTS, DIES AND CURRENCY: *essays dedicated to the memory of Albert Baldwin*, edited by R. A. G. Carson (London: Methuen, 1971). 'Albert Baldwin: two appreciations', by Philip Whitting and Douglas Liddell, pp. 1–8.
'A hoard of trachea of John II and Manuel I from Cyprus', by P. J. Donald and Philip Whitting, pp. 75–84.

NUMISMATIC CHRONICLE, 6th series, vol. 11, 1951. [Class II of the anonymous Byzantine issues], Proceedings, p. 3.
Vol. 12, 1952. 'A follis of Heraclius with two Sicilian countermarks', pp. 131–3.
[Two solidi of Heraclius], Proceedings, p. ix.
Vol. 14, 1954. 'The bronze coinage of the Byzantine Empire from John Zimisces to Alexius I', Proceedings, p. xxi.
Vol. 15, 1955. 'The anonymous Byzantine bronze', pp. 89–99, pls. ix–x.
'An overstriking of Justinian I at Rome', pp. 238–40; Proceedings, p. xiv.
'The debasement of the Byzantine nomisma in the eleventh century', Proceedings, pp. xvii–xviii.
Vol. 16, 1956. [Four nomismata of Manuel I], Proceedings, p. 9.
[A bronze coin of the Il-Khanid Sulaiman (1339–43)], Proceedings, p. 10.
Vol. 17, 1957. [The profile bust issue of Justinian I at Rome; and, Justinian II dating], Proceedings, p. 1.
[Hyperpers and half-hyperpers of Manuel II and John VIII], Proceedings, p. 11.
Vol. 18, 1958. 'Two suspected Byzantine forgeries', pp. 177–8.
[A miliaresion of Constantine VII and Romanus II; and, A series of Constans II solidi], Proceedings, p. 4.
Vol. 19, 1959. [A new transitional Byzantine issue of A.D. 582], Proceedings, p. 5. [See 1960.]
[Sixty solidi of Phocas and Heraclius], Proceedings, p. 6.
Vol. 20, 1960. 'A new transitional Byzantine issue of A.D. 582', pp. 133–5.
[Justin II with a stubby beard; and, OB*+* engraved over CONOB], Proceedings, p. 5 (Byzantine symposium).
7th series, Vol. 1, 1961. 'Reviews: *The numismatic iconography of Justinian II*, by J. D. Breckenridge', pp. 253–5.
Vol. 2, 1962. 'The silver coinage at the end of the Byzantine Empire', Proceedings, p. 5.
Vol. 3, 1963. 'Reviews: *Coins – ancient, mediaeval & modern*, by R. A. G. Carson', pp. 279–81.
Vol. 6, 1966. 'A seventh-century hoard at Carthage', pp. 225–33, pl. xix.

Vol. 8, 1968. [The effigy of the Georgian king David the Builder], Proceedings, p. x.

Vol. 12, 1972. 'Reviews: *Byzantine coins (Archaeological exploration of Sardis, Monograph no. 1)*, by George E. Bates', pp. 334-5.

Vol. 17, 1977. 'The Stockwell bequest and presidential badge', author M. M. A[rchibald]; [biographical details about Doris Stockwell provided by P. D. Whitting], Proceedings, pp. xxviii-xxx, pl. 16.

NUMISMATIC CIRCULAR, vol. [56], 1948. 'A follis of Justinian II, second reign (705-711), overstruck by Theodosius III of Adramytium (716-717)...in the collection of Mr P. D. Whitting', [signed] L. F[orrer], col. 541.

Vol. [57], 1949. 'An XIth-century solidus', cols. 217-18.

'An unusual pentenummia of Tiberius II', col. 348.

Vol. 58, 1950. 'Byzantine solidi of the VIIth century', cols. 669-70.

Vol. 60, 1952. 'A bronze of Isaac II', col. 512.

Vol. 65, 1957. 'Reviews: *Lightweight solidi and Byzantine trade during the sixth and seventh centuries*, by Howard L. Adelson', cols. 423-4.

Vol. 68, 1960. 'Reviews: *A handbook of the coinage of the Byzantine Empire*, by Hugh Goodacre', pp. 53-4. [Reprinted from the *Journal of Hellenic Studies*, 1959.]

'Review: *The numismatic iconography of Justinian II*, by J. D. Breckenridge', p. 4.

Vol. 69, 1961. 'Obituary: R. G. Bartlett, 1921-1961', p. 145.

Vol. 70, 1962. 'Obituary: Dr F. Parkes Weber', p. 158.

Vol. 71, 1963. [A mould for coining required for William III, 1688], p. 77.

Vol. 72, 1964. 'An Heraclius die identity', p. 134.

Vol. 73, 1965. 'Another Heraclius die identity', pp. 129-30.

'Reviews: *Carthaginian gold and electrum coins*, by G. K. Jenkins and R. B. Lewis', p. 213.

Vol. 74, 1966. 'A hoard of early Heraclius folles', by R. N. Bridge and P. D. Whitting, pp. 131-2, 183.

Vol. 75, 1967. 'Two coins of John IV of Nicaea', by P. D. Whitting and P. J. Donald, p. 32.

'Another Heraclius die identity', p. 63. [Not the same as in 1965.]

'A VIIth-century hoard from Cyprus', by P. J. Donald and P. D. Whitting, pp. 162-5; addendum, p. 204.

Vol. 76, 1968. 'A new profile-type follis of Justinian I', by P. D. Whitting and P. J. Donald, p. 260.

'Conscious imitation on Byzantine coins', p. 332.

'A late VIIth-century hoard', pp. 370-1.

Vol. 79, 1971. 'A new Tiberius II lightweight solidus', pp. 102-3.

'A late Palaeologan hoard', pp. 156-7.

'A recent hoard from Cilician Armenia', p. 202.

Vol. 80, 1972. 'Miliaresia of Andronicus II and Michael IX', pp. 270-4, 324-6.

'Reviews: *La monetazione aurea delle zecche minori bizantine dal vi al ix secolo*, by D. Ricotti Prina', pp. 322–3.
Vol. 81, 1973. 'A Theodosius II forgery from Hungary', p. 6.
'Review: *Welsh tokens of the 17th century*, by George C. Boon', pp. 374–5.
'Some unusual Byzantine copper of the sixth century', by P. J. Donald and P. D. Whitting, p. 466.
Vol. 82, 1974. 'Review: *The Billon Trachea of Michael VIII Palaeologos, 1258–1282*, by S. Bendall and P. J. Donald', p. 291.

ROYAL INSTITUTE OF BRITISH ARCHITECTS. *Journal of the RIBA*, 3rd series, vol. 44, 1936/7. 'A visit to St Luke's of Phocis', pp. 18–22.

ST PAUL'S SCHOOL. *Folio*, vol. 5 (no. 1), Christmas 1957. 'Harmless necessary cat', p. 4 [A review of *Cats* by Brian Vesey-Fitzgerald.]
The Pauline, vol. 51, 1933 – 'Book reviews: *Public School religion*, edited by Arnold Lunn', pp. 121–3.
Vol. 54, 1936. 'Book reviews: *The Klephtic ballads in relation to Greek history*, by J. W. Baggally', pp. 193–4.
Vol. 73, 1955. 'Letters Patent for Samuel Pepys', pp. 17–19.
'Marlborough [the Duke] and cricket', p. 42.
'A proposed scholarship at Wadham College [in 1611]', pp. 61–2.
Vol. 76, 1958. 'Book review: *A tale of old Yanina*, by D. R. Morier; edited by J. W. Baggally', pp. 69–70.
Vol. 77, 1959. 'Book reviews: *Regimental fire!*, by R. F. Johnson', pp. 102–3.
Vol. 78, 1960. 'Book reviews: *The Prince Consort*, by Frank Eyck', pp. 62–3.
Vol. 79, 1961. 'G. K. Chesterton and the History VIII', pp. 238–9.

ST PAUL'S SCHOOL CHESTERTON SOCIETY. *The Debater*, vol. 7 (no. 36), summer 1933. 'Per ardua ad astra', pp. 9–11.
Vol. 8 (no. 40), winter 1934. 'A socialist plan: *England's political future*, [by] Lord Allen of Hurtwood', pp. 4–6.
Vol. 8 (no. 41), spring 1935. 'Revision of unemployment', pp. 19–21.
Vol. 9 (no. 43), winter 1935. '[Reviews]: *Challenge to schools: a pamphlet on public school education*, by Arthur Calder-Marshall', p. 13.
Vol. 11 (no. 3), Christmas 1946. 'Numerical reform', [by] P. D. Zarathrustra [i.e. P. D. Whitting?], p. 15.
Vol. 12 (no. 2), summer 1950. [G. Terence White], p. 17 (Correspondence).

SEABY'S COIN AND MEDAL BULLETIN. 1949. 'The anonymous Byzantine bronze', by P. D. Whitting and C. H. Piper, pp. 328–9.
1950. 'Another Byzantine anonymous', by P. D. Whitting and C. H. Piper, pp. 162–3.
'Byzantine anonymous again', by P. D. Whitting and C. H. Piper, pp. 529–31.

1951. 'Re-arranging some Byzantine anonymous types', by P. D. Whitting and C. H. Piper, pp. 143–5.
'Dating Byzantine anonymous issues', [by] P. D. Whitting and C. H. Piper, pp. 192–3.
'A common Byzantine anonymous overstrike', [by] P. D. Whitting and C. H. Piper, pp. 359–61.
1952. 'Overstrikes in the Byzantine anonymous bronze series', by P. D. Whitting and C. H. Piper, pp. 377–81.
1953. 'The Byzantine anonymous series', p. 148. [A correction to 1950, pp. 529–31.]

SEABY COIN & MEDAL BULLETIN. 1989. 'Beginning with the Byzantines'; [edited by Peter Clayton], pp. 131–3.

SPECIOUS TOKENS AND THOSE STRUCK FOR GENERAL CIRCULATION, 1784–1804, [by] R. C. Bell (Newcastle upon Tyne: Corbitt & Hunter, 1968). 'Preface', [by] Philip Whitting, pp. ix–x.

UNIVERSITY OF BIRMINGHAM HISTORICAL JOURNAL, vol. 12 (no. 2), 1970. 'Iconoclasm and the Byzantine coinage', pp. 158–63, plate.

Appendix: OBITUARIES

Independent, 17 Dec. 1988, by Anthony Bryer, p. 11; drawing by Edward Halliday, 1941.
The Times, 17 Dec. 1988, p. 12; 31 Dec. 1988, by Kenneth Baker, p. 12.
Guardian, 20 Dec. 1988. 'Teaching the historians', by Christopher Thorne, p. 25.
Daily Telegraph, 21 Dec. 1988, p. 17.
Seaby Coin & Medal Bulletin, 1989, by Peter A. Clayton, pp. 3–5.
Local Historian, vol. 19, 1989/90, p. 44 (Notes on news, compiled by Philip Morgan).
The Pauline, vol. 107, 1989, by Karl Leyser, pp. 2–3: photo; reminiscences by Lord Beloff, Sir Andrew Shonfield, Kenneth Baker, p. 54.
Spink Numismatic Circular, vol. 97, 1989, by David Sellwood, p. 114.
The Historian, no. 22, Spring 1989, p. 23.

XV. PRISCILLA METCALF PhD
1915–1989

By CAROLINE BARRON

Dr Priscilla Metcalf came to England in the 1950s. I first met her at my parents' home where she was a welcome guest at supper on Sunday nights. Although she was then in her early forties she was embarking upon a new direction in her life: she was determined to live and work in England, to become a British citizen and to study for a doctorate in architectural history. It was not easy to live in England, it took a long time to become a citizen and postgraduate work under Professor Nikolaus Pevsner at the Courtauld Institute meant long hours of work without, of course, earning any money. But Priscilla was never daunted by what was difficult, and with slow determination she achieved all her goals. I was too young at the time to think much about why she wanted to be British – was it dislike of America or love of England? – but at the end of her life, when the great work of scholarship and pietás in editing Pevsner's *Cathedrals of England* was completed, she turned her attention back to her American family and to her American roots. Somewhat surprisingly, Priscilla had grown up in Los Angeles, but most of her ancestors were New Englanders from Rhode Island, and businessmen from Rochester, New York, and their spirit of thrift and enterprise characterised Priscilla's approach to her life in England.

Priscilla had the questing intellect and demanding academic standards characteristic of women educated in the best American colleges. More unusually, perhaps, she wrote with careful clarity, and with wit and brevity. Her pen was her scalpel and she used it with precision and dexterity. She could not be hurried: in part because she had high standards and in part because she savoured what she did. Although she had to live by her writing, yet she never allowed herself to cut corners for purely pecuniary ends. In order to do the job properly she would save money in other ways. And Priscilla never had much money, but yet she made a virtue out of this comparative poverty. She learnt, in ways that those, who have more, often fail to do, how to appreciate what she had, to value it and to treasure it. She chose gifts carefully for her friends, perhaps a book that she had reviewed, an appropriate print that she had found, or a cherished piece of her family silver. In return she appreciated gifts she received, and shared them – a bottle of special wine or some home-made cookies or jam. She took nothing for granted and when she visited our house she would notice and appreciate things which we, in our busy careless way, had come to take for granted or failed to see. In her capacity to take pleasure in otherwise unconsidered trifles, Priscilla not only enriched her own life but also added new dimensions to the lives of her numerous friends.

Partly because she was poor, and partly because she enjoyed it, Priscilla walked a great deal, around and across the villages of suburban London. She savoured the varieties of architectural styles, the ironwork, pub decoration, shop signs, street names and gardens. When she didn't walk she took a bus and sat, ever observant as she surveyed the slowly-changing vistas from the vantage point of the top deck. A fellow-passenger, glancing casually across the aisle at Priscilla, might be forgiven for thinking that she was, perhaps, an affectionate grandmother, possibly of German origin. Indeed her rosy cheeks, rounded figure and chuckling laugh might well suggest such a character. But closer scrutiny would have revealed a distinct and stylish dress-sense and a gentleness which never degenerated into softness or sentimentality, as grandmothers are, perhaps, prone to do! In spite of a certain deafness (a wartime legacy), Priscilla's voice was ever 'gentle and low' and she laughed easily and with real enjoyment. Her New England orderliness, and sense of purpose, gave structure to her life and enabled her to make the most of her gifts for intellectual enquiry, for observation, for good writing and for friendship, for she knew well how to cherish her friends. Above all, I shall remember her as a remarkable person who knew the true value of things and people, and in a largely materialistic world carefully chose for herself a path which was distinctive, unworldly and happy. It is given to few people to hold onto their ideals so triumphantly.

* * *

Lines on a Westminster Abbey Mouse at large in the North Transept during a Prelude & Fugue on the Organ

> Oh mouse called out by trumpet tones
> To come and scan these lettered stones
> Up-roused as diapasons roar
> To race your shadow on the floor
> And dart past Gladstone's marble toes
> While Canning and Disraeli doze.
>
> Did once a thousandth great-aunt scoot
> Before the grand old man's own boot?
> Or did a millionth uncle flit
> Across the path of elder Pitt?
> Ancestral paws flee Purcell, soft
> Ascending to the organ loft?
>
> No sacrist here sets traps for mice
> So rare are mice in paradise.
> No Abbey tabbys watchful stalk
> And slow th'unseeing sightseers walk
> While Baedeker directs their feet –
> But one small footnote pounds the beat.

Priscilla Metcalf

BIBLIOGRAPHY

Books: Alone

Victorian London. London, Cassell, 1972. (15), 190 pp, illus, map.

The Halls of the Fishmongers' Company. An architectural history of a riverside site. Chichester, Phillimore, 1977. xii, 214 pp, illus, col. front, facsims, plans.

The Park Town Estate and the Battersea triangle. A peculiar piece of Victorian London property development and its background. (LTS

Publication No. 121). London, London Topographical Society, 1978. 61 pp, illus, maps, plans.

James Knowles, Victorian editor and architect. Oxford, Clarendon Press, 1980. xvi, 382 pp, illus, front. port.

A short history of the Fishmongers' Hall. London, Worshipful Company of Fishmongers, (1982?). 5 pp. Based on the larger work of 1977.

Books: Joint

Nikolaus Pevsner & Priscilla Metcalf. *The Cathedrals of England.* With contributions by various hands. 2 vols. Harmondsworth, Viking, 1985. *Southern England.* 381 pp, illus, map, plans. *Midland, Eastern and Northern England.* 399 pp, illus, map, plans. Based on the relevant entries in the "Buildings of England" series.

Vanessa Harding & Priscilla Metcalf. *Lloyd's at home.* Part One: *The background.* Part Two: *The buildings.* Colchester, Lloyd's of London Press, for the Corporation of Lloyd', 1986. 168 pp, illus (inc. colour), facsims, maps, plans.

Articles

A lean, pale, sallow, shabby, striking young man. In *Listener*, Vol. 86, 30 Dec. 1971, pp. 904–905, port. [Irwin Russell, 1853–79, poet. Priscilla Metcalf's grandmother's brother, concerned with the promotion of black American poetry.]

At home in Westminster. In *Architectural Review* Vol. 155, 1974, pp. 135–138, illus.

Seven centuries in White Hart Court. In *Guildhall Studies in London History*, Vol. 4, 1979, pp. 1–18, illus.

Living over the shop in the City of London. In Architectural History, Vol. 27, 1984, pp. 96–103, illus.

Reviews & Articles in Times Literary Supplement
Reviews

Philip Howard. *London's River.* 1975, p. 1341.
Ralph Hyde. *Printed maps of Victorian London 1851–1900.* 1975, p. 904.
Geoffrey Trease. *London.* 1975, p. 1272.
Anthony Bird. *Paxton's Palace.* 1976, p. 1131.
Survey of London. Vol. 38. 1976, p. 203.
Gerald Cobb. *English Cathedrals.* 1980, p. 818.
Anthony New. *A guide to the cathedrals of Britain.* 1980, p. 818.

Articles

Viewpoint (on the River Thames, etc.). 1973, p. 150.
Viewpoint (on Inigo Jones etc.). 1973, p. 1000.

Compiled by David Webb

LONDON TOPOGRAPHICAL SOCIETY

The London Topographical Society, founded as the Topographical Society of London in 1880, is mainly a publishing Society and is registered as a charity. Its purpose is to make available facsimiles of maps, plans and views illustrating the history and topography of London, and to publish papers relating to them.

Most of the publications are maps, plans and views, but there are also twenty-six volumes of the Society's journal, the *London Topographical Record*, which contains articles relating to these maps and plans, and to London topography in general. Several other books have been published, including a facsimile of Mills and Oliver's *Surveys of the Building Sites in the City of London after the Great Fire of 1666* (5 volumes, with full indexes) and the series of four volumes, *The A to Zs of Elizabethan, Georgian, Regency and Victorian London*.

The annual subscription is £10.00 for which members receive free the publication for the year. From time to time additional items are published and members can obtain these and earlier publications at a preferential rate.

The Society's official address is 36 Old Deer Park Gardens, Richmond, Surrey; the Chairman (17 Blandford Road, W.5, 01-567-9744) and the Editor (3 Meadway Gate, N.W.11, 01-455-2171) are always ready to try to answer queries and to help members. Publications may be obtained, by appointment, from the Bishopsgate Institution, 230 Bishopsgate, E.C.2.

LONDON TOPOGRAPHICAL SOCIETY

Patron	His Royal Highness The Duke of Edinburgh, KG, KT
President	Ralph Merrifield, BA, FSA, FMA
Vice-Presidents	E. S. de Beer, CBE, MA, DLitt, FBA, FSA
	Stephen Marks, MA, FSA, RIBA
	Miss Irene Scouloudi, MSc, FSA, FRHistS
Chairman	Peter Jackson, FSA
Hon. Treasurer	Roger Cline, MA, LL B
Hon. Editor	Mrs Ann Saunders, PhD, FSA
Publications Secretary	Simon Morris, MA
Membership Secretary	Trevor Ford, Dip RAM
Hon. Secretary	Patrick Frazer, MA, MSc
Hon. Auditor	Hugh Cleaver, MA, ACA
Members of Council	Iain Bain
	Peter Barber, MA
	Felix Barker
	Victor Belcher, MA
	Mrs Penelope Hunting, PhD
	Ralph Hyde, FLA, FSA
	David J. Johnson, MA, FSA, FRHistS
	John F. C. Phillips, BA
	Miss Caroline Ryan, BA
	Miss Elspeth Veale, PhD
	Miss Helen Wallis, DPhil, DLitt, FSA
	David R. Webb, BA, FLA
	Mrs Rosemary Weinstein, BA, FSA, FSA(Scot)

RULES

I. The London Topographical Society is a publishing Society: Its purpose is to assist the study and appreciation of London's history and topography by making available facsimiles of maps, plans and views and by publishing research.

II. The affairs of the Society shall be conducted by a Council, consisting of the President, two or more Vice-Presidents, Honorary Treasurer, Secretary, and not more than twenty-one elected Members of the Society.

III. The Subscription shall be not less than one guinea, payable in advance on the 1st January.

IV. The names of those wishing to become Members shall be submitted to the Council for approval.

V. There shall be each year a General Meeting of the Society, at which the Council elected for the preceding year shall report upon the work of the Society during that year.

VI. At each Annual Meeting all the members of the Council shall retire from office, and shall be eligible for re-election.

VII. No Member whose Subscription for the preceding year remains unpaid shall be eligible for election to the Council.

VIII. A certified Cash Statement shall be issued to all Members with the Annual Report of the Council.

IX. The Council shall have power to fill up occasional vacancies in their number during the year, and to elect any Member of the Society to serve on any Committee or Sub-Committee of the Council.

X. A publication of the Society shall be issued each year to all members whose subscriptions have been paid. No Member whose subscription is in arrears shall be entitled to receive such publication. Occasional additional publications may be issued at a reduced rate to paid-up Members.

XI. No alteration shall be made in these Rules except at an Annual General Meeting, or at a Special General Meeting, called upon the requisition of at least five Members. One month's previous notice of the change to be proposed shall be given in writing to the Secretary, and the alteration proposed must be approved by at least three-fourths of the Members present at such Meeting.

London Topographical Society

REPORT FOR THE YEARS 1985–1989

The separate reports of the Council for each year were distributed to members before the subsequent Annual General Meetings, appearing in *London Topographical News*, nos. 22, 24, 26, 28 and 30.

Ralph Merrifield served throughout the period as President, as did Peter Jackson as Chairman, Dr Ann Saunders as Honorary Editor, Patrick Frazer as Honorary Secretary, Trevor Ford as Membership Secretary and Simon Morris as Publications Secretary.

Roger Cline was elected as Honorary Treasurer in 1985, following the retirement of Anthony Cooper after 10 years of service. In 1988 Caroline Ryan was elected as joint Publications Secretary, to help with the heavy demand for the Society's publications. Hugh Cleaver was coopted as Honorary Auditor in 1988, after Allan Tribe, who had audited the Society's accounts since 1975, decided to stand down.

Two Vice-Presidents, Professor Grimes (Chairman from 1961 to 1974) and Dr Philip Whitting (Honorary Secretary from 1958 to 1966) died during 1988. Stephen Marks (Honorary Secretary from 1966 to 1983) and Miss Irene Scouloudi (Council member since 1947) were appointed as Vice-Presidents.

Members received the following publications for their subscriptions: *London Topographical Record* xxv edited by Ann Saunders (publication no. 132 for 1985), James King's *Kentish Town Panorama c1850* with explanatory booklet by John Richardson (133, 1986), *The London Surveys of Ralph Treswell* edited by John Schofield (135, 1987), *Hugh Alley's Caveat – the Markets of London in 1598* edited by Ian Archer, Caroline Barron and Vanessa Harding (137, 1988), and *Good and Proper Materials* edited by Hermione Hobhouse and Ann Saunders (140, 1989).

Apart from the publications issued against subscriptions, several extra publications could be purchased by members, viz. *The A to Z of Regency London* with an introduction by Paul Laxton and index by Joseph Wisdom (131, 1985), *Satellite View of London 1984* (134, 1986), *The A to Z of Victorian London* with notes by Ralph Hyde (136, 1987), Hollar's *Prospect of London and Westminster taken from Lambeth c1665 and c1707* (138, 1988), and Barker's *Panorama of London from the Roof of the Albion Mills 1792* with introduction by Ralph Hyde and keys by Peter Jackson (139, 1988).

The Society's newsletter, *London Topographical News*, continued to be published twice a year. Dr Penelope Hunting became newsletter editor from issue no. 29 (November 1989), succeeding Stephen Marks, who had been editor since its introduction in 1975.

Annual General Meetings were held at Staple Inn Hall (1985), St Bartholomew's Hospital (1986), Crosby Hall (1987), the Honourable Artillery Company (1988), and the Royal Geographical Society (1988). All the meetings were well attended, with an average of about 200 members and their guests.

To provide funds for the Society's ambitious programme of publications, in 1984 members had approved an increase in the annual subscription from £5 to £10 which became effective in 1986. Paid up membership rose from 686 at the end of 1984 to 788 at the end of 1989.

The Society's finances remained strong, following the great success of the Rhinebeck panorama in 1981. Sales of publications exceeded subscription income in most years during the period, and there was a surplus of income over expenditure in every year.

LIST OF PUBLICATIONS

Dimensions are given in inches to the nearest eighth of an inch, with centimetres in brackets to the nearest half-centimetre; height precedes width. In the case of items in several sheets dimensions may be approximate only. Borders, original titles and other wording are included in the dimensions but added titles are excluded.

Dates in brackets are those of publication and do not necessarily coincide with the year for which the publication was issued. Recently extra publications have been available by purchase and are indicated accordingly.

1. Van den Wyngaerde's View of London, *c.* 1550 (Topographical Society of London, 1881–2): $20\frac{3}{4} \times 116\frac{1}{4}$ in. (52·5 × 295·5 cm), on 7 sheets $31\frac{1}{2} \times 23\frac{3}{8}$ in. (80 × 59·5 cm), with sheet of text.
2. Plan of London, *c.* 1560, attributed to Hoefnagel (T.S.L. 1882–3): from Braun and Hogenberg's *Civitates Orbis Terrarum*, second state showing Royal Exchange, $13\frac{1}{4} \times 19\frac{3}{8}$ in. (35·5 × 49 cm), on sheet $22\frac{1}{4} \times 29\frac{1}{4}$ in. (56·5 × 74·5 cm).
3. *Illustrated Topographical Record of London*, first series (1898), drawings by J. P. Emslie of changes and demolitions, 1880–7: $11\frac{1}{2} \times 8\frac{7}{8}$ in. (29 × 22·5 cm), sewn, paper wrapper.
4. Visscher's View of London, 1616 (T.S.L. 1883–5): $16\frac{5}{8} \times 85\frac{1}{4}$ in. (42 × 216·5 cm), on 4 sheets $23\frac{1}{4} \times 31\frac{3}{4}$ in. (59 × 78 cm). (See 'Notes on Visscher's View of London, 1616', by T. F. Ordish, *L.T.R.* VI, 39.)
5. Porter's 'Newest and Exactest Mapp of London and Westminster', *c.* 1660 (1898): $11\frac{1}{4} \times 30\frac{1}{4}$ in. (28·5 × 77 cm), on 2 sheets $22\frac{1}{4} \times 29\frac{1}{4}$ in. (56·5 × 74·5 cm).
6. *Illustrated Topographical Record of London*, second series (1899), drawings by J. P. Emslie of changes and demolitions, 1886–7: $11\frac{1}{2} \times 8\frac{7}{8}$ in. (29 × 22·5 cm), sewn, paper wrapper.
7. Norden's Maps of London and Westminster, 1593, from the *Speculum Britanniae* (1899): $6\frac{3}{4} \times 9\frac{5}{8}$ in. (17 × 24·5 cm) and $6\frac{1}{8} \times 9\frac{7}{8}$ in. (15·5 × 25 cm), on one sheet $29\frac{1}{4} \times 22\frac{1}{8}$ in. (74·5 × 56 cm). (See 'Notes on Norden and his Map of London', by H. B. Wheatley, *L.T.R.* II, 42.)
8. Kensington Turnpike Trust Plans, 1811, by Salway, of the road from Hyde Park Corner to Counter's Bridge (1899–1903): $20\frac{5}{8}$ in. × 56 ft $1\frac{5}{8}$ in. (52·5 × 1711 cm), in colour, 30 sheets and title-page 24 × 27 in. (61 × 69 cm). (See 'Notes on Salway's Plan', by W. F. Prideaux, *L.T.R.* III, 21, and V, 138.)
9. *Illustrated Topographical Record of London*, third series (1900), drawings by J. P. Emslie of changes and demolitions, 1888–90: $11\frac{1}{2} \times 8\frac{7}{8}$ in. (29 × 22·5 cm), sewn, paper wrapper.

10. Comparative Plan of Whitehall, 1680/1896: modern ground plan superimposed on Fisher's plan of 1680 as engraved by Vertue (1900): one sheet $26\frac{5}{8} \times 22$ in. (67·5 × 56 cm).
11. *Annual Record* I, ed. T. F. Ordish (1901): $8\frac{3}{4} \times 5\frac{7}{8}$ in. (22·5 × 15 cm), quarter cloth; continued as *London Topographical Record*.
12. Hollar's West-Central London, *c.* 1658, a bird's eye-view (1902): $13\frac{1}{8} \times 17\frac{1}{4}$ in. (33·5 × 44 cm) on $17 \times 23\frac{1}{4}$ in. (43 × 59 cm). (See 'Hollar's Map', by W. R. Lethaby and R. Jenkins, *L.T.R.* II, 109).
13. *London Topographical Record*, II, ed. T. F. Ordish (1903): $9 \times 5\frac{7}{8}$ in. (23 × 15 cm), quarter cloth.
14. Kip's View of London, Westminster and St James's Park, 1710 (1903): *c.* 53×82 in. (134 × 208 cm) on 12 sheets $22\frac{1}{8} \times 24$ in. (50 × 61 cm).
15. Morden and Lea's Plan of London, 1682, also known as Ogilby and Morgan's Plan (1904): 300 ft to 1 in., $59\frac{1}{4} \times 93\frac{3}{8}$ in. (150·5 × 238 cm) on 9 sheets $22 \times 30\frac{1}{8}$ in. (56 × 76·5 cm) and 3 sheets $22 \times 14\frac{1}{8}$ in. (56 × 38·5 cm). (See 'Morden and Lea's Plan of London', by W. L. Spiers, *L.T.R.* V, 117.)
16. *London Topographical Record*, III, ed. T. F. Ordish (1906): $9 \times 5\frac{7}{8}$ in. (23 × 15 cm), quarter cloth.
17. Map of Elizabethan London, formerly attributed to Ralph Agas (1905): $28\frac{1}{8} \times 72$ in. (71·5 × 183 cm) on 8 sheets $23\frac{1}{2} \times 17\frac{1}{4}$ in. (59·5 × 44 cm).
18. Faithorne and Newcourt's Map of London, 1658 (1905): map *c.* $32\frac{1}{2} \times 71$ in. (82·5 × 180·5 cm) on 6 sheets $20 \times 25\frac{1}{4}$ in. (51 × 64 cm) and 2 sheets $20 \times 12\frac{1}{2}$ in. (51 × 31·5 cm), and title on 4 pieces.
19. Hollar's Long View of London, 1647 (1906–7): 18×92 in. (46 × 233·5 cm) in 7 pieces on 6 sheets $25\frac{3}{8} \times 19$ in. (64·5 × 48·5 cm).
20. *London Topographical Record*, IV (1907): $9 \times 5\frac{7}{8}$ in. (23 × 15 cm), quarter cloth.
21. Wren's Drawings of Old St Paul's (1908): (i) plan of old cathedral before the Great Fire, $18\frac{7}{8} \times 14\frac{1}{2}$ in. (48 × 37 cm); (ii) section of Wren's scheme for rebuilding, $18\frac{1}{2} \times 12$ in. (47 × 30·5 cm); on 2 sheets $29\frac{3}{8} \times 22\frac{1}{2}$ in. (74·5 × 57 cm). (See 'Wren's Drawings of Old St Paul's…', by W. R. Lethaby, *L.T.R.* V, 136.)
22. and 26. Hollar's 'Exact Surveigh', 1667 (1908, 1909): $21\frac{3}{8} \times 32\frac{1}{2}$ in. (54·5 × 82·5 cm) on 2 sheets $25\frac{3}{8} \times 19$ in. (64·5 × 48·5 cm).
23. *London Topographical Record*, V (1908): $9 \times 5\frac{7}{8}$ in. (23 × 15 cm), quarter cloth.
24. The Palace of Whitehall, View from the River, 1683 (1909): $14 \times 24\frac{1}{2}$ in. (35·5 × 62·5 cm) on sheet 24×35 in. (61 × 89 cm). (See 'View of the Palace of Whitehall', by W. L. Spiers, *L.T.R.* VII, 26.)
25. *London Topographical Record*, VI (1909): $9 \times 5\frac{7}{8}$ in. (23 × 15 cm), quarter cloth.

26. See 22.
27. Seven London Views by Deceased Artists (1910): 14 × 11 in. (35·5 × 28 cm). (See 'Notes on London Views', by P. Norman, *L.T.R.* VIII, 94.)
28. *London Topographical Record*, VII, ed. H. G. Head (1912): 9 × 5$\frac{7}{8}$ in. (23 × 15 cm), quarter cloth.
29. Seven More London Views by Deceased Artists (1911): 14 × 11 in. (35·5 × 28 cm). (See 'Notes on London Views', by P. Norman, *L.T.R.* VIII, 94.)
30. Roads out of London, from Ogilby's *Britannia*, 1675, with descriptive letterpress (1911): 15$\frac{1}{8}$ × 11$\frac{3}{8}$ in. (38·5 × 29 cm), sewn, paper wrapper.
31. Jonas Moore's Map of the River Thames from Westminster to the Sea, 1662 (1912): part only, 20$\frac{7}{8}$ × 23$\frac{5}{8}$ in. (52 × 60 cm) on one sheet 23$\frac{1}{4}$ × 33 in. (59 × 84 cm). (See 'A Seventeenth Century Map of London and the Thames', by M. Holmes, *L.T.R.* XX, 26.)
32. *London Topographical Record*, VIII, ed. H. G. Head (1913): 9 × 5$\frac{7}{8}$ in. (23 × 15 cm), quarter cloth.
33. Seven Drawings of London Bridge by E. W. Cooke (1913): 14 × 11 in. (35·5 × 28 cm). (See 'Drawings of Old and New London Bridge by E. W. Cooke', by P. Norman, *L.T.R.* IX, 1.)
34, 36, 37, 41, 42, 43 and 44. Rocque's Survey of London, 1746 (1913–19): 6 ft. 8 in. × 12 ft. 8$\frac{1}{2}$ in. (203 × 387·5 cm), with key, on 49 sheets 17 × 22$\frac{1}{2}$ in. (43 × 57 cm). (See 'Rocque's Plan of London', by H. B. Wheatley, *L.T.R.* IX, 15.)
35. *London Topographical Record*, IX, ed. H. G. Head (1914): 9 × 5$\frac{7}{8}$ in. (23 × 15 cm), quarter cloth.
36. See 34.
37. See 34.
38. *London Topographical Record*, X, ed. H. G. Head (1916): 9 × 5$\frac{7}{8}$ in. (23 × 15 cm), quarter cloth.
39. A Plan of Ebury Manor, *c.* 1663–70 (1915): in colour, one sheet 30$\frac{1}{8}$ × 19$\frac{7}{8}$ in. (76·5 × 20·5 cm), and sheet of text.
40. *London Topographical Record* XI, ed. H. G. Head (1917): 9 × 5$\frac{7}{8}$ in. (22·5 × 15 cm), quarter cloth.
41. See 34.
42. See 34.
43. See 34.
44. See 34.
45. A View of London Bridge by John Norden, 1597 (1919): 15$\frac{1}{4}$ × 20$\frac{1}{8}$ in. (38·5 × 51 cm) on one sheet 18$\frac{7}{8}$ × 25$\frac{1}{4}$ in. (48 × 64 cm).
46. *London Topographical Record*, XII, ed. H. G. Head (1920): 9 × 5$\frac{7}{8}$ in. (22·5 × 15 cm), quarter cloth.
47. A View of London Bridge from both sides, by Sutton Nicholls, *c.* 1710 (1921): 11 × 17$\frac{1}{4}$ in. (28 × 44 cm) on sheet 14$\frac{7}{8}$ × 22 in. (37·5 × 56 cm).

LIST OF PUBLICATIONS

48. Tallis's Plan of Bond Street (1921): 12 pages $5\frac{3}{4} \times 9$ in. (14·5 × 23 cm). (See 'Tallis's Street Views of London', by E. B. Chancellor, *L.T.R.* XII, 67; see also publication 110.)
49. Matthew Merian's View of London, 1638 (1922): $8\frac{3}{4} \times 27\frac{1}{2}$ in. (22 × 70 cm) on sheet $13\frac{1}{2} \times 31\frac{1}{2}$ in. (34·5 × 80 cm), folded.
50. Seven Unpublished Drawings by Hollar: from the Pepysian Library, Cambridge (1922): 4 sheets 14 × 11 in. (35·5 × 28 cm).
51. *London Topographical Record*, XIII, ed. H. G. Head (1923): $8\frac{7}{8} \times 5\frac{7}{8}$ in. (22·5 × 15 cm), quarter cloth.
52. and 53. Views of Westminster, 1801–1815, by William Capon, with Capon's descriptions annotated by P. Norman (1923–4): 16 views, 1 in colour, and map, 5 sewn sections 14 × 11 in. (35·5 × 28 cm).
53. See 52.
54. A London Plan of 1585 (1925): $22\frac{1}{2} \times 31$ in. (57 × 98 cm) on sheet $25\frac{1}{8} \times 35\frac{7}{8}$ in. (63·5 × 91 cm), folded. See 55.
55. *The Early History of Piccadilly, Leicester Square, Soho and their Neighbourhood*, by C. L. Kingsford (1925): written to explain the map of 1585 (see 54), $8\frac{7}{8} \times 5\frac{7}{8}$ in. (22·5 × 15 cm), quarter cloth, uniform with the *Record*.
56. Drawings of Buildings in the Area described in *The Early History of Piccadilly*...(1926): 11 drawings on 7 sheets 14 × 11 in. (35·5 × 28 cm), and plan of West London, c. 1710, on double sheet folded. See 55.
57. *London Topographical Record*, XIV, ed. W. H. Godfrey (1928): $8\frac{7}{8} \times 5\frac{7}{8}$ in. (22·5 × 15 cm), quarter cloth.
58. Plan of Nevill's Alley, Fetter Lane, 1670 (1928): $20\frac{1}{4} \times 25$ in. (51·5 × 63·5 cm). (See 'Nevill's Court, Fetter Lane', by W. G. Bell, *L.T.R.* XV, 87.)
59. Seven Views of the Inns of Court and Chancery, with notes by J. B. Williamson (1928): 14 × 11 in. (35·5 × 28 cm), 2 sewn sections.
60. *London Topographical Record*, XV, ed. W. H. Godfrey (1931): $8\frac{7}{8} \times 5\frac{7}{8}$ in. (22·5 × 15 cm), quarter cloth.
61. Area east of St Katherine's Dock, c. 1590, from a tracing by M. B. Honeybourne of a plan in the Public Record Office (1929): $27\frac{1}{4} \times 21\frac{1}{4}$ in. (69 × 54 cm) on sheet $29\frac{1}{4} \times 23\frac{3}{8}$ in. (74 × 59·5 cm), folded.
62. *London Topographical Record*, XVI, ed. W. H. Godfrey (1932): $8\frac{7}{8} \times 5\frac{7}{8}$ in. (22·5 × 15 cm), quarter cloth.
63. Hollar's View of Greenwich, 1637 (1930): $5\frac{3}{4} \times 33$ in. (14·5 × 84 cm) on sheet $7\frac{3}{4} \times 35\frac{3}{4}$ in. (19·5 × 90·5 cm), folded.
64. A Plan in the Public Record Office of property on the south-east side of Charing Cross, 1610 (1930): 19 × 26 in. (48 × 66 cm) on sheet $22\frac{5}{8} \times 29\frac{1}{8}$ in. (57·5 × 74 cm).
65. Plan of the Manor of Walworth and Parish of Newington, Surrey, 1681 (1932): two-thirds scale, $13\frac{3}{4} \times 18$ in. (35 × 45·5 cm) on

sheet $18\frac{1}{8} \times 23$ in. (46×58 cm). (See 'Thomas Hill's Maps...' by I. Darlington, *L.T.R.* XXI, 37.)

66. Plan of the Duke of Bedford's Estates, 1795 (1933), from Bloomsbury to the river: $23\frac{1}{8} \times 40\frac{7}{8}$ in. (59×103.5 cm) on sheet $28\frac{3}{8} \times 43\frac{1}{8}$ in. (72×109.5 cm), folded. (See 'Duke of Bedford's Estate Map, 1295' by E. Jeffries Davis, *L.T.R.* XVIII, 134.)

67. Plan of the Parish of St Mary, Kensington, 1822 (1934): $28 \times 34\frac{5}{8}$ in. (71×88 cm) on sheet $31\frac{5}{8} \times 37\frac{1}{4}$ in. (81.5×94.5 cm), folded, with sheet of notes.

68. Eight Views of Kensington, from originals in Kensington Public Library (1934): 8 sheets 11×14 in. (28×35.5 cm).

69. *London Topographical Record*, XVII, ed. W. H. Godfrey (1936): $8\frac{7}{8} \times 5\frac{7}{8}$ in. (22.5×15 cm), quarter cloth.

70. Four drawings by Philip Norman (1936): (i) Vine Tavern, Mile End; (ii) nos. 5 and 7, Aldgate; (iii) nos. 10 and 11, Austin Friars; (iv) St Magnus' Church, London Bridge: 4 sheets 14×11 in. (35.5×28 cm), in folder.

71. A further four drawings by Philip Norman (1937): (v) no. 13, Leather Lane; (vi) Staircase in the Old Bell Inn, Holborn; (vii) Old houses, Chelsea; (viii) Backs of old houses, Cheyne Walk: 4 sheets 14×11 in. (35.5×28 cm), in folder with 70.

72. Clothworker's Company, Survey of Properties in 1612 and 1728 (1938): (i) Clothworkers' Hall, 1612; (ii) St James's in the Wall, 1612; (iii) St James's in the Wall, 1728: 3 sheets $22\frac{1}{2} \times 17\frac{1}{2}$ in. (57×44.5 cm) in folder. (See 'The Clothworkers' Company: Book of Plans...' by W. H. Godfrey, *L.T.R.* XVIII, 51.)

73. Clothworkers' Survey (1939): (iv) Neighbourhood of the Fleet Prison, 1612: (v) Neighbourhood of the Fleet Prison, 1728; 2 sheets $22\frac{1}{2} \times 17\frac{1}{2}$ in. (57×44.5 cm), in folder.

74. Clothworkers' Survey (1940): (vi) Richard Fishburne's House, Throgmorton Street, 1612: $22\frac{1}{2} \times 35$ in. (57×89 cm), folded; (vii) Fox Court, Nicholas Lane, 1612: $22\frac{1}{2} \times 17\frac{1}{2}$ ins (57×44.5 cm), in folder.

75. Clothworkers' Survey (1941): (viii) Sir Edward Darcy's House, Billiter Street, 1612, $22\frac{1}{2} \times 34\frac{7}{8}$ in. (57×88.5 cm), folded; (ix) Fox Court, Nicholas Lane, 1728: $22\frac{1}{2} \times 17\frac{1}{2}$ in. (57×44.5 cm), in folder.

76. *London Topographical Record*, XVIII, ed. W. H. Godfrey (1942): $8\frac{7}{8} \times 5\frac{7}{8}$ in. (22.5×15 cm), quarter cloth.

77. Van den Wyngaerde's View of the City of London between Fleet River and London Bridge, *c.* 1550 (1944): $5\frac{1}{8} \times 50\frac{1}{8}$ in. (13×127.5 cm) in 3 sections on 1 sheet $23\frac{1}{8} \times 18\frac{5}{8}$ in. (58.5×47 cm).

78. View of London from Southwark, attributed to Thomas Wyck (1616–77), at Chatsworth House (1945): original size $20\frac{1}{4} \times 30\frac{1}{2}$ in. (51.5×87.5 cm) reproduced $15\frac{5}{8} \times 22\frac{3}{4}$ in. (38.5×57.5 cm) on sheet 20×30 in. (51×76.5 cm).

79. *Survey of Building Sites in the City of London after the Great Fire of 1666* by Mills and Oliver, vol. I. (i.e. part i of Mills I) (1946, extra publication): reduced facsimile, $8\frac{7}{8} \times 5\frac{5}{8}$ in. (22·5 × 14·5 cm), quarter cloth, uniform with the *Record*.
80. *London Topographical Record*, XIX, ed. W. H. Godfrey (1947): $8\frac{7}{8} \times 5\frac{7}{8}$ in. (22·5 × 15 cm), quarter cloth.
81. Whitehall Palace, a seventeenth-century painting at Kensington Palace (1948): original size 16 × 37 in. (40·5 × 94 cm) reproduced $9\frac{1}{2} \times 22\frac{1}{4}$ in. (24 × 56·5 cm) on sheet 20 × 30 in. (51 × 76·5 cm).
82. A Prospect of the City of London from the South-east, 1945, by Cecil Brown (1949): original size 6 ft 6 in. × 9 ft (183 × 274·5 cm) reproduced $14\frac{7}{8} \times 20\frac{7}{8}$ in. (37·5 × 53 cm) on sheet $16\frac{5}{8} \times 23$ in. (42 × 58·5 cm).
83. *Le Guide de Londres*, 1693, by F. Colsoni, edited by W. H. Godfrey (1951): $8\frac{7}{8} \times 5\frac{7}{8}$ in (22·5 × 15 cm), quarter cloth, uniform with the *Record*.
84. Seventeenth-century Plans of the Properties belonging to St Bartholomew's Hospital (1950–1): (i) The Grey Friars, *c*, 1617 (see 'The Precinct of the Greyfriars' by M. B. Honeybourne, *L.T.R.* XVI, 9); (ii) Properties adjoining Hosier Lane; (iii) St Nicholas Flesh Shambles: on 2 sheets $19\frac{1}{2} \times 29$ in (49·5 × 74 cm) folded, in folder.
85. *London Topographical Record*, XX, ed. W. H. Godfrey (1952): $8\frac{7}{8} \times 5\frac{7}{8}$ in. (22·5 × 15 cm), quarter cloth.
86. *Berkeley Square to Bond Street, The early history of the neighbourhood*, by B. H. Johnson (1952): $8\frac{7}{8} \times 5\frac{7}{8}$ in. (22·5 × 15 cm), full cloth, published by John Murray in association with the London Topographical Society.
87. Seventeenth-century Plans of Properties belonging to St Bartholomew's Hospital (1953–4): (iv) south-west portion of the hospital adjoining the City Wall Ditch; (v) North portion adjoining Smithfield from Duck Lane to the Hospital Church; (vi) South portion adjoining Little Britain and the City Wall and Ditch; (vii) property north of Chick Lane: on 2 sheets $19\frac{1}{2} \times 29$ in. (49·5 × 74 cm), folded in folder with 84.
88. Plan of the precinct of St Bartholomew's Hospital, *c*. 1617 (1955): in colour, $15\frac{7}{8} \times 10\frac{1}{4}$ in. (40·5 × 26 cm) on sheet $19\frac{1}{2} \times 14\frac{1}{2}$ in. (49·5 × 37 cm), in folder with 84. (See 'The Fire of London and St Bartholomew's Hospital' by G. Whitteridge, *L.T.R.* XX, 47.)
89. *Survey of Building Sites in the City of London after the Great Fire of 1666*, by Mills and Oliver, vol. II (i.e. part ii of Mills I) (1956, extra publication): reduced facsimile, $8\frac{7}{8} \times 5\frac{5}{8}$ in. (22·5 × 14·5 cm), quarter cloth, uniform with the *Record*.
90. Plan of the precinct (eastern part) of the Hospital of St Katherine by the Tower, 1685, part of a survey by John Ogilby (1957); 1 sheet 28 × 25 in. (71·5 × 63·5 cm).

LIST OF PUBLICATIONS

91. *London Topographical Record*, XXI, ed. W. H. Godfrey (1958): $8\frac{7}{8} \times 5\frac{7}{8}$ in. (22·5 × 15 cm), quarter cloth.
92. The City of London, showing Parish Boundaries prior to the Union of Parishes Act, 1907, on the 1:2500 Ordnance Survey map, 1st edition, 1876 (1959): in colour, $24\frac{3}{4} \times 44\frac{1}{8}$ in. (63 × 112 cm) on sheet 27 × 46 in. (68·5 × 117 cm).
93. A Map of London under Richard II, from original sources, by M. B. Honeybourne (1960): $27\frac{3}{8} \times 41$ in. (69·5 × 104 cm) on sheet $31\frac{7}{8} \times 45$ in. (81 × 114·5 cm).
94. A View of London, 1600, by John Norden, from the engraving in the de la Gardie Collection in the Royal Library, Stockholm (1961): $18\frac{1}{8} \times 47\frac{7}{8}$ in. (46 × 121·5 cm) on sheet $24\frac{1}{8} \times 50$ in. (61 × 127 cm).
95 and 96. A Survey of the Parliamentary Borough of St Marylebone, including Paddington and St Pancras, 1834, engraved by B. R. Davis (1962–3): slightly reduced scale, $40\frac{1}{2} \times 34\frac{1}{8}$ in. (103 × 86·5 cm) on 2 sheets $22 \times 36\frac{1}{2}$ in. (50·5 × 92·5 cm).
96. See 95.
97, 98 and 99. *Survey of Building Sites in the City of London after the Great Fire of 1666*, by Mills and Oliver, vols. III, IV and V (i.e. Mills II and Oliver I and II) (1963, extra publication): reduced facsimile, $9\frac{1}{2} \times 6\frac{1}{8}$ in. (24 × 15·5 cm), quarter cloth.
98. See 97.
99. See 97.
100. *The Map of Mid-sixteenth Century London, An investigation into the relationship between a copper-engraved map and its derivatives*, by S. P. Marks (1964): $11\frac{1}{2} \times 8\frac{7}{8}$ in. (29 × 22·5 cm), quarter cloth.
101. *Survey of Building Sites in the City of London after the Great Fire of 1666*, by Mills and Oliver, vol. II (i.e. Mills I) (1965, extra publication): reduced facsimile, $9\frac{1}{2} \times 6\frac{1}{8}$ in. (24 × 15·5 cm), quarter cloth, uniform with 97, 98 and 99; a new edition of 79 and 89.
102. *London Topographical Record*, XXII, ed. M. B. Honeybourne (1965): $8\frac{7}{8} \times 5\frac{7}{8}$ in. (22·5 × 15 cm), quarter cloth.
103. *Survey of Building Sites in the City of London after the Great Fire of 1666*, by Mills and Oliver, vol. I, Introduction and Indexes (1967, extra publication): $9\frac{1}{2} \times 6\frac{1}{4}$ in. (24 × 16 cm), quarter cloth, uniform with 97, 98, 99 and 101.
104. Hollar's 'Exact Surveigh', 1667 (1966, extra publication): $21\frac{1}{2} \times 32\frac{1}{2}$ in. (54·5 × 82·5) on sheet $25 \times 34\frac{3}{4}$ in. (63·5 × 88 cm), replacing 22 and 26.
105. Grand Architectural Panorama of London, Regent Street to Westminster, by R. Sandeman, 1849 (1966): $4\frac{3}{4}$ in. × 22 ft 6 in. long (12 × 686 cm) in a small case $6 \times 7\frac{1}{8}$ in. (15 × 18 cm).
106. Horwood's Plan of London, 1792–9 (1966, extra publication, a memorial to the work of the London Survey Committee): 7 ft

3 in. × 13 ft 4 in. (221 × 406 cm), iv + 32 sheets 24½ × 23 in. (62·5 × 58·5 cm); includes variant plates from sheets A1 and B1. (See 'Richard Horwood's Plan of London: a guide to additions and variants, 1792–1819' by Paul Laxton, *L.T.R.* xxvi, 214.)

107. The Banqueting House with the Whitehall and Holbein Gates, by Inigo Jones, for a masque by Ben Jonson performed in 1623 (1967): 14 × 24¼ in. (37 × 61·5 cm) on sheet 19 × 26⅞ in. (48 × 68 cm); with 'A Prospect of Whitehall by Inigo Jones' by J. Harris, from *The Burlington Magazine*, February 1967.

108. *Index to Rocque's Plan of the Cities of London and Westminster and the Borough of Southwark, 1747* (1968, extra publication): facsimile, 11½ × 8⅞ in. (29 × 22·5 cm), full cloth, uniform with 100.

109. *The London Panoramas of Robert Barker and Thomas Girtin, c. 1800*, by H. J. Pragnell (1968): 11½ × 8⅞ in. (29 × 22·5 cm), quarter cloth, uniform with 100. (See 139.)

110. *John Tallis's London Street Views, 1838–1840 and 1847*, with introduction by Peter Jackson (1969, extra publication): 8 × 11¼ in. (20 × 28·5 cm), full cloth; published by Nattali and Maurice in association with the London Topographical Society. See No. 48.

111. Map of Chelsea by F. P. Thompson, 1836 (1969): 27¾ × 41⅝ in. (70·5 × 104·5 cm) on two sheets 29⅞ × 22½ in. (76 × 57·5 cm).

112. Hollar's Long View of London from Bankside, 1647 (1970, extra publication): 18½ × 93 in. (47 × 236·5 cm) on 7 sheets 25 × 19 in. (63·5 × 48 cm), replacing 19.

113. A Selection of Drawings of Old and New London Bridge, *c.* 1830, by E. W. Cooke (1970): 14 × 11 in. (35·5 × 28 cm) in folder.

114. Langley and Belch's 'New Map of London', 1812 (1971): in colour, 20¾ × 30¾ in. (52·5 × 78 cm) on sheet 25 × 35 in. (63·5 × 89 cm).

115. *London Topographical Record*, XXIII, ed. M. B. Honeybourne (1972): 8¾ × 5¾ in. (22 × 14·5 cm), quarter cloth.

116. Map of the Railways proposed by the Bills of the Session of 1863 in the Metropolis and its vicinity (1973): 24¼ × 23⅞ in. (61·5 × 60·5 cm) on sheet 28 × 24⅞ in. (71 × 63 cm). (See 'Parliament and the Railways', by David J. Johnson, *L.T.R.* XXIV, 147.)

117. *The Public Markets of the City of London surveyed by William Leybourne in 1677*, by Betty R. Masters (1974): 11½ × 8⅞ in. (29 × 22·5 cm), quarter cloth.

118 and 119. Thomas Milne's Land Use Map of London and Environs in 1800, with an introduction by Dr G. B. G. Bull (1975–6): north and south sections each 20⅛ × 40½ in. (51 × 103 cm) on 3 sheets, iii + 6 sheets 24 × 16½ in. (61 × 42 cm) in folder, available in colour and in black and white.

119. See 118.

LIST OF PUBLICATIONS

120. *The Artillery Ground and Fields in Finsbury*, Two maps of 1641 and 1703 reproduced with a commentary by James R. Sewell (1977): $14\frac{3}{4} \times 11\frac{3}{8}$ in. ($37 \cdot 5 \times 29$ cm), sewn in card wrapper.
121. *The Park Town Estate and the Battersea Tangle, A peculiar piece of Victorian London property development and its background*, by Priscilla Metcalf (1978): $11\frac{1}{8} \times 9$ in. ($28 \times 22 \cdot 5$ cm), Linson.
122. *The A to Z to Elizabethan London*, compiled by Adrian Prockter and Robert Taylor with introductory notes by John Fisher (1979): $12\frac{1}{4} \times 8\frac{5}{8}$ in. (31×22 cm), Linson, published concurrently by the Society and by Harry Margary in association with Guildhall Library.
123. *London Topographical Record*, XXIV, ed. Ann Saunders (1980, centenary volume): $8\frac{3}{4} \times 5\frac{3}{4}$ in. ($22 \times 14 \cdot 5$ cm), quarter cloth.
124. London from the North, by J. Swertner, 1789 (1980, extra centenary publication): $19 \times 31\frac{1}{2}$ in. ($48 \cdot 5 \times 80$ cm) on sheet $23 \times 32\frac{3}{4}$ in. ($58 \cdot 5 \times 83$ cm).
125. The 'Rhinebeck' Panorama of London, *c.* 1810, with an introduction by Ralph Hyde and keys by Peter Jackson (1981): in colour, reduced, $18\frac{3}{4} \times 107\frac{1}{4}$ in. ($47 \cdot 5 \times 272 \cdot 5$ cm) on iii + 4 sheets $24\frac{5}{8} \times 18\frac{3}{4}$ in. ($62 \cdot 5 \times 47 \cdot 5$ cm) in folder. (See 'A London Panorama, *c.* 1800 Resurrected', by Ralph Hyde, *L.T.R.* XXIV, 211.)
126. *The A to Z of Georgian London*, with introductory notes by Ralph Hyde (1982): $12 \times 8\frac{5}{8}$ in. ($30 \cdot 5 \times 22$ cm), Linson, uniform with 122, published concurrently by the Society and by Harry Margary in association with Guildhall Library.
127. *Robert Baker of Piccadilly Hall and His Heirs*, by Francis Sheppard (1982, extra publication): $11\frac{1}{8} \times 8\frac{3}{4}$ in. (28×22 cm), Linson, uniform with 121.
128. A Survey of Hatton Garden by Abraham Arlidge 1694 (1983): in colour, $31\frac{1}{4} \times 26$ in. ($79 \cdot 5 \times 66$ cm) on sheet $36\frac{3}{8} \times 27\frac{3}{4}$ in. ($92 \cdot 5 \times 70 \cdot 5$ cm) with sheet of text. (See 'The Survey of Hatton Garden in 1694 by Abraham Arlidge', by Dr Penelope Hunting, *L.T.R.* XXV, 83.)
129. A Plan of the Tower of London in 1682 (1983): in colour, $25\frac{3}{4} \times 25\frac{3}{4}$ in. (63×63 cm) on sheet $30 \times 27\frac{1}{2}$. (76×70 cm) with sheet of text. (See 'Five Seventeenth-Century Plans of the Tower of London', by Geoffrey Parnell, *L.T.R.* XXV, 63.)
130. Charles Booth's Descriptive Map of London Poverty 1889, with an introduction by Dr David A. Reeder (1984): in colour, $36\frac{1}{2} \times 46\frac{1}{4}$ in. ($93 \times 117 \cdot 5$ cm) on ii + 4 sheets averaging 20×25 in. (51×64 cm) in folder.
131. *The A to Z of Regency London*, with introduction by Paul Laxton and index by Joseph Wisdom (1985, extra publication): $12 \times 8\frac{5}{8}$ in. ($30 \cdot 5 \times 22$ cm), Linson, uniform with 122, published concurrently by the Society and by Harry Margary in association with Guildhall Library.

132. *London Topographical Record*, xxv, ed. Ann Saunders (1985): $8\frac{3}{4} \times 5\frac{3}{4}$ in. (22 × 14·5 cm), quarter cloth.
133. The Kentish Town Panorama, by James Frederick King c1850, with explanatory booklet by John Richardson (1986): 6 in. × 39 ft. 8 in. (15·5 × 1209 cm) on 26 sheets $9\frac{3}{8} \times 23\frac{1}{2}$ in. (24 × 59·5 cm) in folder.
134. Satellite View of London, taken by Landsat on 21 October 1984 (1986, extra publication): in colour, $20\frac{1}{4} \times 23\frac{1}{2}$ in (51·5 × 59·9 cm) on sheet.
135. *The London Surveys of Ralph Treswell*, ed. by John Schofield, illustrated with all of Treswell's London plans, several in colour (1987): $11\frac{1}{8} \times 9$ in. (28 × 23 cm).
136. *The A to Z of Victorian London*, with notes by Ralph Hyde (1987, extra publication): $12\frac{1}{8} \times 8\frac{3}{4}$ in. (31 × 22 cm), Linson, uniform with 122, published concurrently by the Society and Harry Margary in association with Guildhall Library.
137. *Hugh Alley's Caveat – The Markets of London in 1598*, ed. Ian Archer, Caroline Barron and Vanessa Harding (1988): $9\frac{1}{2} \times 10\frac{1}{4}$ in. (24 × 26 cm).
138. Hollar's Prospect of London and Westminster taken from Lambeth, in 2 versions c1665 and c1707 (1988, extra publication): $12\frac{3}{4}$ in. × 10 ft. $6\frac{1}{2}$ in (32·5 × 322·4 cm) on 8 sheets 14 in. × 11 ft. (35·6 × 336 cm) in folder. (See 'Some Notes on Hollon's *Prospect*...' by Peter Jackson, *L.T.R.* xxvi, 134.)
139. Barker's Panorama of London from the Roof of the Albion Mills, 1792, with an introduction by Ralph Hyde and keys by Peter Jackson (1988, extra publication): in colour, $16\frac{3}{4} \times 130\frac{1}{2}$ in. (42.5 × 343 cm) on iii+6 sheets $19\frac{1}{4} \times 23\frac{1}{2}$ in. (49 × 59·5 cm) in folder, published concurrently by the Society and Guildhall Library. (See 109.)
140. *Good and Proper Materials, The Fabric of London since the Great Fire*, papers given at a conference organised by the Survey of London, ed. Hermione Hobhouse and Ann Saunders (1989): $10\frac{3}{4} \times 8\frac{3}{4}$ in. (27·5 × 22 cm), paperback.

LIST OF MEMBERS 1990
Institutional members are listed separately

His Royal Highness The Duke of Edinburgh, KG, KT, *Patron*

Adams, Mrs Anna   Roots, Dartnell Avenue, West Byfleet, Surrey KT14 6PJ

Adams, Mr Bernard P. F.   24 South Park Court, Park Road, Beckenham, Kent BR3 1PH

Addis-Smith, Mrs Sonia W.   Cross End House, Thurleigh, Bedfordshire MK44 2EE

Addison, Mr Robert J., F.C.A.   Eastering, 13 Austin Avenue, Bickley, Kent BR2 8AJ

Aickin, Dr Robert M.   69 Lauriston Road, London E9 7HA

Ainsworth-Smith, Revd I. M., M.A.   Knutsford Cottage, North Street, Milverton, Taunton, Somerset TA4 1LG

Aldous, Mr Anthony M.   12 Eliot Hill, London SE13 7EB

Allen, Mr G. R.   183 Putnoe Street, Bedford MK41 8JR

Allin, Mr Paul V., M.Sc.   15 Rosemead Gardens, Hutton, Brentwood, Essex CM13 1HZ

Allinson, Mr Tom   19 Breton Road, Sault Ste Marie, Ontario P6B 5S9, Canada

Anderson, Mr T. W. M., MC., T.D., B.A.   69 Monkhams Lane, Woodford Wells, Essex IG8 0NN

Anwyl-Harris, Mr Peter D.   13 Caroline Terrace, Belgravia, London SW1W 8JS

Archer, Mr Ian, M.A.   Downing College, Cambridge CB2 1DQ

Ash, Mr H. J.   Cedar Lodge, 39 Church Road, Newick, East Sussex BN8 4JX

Ashdown, Mr John H., F.S.A.   53 Bainton Road, Oxford OX2 7AG

Austin, Mrs Constance R.   Landor House, 24 Glebe Avenue, Enfield, Middlesex EN2 8NY

Austin, Mr Michael, A.I.M.L.S.   Landor House, 24 Glebe Avenue, Enfield, Middlesex EN2 8NY

Aylward, Mr Roger   81 Broxholm Road, London SE27 0BJ

Ayres, Mr R. T.   Flat B, 6 Stanley Street, Bedford MK41 7RF

Backman, Mr P. R.   48 Hallswelle Road, London NW11 0DJ

Bacon, Mr Timothy R.   67 Britannia Road, London SW6 2JR

Bailey, Mr K. A.   115 High Street North, Stewkley, Leighton Buzzard, Bedfordshire LU7 0EX

Bailey, Mr Simon   24 Albion Drive, London E8 4ET

Bain, Mr Iain S., *Member of Council*   New Cottage, Newnham, Baldock, Hertfordshire

Baker, Mr James W.   231 Sandwich Street, Plymouth, Massachusetts 02360, U.S.A.

Bales, Mr Kevin B.   3 Manor Place, London SE17 3BD
Ball, Dr Enid   12 Warren Way, Digswell, Hertfordshire AL6 0AD
Bamfield, Mr K. F.   24 Thorndon Gardens, Ewell, Epsom, Surrey
Bankes, Mr A. G. K.   64 Kingsley Way, London, N2 0EW
Bankes, Mrs Juliet M.   96 Elms Crescent, Clapham Park, London SW4 8QU
Banks, Miss Elizabeth   21 The Little Boltons, London SW10 9LJ
Bar, Mr Norman C.   48 Granville Road, Finchley, London N12
Barber, Mr Peter M., *Member of Council*   16 Tivoli Road, London N8 8RE
Barker, Mr B. Ashley   39 Kensington Park Gardens, London W11 2QT
Barker, Mr R. F. R., *Member of Council*   4 Lindsey House, Lloyd's Place, Blackheath, London SE3
Barkley, Mr Harold   4 Trafalgar Road, Cambridge CB4 1EU
Barnett, Mr J. A.   40 York Terrace West, London NW1 4QA
Barrett, Mr A. C., C.B.E., F.C.I.O.B., F.R.S.H.   102 Chepstow Road, London W2 5QW
Barriff, Mr Stephen J.   4 Montacute Road, Bushey Heath, Hertfordshire WD2 1PJ
Barron, Dr Caroline   Royal Holloway & Bedford College, Egham Hill, Egham, Surrey TW20 0EX
Barter Bailey, Mrs Sarah E., F.S.A.   3 Cambridge Mansions, Cambridge Road, London SW11 4RU
Barton, Mrs Gwen C.   The Anchorage, 11 Alpine Road, Hove, East Sussex, BN3 5HG
Barty-King, Mr Hugh D. O.   Holgate House, Ticehurst, Wadhurst, East Sussex TN5 7AA
Bates, Revd James B.   7 Ashley Terrace, St Brannock's Park Road, Ilfracombe, Devon EX34 8JG
Baxter, Mr Alan J.   Alan Baxter Associates, 14/16 Cowcross Street, London EC1M 6DR
Beacham, Mr J. W.   Draycott, The Ridgway, Pyrford, Woking, Surrey GU22 8PN
Beard, Mr James H.   40 Littlefield Road, Edgware, Middlesex HA8 0TD
Bebbington, Mr John G.   Newlands, 4 Bishopsmead Close, East Horsley Leatherhead, Surrey KT24 6RY
Beckett, Mr L. J.   7 Braeside Close, Sevenoaks, Kent TN13 2JL
Beer, Dr E. S. de., C.B.E., D.Litt., F.B.A., F.S.A., *Vice-President*   Stoke House, Stoke Hammond, Milton Keynes, Buckinghamshire MK17 9NB
Belcher, Mr Victor R., M.A., *Member of Council*   55 Gore Road, London E9 7HN
Bennell, Mr John E. G., M.Litt.   92 Buriton Road, Harestock, Winchester, Hampshire SO22 6JF
Bennett, Mr A. Francis   13 Seagrave Road, London SW6 1RP
Bensley, Mr John S.   60 The Ryde, Hatfield, Hertfordshire AL9 5DL

LIST OF MEMBERS 329

Bentley, Mr David, B.A. (Hons.)   48A Parkholme Road, London E8 3AQ
Berry, Mr Robert E. H.   7A Heyford Avenue, London SW8 1EA
Bettley, Mr James   155 Hartington Road, London SW8 2EY
Biffin, Mr Edward H., M.A.   8 Orchard Row, Church Road, Grafham, Huntingdon, Cambridgeshire PE18 0BD
Bimson, Miss Mavis, F.S.A.   32 Upper Park Road, London NW3 2UT
Binfield, Dr J. C. G.   Department of History, University of Sheffield, Sheffield S10 2TN
Binney, Mr Marcus, O.B.E.   Domaine des Vaux, St Lawrence, Jersey CI
Bishton, Mr Colin D.   82 Exford Road, Lee, London SE12 9HA
Black, Mr Edward J.   43 Cromwell Road, Wembley, Middlesex HA0 1JS
Black, Mr G. D.   106 Lauderdale Mansions, Lauderdale Road, Maida Vale, London W9 1LY
Blake, Mr Paul A.   18 Rosevine, London SW20 8RB
Bloice, Mr Brian J.   220 Woodmansterne Road, Streatham, London SW16 5UA
Bloor, Ms Elizabeth, M.A.   139 Malden Road, Kentish Town, London NW5 4HS
Blowers, Mrs Shirley   Chelsea House, Princess Street, Ipswich IP1 1QT
Boardman, Dr Brigid M. U., B.A., M.A., Ph.D.   54 St James Park, Bath, Avon, BA1 2SX
Boast, Miss Mary   29 Ruskin Park House, Champion Hill, London SE5 8TQ
Bodnar, Mr Andrew   18 Briarwood Road, London, SW4 9PX
Bond, Mr J. P.   Newburn, Well Place, Cheltenham, Gloucestershire GL50 2PJ
Bonnell, Mrs Fraser C.   777 South Orange Grove Blvd No 2, Pasadena, California CA 91105, U.S.A.
Booth, Mr Philip A.   Sheffield University, Department of Town and Regional Planning, 8 Claremont Place, Sheffield S10 2TB
Bowen-Bravery, Mr David   7 Revell Close, Fetcham, Leatherhead, Surrey KT22 9PT
Bowers, Mr J Michael   39 Cloudesley Road, Islington, London N1 0EL
Bowlt, Mrs Eileen M., B.A.   7 Croft Gardens, Ruislip, Middlesex HA4 8EY
Bowyer, Mr L. F.   110 Andrewes House, Barbican, London EC2Y 8AY
Bradshaw, Mr Christopher   16 Ornan Road, London NW3 4PX
Bresslaw, Mr Bernard   6 Osborne Gardens, Little Heath, Potters Bar, Hertfordshire EN6 1RZ
Brett, Mr M. P.   37 Glanville Road, Bromley, Kent BR2 9LN
Briginshaw, Dr Anthony   Department of Mathematics, The City University, London EC1V 0HB
Brittain, Mr J. H.   Edge Hill, 46 Cromwell Avenue, Billericay, Essex CM12 0AG
Britten, Mr Norman W., F.C.A.   Suite 776, Lloyd's, Lime Street, London EC3M 7DQ

Brooke, Mrs Patricia K.   Littleshaws, 29 Blacksmiths Hill, Sanderstead, Surrey CR2 9AZ
Brown, Mr Bernard J.   39 Lower Camden, Chislehurst, Kent BR7 5HY
Brown, Mr David G.   168 Wightman Road, London N8 0BD
Bruce, Mrs M. E.   Flat 6, 15–16 Cephas Avenue, London E1
Brushfield, Mr John M., M.C.A.M., M.Inst.M   30 Herald's Place, Renfrew Road, Kennington, London SE11 4NP
Buckley, Mr R. A.   164 Sutherland Avenue, London W9 1HR
Buckley, Mr Richard W   19 Corringham Road, Wembley, Middlesex HA9 9PX
Bulmer-Thomas, Mr Ivor, M.A., F.S.A.   12 Edwardes Square, London W8 6HG
Butler, Mr Roger   66 Widmore Road, Bromley, Kent BR1 3BD
Byron, Mr Richard E.   28 Bramerton Street, London SW3
Cain, Mr Piers W. A., B.A. (Hons)   94 Byron Way, Northolt, Middlesex UB5 6AZ
Capel, Mr R. I.   91 Thomas More House, Barbican, London EC2
Caplin, Miss Esther T., M.A.   57 Erskine Hill, London NW11 6RY
Carlin, Dr Martha   Smith College, Department of History, Northampton MA 01063, U.S.A.
Carnaby, Mr John J., Engl Tech., M.I.P.   3 Lakeside Crescent, East Barnet, Hertfordshire EN4 8QH
Carter, Mr N. R. W., F.R.I.C.S., DipTP   40 Westwood Park, Forest Hill, London SE23 3QH
Catford, Mr K. E.   Mere Hall, Noctorum Lane, Birkenhead, Merseyside L43 9TZ
Caustin, Mr C. E.   59B Clapham Common South Side, London SW4 9DA
Cerasano, Prof. Susan P.   Department of English, Colgate University, Hamilton, New York NY 13346, U.S.A.
Chaffin, Mr David E.   17 Princes Street, Stirling, Scotland FK8 1HQ
Chaplin, Mr Norman W.   24 Ladbroke Gardens, London W11 2PY
Chapman, Dr Hugh P. A.   39 Mundania Road, London SE22
Chard, Prof Timothy   509 Mountjoy House, Barbican, London EC2
Charlton, Mr S. A.   367 St John Street, London EC1V 4LB
Cherrill, Mrs Mavis   18 Blenheim Gardens, Sanderstead, Surrey CR2 9AA
Cherry, Mrs Bridget K.   58 Lancaster Road, London N4 4PT
Chilton, Mr Charles F. W.   31 Crediton Hill, West Hampstead, London NW6 1HS
Clarke, Mrs Patricia A.   31 Lynton Road, Harrow, Middlesex HA2 9NJ
Cleaver, Mr Hugh, *Honorary Auditor*   91 Malford Grove, South Woodford, London E18
Clifton, Dr Gloria C., B.A., Ph.D.   55 The Ridgway, Sutton, Surrey SM2 5JX
Cline, Mr Roger L., M.A., LL.B., *Honorary Treasurer*   34 Kingstown Street, London NW1 8JP

LIST OF MEMBERS

Clingman, Mr S. N.   46 Church Court, Finchley, London N3
Clout, Mr Martin S., L.T., F.T.C.L.   Whitehouse Farm, Hooe Road, Ninfield, Battle, East Sussex TN33 9EH
Clute, Mrs Judith R.   221 Camden High Street, London NW1 7BU
Cockings, Mr James H., B.A., M.Sc.   78 Prebend Street, Islington, London N1 8PR
Collier, Mrs J. Rosalind, B.A.   44 Kitchener Road, London N2 8AS
Collier, Mr L. H.   46 Vallance Road, Wood Green, London N22 4UB
Connah Jnr, Mr Douglas D.   410 West Lombard Street, Baltimore, Maryland MD 21201, U.S.A.
Cook, Mr Andrew S.   66 Compton Road, Winchmore Hill, London N21 3NS
Coombes, Mr A. J.   24 Horsham Road, Dorking, Surrey RH4 2JA
Coombes, Mr J. Michael   Durrant Cross, 2A Durrant Road, Bournemouth, Hampshire BH2 6LE
Cooper, Mr Anthony, F.R.I.B.A., *Member of Council*   6 Waterside Place, Princess Road, London NW1 8JT
Cooper, Mr Robert M.   120 Wheelers Lane, Kings Heath, Birmingham B13 0SG
Corfield, Dr Penelope J., M.A., Ph.D.   Department of History, Royal Holloway & Bedford New College, Egham, Surrey TW20 0EX
Corner, Mr D. M.   105 Listria Park, London N16 5SP
Cosh, Miss Mary   10 Albion Villas, Thornhill Road, London N1
Cowan, Ms Janet   1 Woodville Gardens, London W5
Cowe, Mr Kenneth W.   38A Warwick Park, Tunbridge Wells, Kent TN2 5TB
Crabb-Wyke, Mr P. V.   46 Pauline Gardens, Billericay, Essex CM12 0LB
Craig, Mr James F.   60 Headley Chase, Brentwood, Essex
Craig, Mr Peter N., M.A. (Cantab)   55 Brompton Park Crescent, London SW6 1SN
Crawford, Mr David   22 The Avenue, Bickley, Bromley, Kent BR1 2BT
Crayford, Mr Robert C.   1 Warren Avenue, Orpington, Kent
Croft, Dr Desmond N.   Bourne House, Hurstbourne Priors, Whitchurch, Hampshire RG28 7SB
Crotty, Mr R. J.   7 Harrowdene Gardens, Sandy Lane, Teddington, Middlesex
Cullis, Mr J. E.   McDaniel & Daw, Well Court, 14–16 Farringdon Lane, London EC1R 3AU
Cumming, Mr Donald L.   1 Marlpit Avenue, Coulsdon, Surrey CR3 2SD
Curtis, Mr B. Thomas   36 Brockswood Lane, Welwyn Garden City, Hertfordshire AL8 7BG
Da Costa, Mr Christopher I., B.A. (Hons)   19 The Green, Reading, Berkshire RG6 2BS
Dagley, Mr D. B.   Daglea Cottage, Mersea Road, Abberton, nr Colchester, Essex

LIST OF MEMBERS

Davidge, Mr Ian   71 Park Lane, Southend-on-Sea, Essex SS1 2SL
Davies, Mr J. G.   60 Embassy House, West End Lane, London NW6 2NB
Day, Mr Martin K.   The Printer on the Rye, Eagle Printing Works, Printer's Yard, Stone Street, Cranbrook, Kent TN17 3HF
De'Ath, Mr David A.   32 Queensway, Caversham Park Village, Reading, Berkshire RG4 0SQ
Dell, Mr David J.   66 Vallance Road, Alexandra Park, London N22 4UB
Delves, Mr H. C.   Moorsfort Cottage, 18 Fife Road, East Sheen, London SW14 7EL
Denny, Mr Stephen W., B.Sc., F.R.G.S.   Wilderness Copse, Fernham Road, Shellingford, Faringdon, Oxfordshire SN7 7PU
Devine, Mrs P. M.   148 Snakes Lane, Woodford Green, Essex IG8 7JB
Diamond, Mr John B., B.Arch., R.I.B.A.   27 Crescent Grove, London SW4
Dibben, Mr A. A., M.A.   18 Clare Road, Lewes, East Sussex BN7 1PN
Draper, Mrs Marie P. G.   39 Vicarage Road, London E10 5EF
Duncan, Mr Douglas H. S.   116 Fairlawn Drive, Berkeley, California 94708, U.S.A.
Mr Thomas P Dungan   1821 Hatfield Road, Huntingtown, MD 20639, U.S.A.
Duttson, Mr B. L.   80 Esmond Road, London W4 1JF
Earl, Mr John   60 Balcaskie Road, Eltham, London SE9 1HQ
Eastment, Mr Mark J.   8 Wilton Avenue, Chiswick, London W4 2HY
Eddowes, Mr John R. B.   Legastat Copying, 57 Carey Street, London WC2
Edwards, Mr Arthur C.   43 Maltese Road, Chelmsford, Essex CM1 2PB
Edwards, Mr H. G.   34 Carstan Square, London E14 9EU
Edwards, Mr R.   12 Blacklands Drive, Hayes End, Hayes, Middlesex UB4 8EU
Eeles, Miss Joanne   27 Charlton Court, High Street South, East Ham, London E6 4JD
Elam, Mr Ronald A.   31 Loxley Road, London SW18 3LL
Elbourne, Miss Judith M.   98 Russell Court, Woburn Place, London WC1H 0LP
Ellen, Mr R. G.   4 Larkfield Road, Farnham, Surrey GU9 7DB
Elliot, Mr James D., B.A., M.A.   21 Harborough Road, Streatham, London SW16 2XP
Engelhart, Mrs Margaret S   72 Brinkerhoff Street, Plattsburgh, NY 12901, U.S.A.
Epstein, Mr Jonathan A.   72 Stanhope Avenue, London N3 3NA
Erler, Ms Mary   Department of English, Fordham University, Bronx, New York 10458, U.S.A.
Esslemont, Mr Peter D.   25 Thomas a Becket Close, Sudbury, Wembley, Middlesex HA0 2SH
Farthing, Mr Cecil H. J., O.B.E., B.A., F.S.A.   129 Iveagh House, King's Road, Chelsea, London SW10 0TY

LIST OF MEMBERS

Fenwick, Mr P. H.   78 Herald Walk, Knight's Manor, Temple Hill, Dartford, Kent DA1 5SS
Field, Mr E. F.   165 Raeburn Avenue, Surbiton, Surrey KT5 9DG
Finch, Mr Harold   9 Bousfield Road, New Cross, London SE14 5TP
Finn, Mr John S.   23 Groombridge Road, London E9 7DP
Finney, Mr Jarlath J.   207 Hampstead Way, London NW11 7YB
Firmager, Mrs Gabrielle M.   72B High Street, Semington, Trowbridge, Wiltshire BA14 6JR
Fleming, Mr R. L.   Flat 4, 5 Royal Avenue, Chelsea, London SW3 4QE
Ford, Mr Trevor T., Dip R.A.M., *Membership Secretary*   151 Mount View Road, Stroud Green, London N4 4JT
Forrest, Mr Jeffrey A.   8 Ponsonby Place, London SW1P 4PT
Forrestier Smith, Mr Peter A.   64 Gordon Road, Carshalton Beeches, Surrey SM5 3RE
Foster III, Mr Fred   5532 Karen-Elaine Drive, Apt 1729, New Carrollton, Maryland MD 2078, U.S.A.
Foulds, Mr William P.   7 The Crescent, Farnborough, Hampshire GU14 7AR
Fox, Mr Howard M.A., F.C.A.   18 Westside, Fortis Green, London N2 9ES
Foxwell, Miss Hilary J.   135 Topshaw Road, London SW17 8SW
Fraser, Mr A.   69 North Grove, London N15 5QS
Fraser, Mr Alan C.   38 Oakland Way, Ewell, Surrey KT19 0EN
Frazer, Mr Patrick A. T., M.A., M.Sc., *Honorary Secretary*   36 Old Deer Park Gardens, Richmond, Surrey TW9 2TL
Freestone, Theresa M.   30 Grange Park, Ealing, London, W5 3PS
French, Mrs Josephine, B.A.   11 Montague Road, Berkhamsted, Hertfordshire HP4 3DS
Frizzell, Miss Helen D. I.   Flat 2, 39 St George's Square, London SW1V 3QN
Fuller, Mr I. R.   62 St Mary's Grove, Chiswick, London W4 3LW
Fulwell, Miss Stella   26 Avenue Road, Leigh-on-Sea, Essex SS1 2DU
Fyffe, Mr C. J.   52 Holmdale Road, London NW6 1BL
Gadbury, Mr Peter   5 Berger Close, Petts Wood, Kent BR5 1HR
Gale, Mr L. Thornton   9404 SE 54th Street, Mercer Island WA 98040, U.S.A.
Gan, Mr Richard L.   The Firs, 25A Tower Road, Strawberry Hill, Middlesex TW1 4PJ
Garrett, Mrs Elizabeth   52 Glebelands, Pulborough, West Sussex RH20 2JJ
Gay, Mr Kenneth D.   201 Alexandra Park Road, London N22 4BJ
Gee, Mrs C. M.   Keats House, Keats Grove, Hampstead, London NW3 2RR
Gentles, Mr Ian   Department of History, Glendon College, 2225 Bayview Avenue, Toronto, Ontario M4N 3M6 Canada
Gestetner, Mr Jonathan   7 Oakhill Avenue, London NW3 7RD

LIST OF MEMBERS

Gibson, Mr Ian   18D Castletown Road, London W14
Gidley-Kitchin, Mr G. C. B.   Wybournes, Kemsing, Sevenoaks, Kent TN15 6NE
Gill, Mr Brian H., M.A.(Edin) DipArchaeol   261 Grove Street, London SE8 3PZ
Ginsburg, Mr Leslie B., AADip., SPDip., R.I.B.A.   65 Warrington Crescent, London W9 1EH
Gladman, Mr Martin I.   14 Linden Road, New Southgate, London N11 1ER
Glover, Col Colin M.   5A Rutland Court, Knightsbridge, London SW7 1BN
Glover, Mrs J. J.   48 Hermitage Lane, London NW2 2HG
Goddard, Mrs Joan   2 Ravens Road, Shoreham-by-Sea, West Sussex
Godfrey, Mrs W. Emil   81 The Causeway, Steventon, Abingdon, Oxfordshire OX13 6SQ
Gollin, Mr G. J., CEng FIMechI   The White House, 24 Ottways Lane, Ashtead, Surrey KT21 2NZ
Good, Mr Charles H.   4100 N Marine Drive, Chicago, Illinois 60613, U.S.A.
Goodway, Dr David J.   York House, Laurel Grove, Keighley, West Yorkshire BD21 2HW
Goodwin, Mrs Gillian T. M., M.A. (Oxon)   29 Chalcot Square, London NW1 8YA
Goodwin, Mr Vernon J.   19 Avenue Road, Woodford Green, Essex IG8 7NU
Gormley, Mr Patrick   20A Greycoat Gardens, Greycoat Street, London SW1P 2QA
Gotlop, Mr Peter F.   2 Pinewood, 287 Nether Street, London N3 1PD
Grafton Green, Mrs Brigid,   88 Temple Fortune Lane, London NW11 7TX
Green, Mr Christopher H   BP House, 1 Albert Road, Melbourne 3004 Australia
Greenfield, Mr D. A.   65 Stainforth Road, Seven Kings, Essex IG2 7EL
Guillen, Mr M. P.   56 Kingston Road, Teddington, Middlesex TW11 9HX
Gurling, Mr P. W.   47 Bouverie Avenue, Salisbury, Wiltshire SP2 8DU
Gurney, Mrs Diana   31 St Mark's Crescent, London NW1 7TT
Hall, Mr A. H., F.S.A., F.L.A.   69 Ingleborough, Cavell Drive, The Ridgeway, Enfield EN2 7PR
Hall, Mr E. G.   Flat 7, 17 Henrietta Street, Bath, Avon BA2 6LW
Hall, Mr John M.   55B Montalt Road, Woodford Green, Essex IG8 9RS
Hallam, Miss Saskia A. W.   99 Vanbrugh Court, Wincott Street, London SE11 4NS
Hammerstone, Mr M. D.   70 Rodenhurst Road, London SW4 8AR
Hammil, Mr L. P.   115 Eleanor Road, London E8 1DN

Hammond, Mr P. W.   3 Campden Terrace, Linden Gardens, London W4 2EP
Hancock, Mrs Denise B.A., LL.B.   The Barn, Wrexon Trull, Taunton, Somerset TA3 7PA
Hanson, Ms Julienne, 23A Richmond Avenue, London N1 0NE
Hanson, Mr Michael P. J.   83 Thetford Road, New Malden, Surrey KT3 5DS
Harbridge, Dr C. C. E.   Crofton, 42 Fourth Avenue, Frinton-on-Sea, Essex CO13 9DX
Harding, Dr Vanessa A.   325 Upper Street, Islington, London N1 2XQ
Harper, Mr Richard P., F.S.A.   48 Western Hill, Durham City DH1 4RJ
Harper Smith, Dr T.   48 Perryn Road, London W3
Harris, Mr Michael R. A.   Extra-Mural Department, London University, 26 Russell Square, London WC1B 5DQ
Harte, Dr Michael J.   Greenham Farm, Wadhurst, East Sussex TN5 6LE
Harte, Mr Negley B.   Department of History, University College London, London WC1E 6BT
Harvey, Mr Gerald P.   8 Briarwood Road, Clapham Park, London SW4 9PX
Hay, Mr Peter L.   Norman Hay plc, Bath Road, Harmondsworth, West Drayton, Middlesex UB7 0BY
Heather, Mr William   23 Pandora Road, London NW6 1TS
Hendy, Mr John LL.B., DipLL, L.L.M.   15 Old Square, Lincoln's Inn, London WC2A 3UH
Hewett, Mr Peter G., M.A.   White Cottage, Church Road, Milford, Godalming, Surrey
Heywood, Mr Neil T.   Owls Wood Cottage, Newsells Village, Royston, Hertfordshire SG8 8DE
Hickin, Mr Michael G   16 Croxted Road, London SE21 8SW
Hidson, Mr Roy   117 St Thomas's Road, Finsbury Park, London N4 2QJ
Hillier, Mr Andrew   9 Rodenhurst Road, London SW4
Hillier, Mr J. F.   36 Watery Lane, Merton Park, London SW20 9AD
Hillman, Ms Judy   16 Chalcot Road, London NW1 8LL
Hird, Miss A. J.   126 Felstead Road, Orpington, Kent BR6 9AF
Hird, Mr Roger L.   25 Rozel Road, London SW4 0EY
Hobbs, Mr Harold C., F.R.S.A., F.R.G.S., FInstLEx   16 Springfield Park Road, Horsham, Sussex RH12 2PW
Hobhouse, Miss Hermione MBE   61 St Dunstan's Road, London W6
Hodges, Mr Richard H.   30 Rectory Park Avenue, Northolt, Middlesex
Hodgkins, Mr J. R., B.Sc(Econ)   34 Hamlet Court Road, Westcliff-on-Sea, Essex SS0 7LX
Hodgson, Mr Godfrey M. T.   27 Polstead Road, Oxford OX2 6TW

Hoey, Mr David J   3510 Clearbrook Road, R. R. No. 8, Abbotsford, British Columbia V2S 6A9 Canada
Holden, Mr C. E. A.   29 Kingswood Road, Shortlands, Bromley, Kent BR2 0HG
Holness, Mrs A   18 Elm Close, Bowerhill, Melksham, Wiltshire SN12 6SD
Holyer, Mr Harold J.   10 Masonsfield, Mannings Heath, Horsham, West Sussex RH13 6JP
Honig, Mr G. J.   Dubbele Buurt 6, 1521 DC Wormerveer, Holland
Honour, Mr David J.   3 Mackenzie Road, Beckenham, Kent BR3 4RT
Hooker, Mr Alan H.   Groombridge Post Office, Phoenix House, Station Road, Groombridge, Tunbridge Wells TN3 9QY
Hopkinson, J. H.   Acre Mead, Whitham Road, Long Bennington, Newark, Nottinghamshire
Hopton, Mr Eric I.   30/31 Chesham Place, London SW1X 8HB
Hornby, Mr J. Roger   27 Cloudesley Square, London N1 0HN
Horne, Mr M. A. C.   175 Exeter Road, South Harrow, Middlesex HA2 9PG
Hughes-Stanton, Mr Corin, 11 Burghley Road, London NW5 1UG
Humby, Mr Michael J.   49 John Trundle Court, Barbican, London EC2Y 8DJ
Hunter, Mr P. B.   311 Willoughby House, Barbican, London EC2Y 8BL
Hunting, Dr P. S., *Member of Council*   40 Smith Street, London SW3
Hurdley, Mr J. R.   17 Bloomfield Terrace, Belgravia, London SW1W 8PG
Hurley, Mr Alan M.   175 South Ashmont Street, Boston, Massachusetts MA 02124, U.S.A.
Hurst, Mr B. L.   The Spinney, 2 Richfield Road, Bushey Heath, Herts WD2 3LQ
Hurst, Mr M. G.   57 Uphill Road, Mill Hill, London NW7 4PR
Hussey, Mr P. J.   9A St Stephens Gardens, Twickenham, Middlesex TW1 2LT
Hyde, Mr Michael R.   Villa Farm, Rushmere St Andrew, Ipswich, Suffolk IP5 7DT
Hyde, Mr Ralph, F.L.A., F.S.A., *Member of Council*   Print Room, Guildhall Library, Aldermanbury, London EC2P 2EJ
Ince, Mr Geoffrey W.   9 The Oval, Hackney Road, London E2
Ingram, Prof William   University of Michigan, Department of English Language and Literature, Ann Arbor, Michigan 48109, U.S.A.
Insall, Mr Donald W.   73 Kew Green, Richmond, Surrey TW9 3AH
Inskip, Mr Henry T., J.P., F.R.I.C.S.   8 Rothsay Gardens, Bedford MK40 3QB
Ishida, Yuzuru   120–38 Okamoto, Kamakura-shi Kanagawa-ken 247 Japan
Jackson, Mr Barry T., M.S., F.R.C.S.   Mapledene, 7 St Matthew's Avenue, Surbiton, Surrey KT6 6JJ

LIST OF MEMBERS

Jackson, Mr D. S.   65 Longley Road, Harrow, Middlesex HA1 4TQ
Jackson, Mr Oliver J. F.   Cobb's Mill, Mill Lane, Hurstpierpoint, West Sussex BN6 9HN
Jackson, Mr Peter C. G., F.S.A., *Chairman*   17 Blandford Road, Ealing, London W5 5RL
James, Mrs Madge   Windyridge, Cornells Lane, Widdington, Saffron Walden, Essex
James, Mrs Robert, 7 Waterside Place, Princess Road, London NW1 8JT
Jay, Mr Barrie S., M.A., M.D., F.R.C.S.   10 Beltane Drive, London SW19 5JR
Jeffcock, Mr David P.   Wellington House, Captains Row, Lymington Hampshire SO41 9RR
Jefferson Smith, Mr P.   22 Iveley Road, Clapham, London SW4 0EW
Jeffery, Dr Paul G., C.B.E., Ph.D   23B Home Park Road, Wimbledon Park, London SW19 7HP
Jenkins, Mr D. K.   68 Clapham Common Northside, London SW4
Jervis, Ms Katherine,   1 Beech Lawn, Epsom Road, Guildford, Surrey GU1 3PE
John-Adrian,   16 Gloucester Court, Swan Street, London SE1 1DQ
Johnson, Mr David J., M.A., FRHistS, *Member of Council*   41 Cranes Park Avenue, Surbiton, Surrey KT5 8BS
Johnson, Mr Ian A.   Green Tiles, Mill Lane, Chalfont St Giles Buckinghamshire
Johnston, John S.   81 Holeburn Road, Glasgow, Scotland, G43 2XN
Johnston, Miss J. L. H., *Honorary Member*   Box 623, Windsor, Nova Scotia, B0N 2T0 Canada
Jollye, Mr S. H.   6 The Glebe, Ipplepen, Newton Abbot, Devon TQ12 5TQ
Jolowicz, Miss Marian R.   9 Akenside Court, 26 Belsize Crescent, London NW3 5QU
Jones, Mr Dermot M. F.   95 Clapham Manor Street, London SW4 6DR
Jones, Prof Keith   57 Marksbury Avenue, Richmond, Surrey
Jones, Mr Norman   Flat 2, 28 Lisle Street, London WC2H 7BA
Jones, Mr P. J.   11 Park Court, Harlow, Essex CM20 2PY
Joseph, Mr Jeremy B.   11 Mallory Avenue, Caversham, Reading, Berkshire RG4 0QN
Jupp, Mrs Denise C.   419 Sidcup Road, Eltham, London SE9 4EU
Jupp, Mr Peter G.   28 Shenley Road, Dartford, Kent DA1 1YE
Kalman, Mr Raymond P., B.A.   21 Florence Mansions, Vivian Avenue, London NW4 3UY
Kaukas, Mr Bernard   13 Lynwood Road, Ealing, London W5 1JQ
Kay, Mr John D., C.B.E., AADipl, A.R.I.B.A.   36 Crescent Grove, London SW4 7AH
Kearon, Mr Peter, H. A.   6 Selwood Place, London SW7
Keene, Mr Derek J.   162 Erlanger Road, Telegraph Hill, London SE14 5TJ

Keir, Mrs Gillian I. Highlands, 17 Battlefield Road, St Albans, Hertfordshire AL1 4DA
Kelly, Miss Averill A., M.A. Flat 8, 34 Phillimore Gardens, London W8 7QF
Kelsall, Mr A. F. 4 Woodlands Avenue, Finchley, London N3 2NR
Kent, Mr J. P. C. Ph.D., F.S.A. 16 Newmans Way, Hadley Wood, Barnet, Hertfordshire EN4 0LR
Kerry, Mr Alan 7 Bermans Way, London NW10 1SD
Kettlewell, Mrs Muriel Banovalum House, Manor House Street, Horncastle, Lincolnshire LN9 5HF
Keynes, Mr Randal H., M.A. 17 Keystone Crescent, London N1 9DS
King, Mr B. D. 17 Wellington House, Eton Road, London NW3 4SY
King, Mr T. T. Ridgemount, Heronway, Hutton Mount, Brentwood, Essex CM13 2LX
Kinney, Mr Laurence 9 Old Park Lane, Farnham, Surrey GU9 0AJ
Kocen, Dr R. S. 127 Willifield Way, London NW11 6XY
Koch, Prof Claude, F. 128 West Highland Avenue, Philadelphia, PA 19118, U.S.A.
Lakin-Thomas, Mr Duane 9 Leys Avenue, Cambridge CB4 2AN
Lambert, Mr Michael J. 23 Gade Tower, Nash Mills, Hemel Hempstead, Hertfordshire HP3 8AE
Lang, Mr Robert G. Department of History, University of Oregon, Eugene, Oregon 97403, U.S.A.
Larkin, Mrs Carol L. 22 The Cedars, Dunstable, Bedfordshire LU6 3JB
Larkin, Mr D. J. W. 21 Afghan Road, London SW11 2QD
Launert, Dr Edmund British Museum (N H), Department of Botany, Cromwell Road, London SW7 5BD
Laurenson, Mr Ian W. Department of English, Monash University, Clayton, Victoria 3168, Australia
Le Maistre, Mr Ian G. 5 Richmond Mansions, Lower Richmond Road, London SW15
Leal, Mr David J. 28 Caterham Road, London SE13 5AR
Leaver, Miss Edith P., F.R.C.S.(Edin), F.R.C.O.G. 147 Wistaston Green Road, Crewe, Cheshire CW2 8RA
Lee, Mrs Antonia 13 Twyford Avenue, London W3 9PY
Lee, Mr C. L. 107 Queens Road, Hertford SG13 8BJ
Lee, Mrs Carol M. 51 Lawford Road, London N1 5BJ
Lee, Miss Olivia D. 7 Hawkridge Close, Chadwell Heath, Romford, Essex RM6 4NR
Lee, Mr Stephen, B.Sc. 94 Cranbourne Avenue, Wanstead, London E11 2BQ
Leese, Mr Peter F. 52 Harvey Road, Croxley Green, Rickmansworth, Hertfordshire WD3 3BT
Leith, Mr Ian 20A Southolm Street, London SW11
Leon, Mrs Aurora 21 Acol Court, Acol Road, London NW6 3AE

Leon, Mr Robert   21 Acol Court, Acol Road, London NW6 3AE
Levey, Miss Santina M., B.A. (Hons), F.M.A.   Flat 3, 45 Newman Street, London W1P 3PA
Levinson, Mr Harry A.   P.O. Box 534, Beverly Hills, California 90213, U.S.A.
Levy, Mrs Eleanor, B.A.   Dickens Lodge, Chigwell Village, High Road, Chigwell, Essex IG7 6QB
Levy, Mr Peter L.   52 Springfield Road, London NW8 0QN
Lewin, Mrs Rosemary   48 Methley Street, London SE11 4AJ
Lewis, Dr A. P.   Tremelling Cottage, The Green Lane, St Erth, Hayle, Cornwall TR27 6HS
Lewis, Mrs Lesley, F.S.A.   38 Whitelands House, Cheltenham Terrace, London SW3 4QY
Lindsay, Mr P. C., CEng M.I.C.E   1 Groveside Close, West Acton, London W3 0DX
Links, Mr J. G., O.B.E.   8 Elizabeth Close, Randolph Avenue, London W9 1BN
Lipton, Mr Stuart A.   10 Bruton Street, London W1X 7AG
Little, Mr R. C.   6 Westmorland Terrace, London SW1V 4AF
Lloyd, Mr Humphrey J., Q.C.   1 Atkin Building, Gray's Inn, London WC1R 5BQ
Lockhart, Mr Donald A., M.A., LL.B.   43 West Hill Road, London SW18 1LL
Logan, Mr Roger I., B.A. (Hons)   18 Belmont Avenue, New Malden, Surrey KT3 6QD
Long, Mr Nicholas J.   58 Crescent Lane, London SW4 9PU
Lord, Mr F. C. W.   114 Hayes Lane, Bromley, Kent BR2 9EP
Lovegrove, Mr J. W.   34 Ashbarn Crescent, Badger Farm, Winchester, Hampshire SO22 4LW
Lucas, Mrs Renee J.   3 Sussex House, Glenilla Road, London NW3 4AR
Lusch, Mr Robert   54 Shorts Gardens, London WC2
Lyons, Mr James R.   30 Castelnau Gardens, Barnes, London SW13 9DU
MacDonald, Mr Hunter   17 Sulgrave Road, London W6 7RD
Macartney, Mrs Sylvia   133 South Park Crescent, Catford, London SE6 1JL
Mackillop, Mr Thomas C., LL.B. (Hons)   53 The Cut, Waterloo, London SE1 8LF
Maddocks, Dr Anne C., B.M. MRCPath   St Mary's Hospital Medical School, Department of Bacteriology, Wright-Fleming Institute, London W2 1PG
Maguire, Mr L. J., M.B.E., T.D.   54 Croham Valley Road, South Croydon, Surrey CR2 7NB
Major, Mr J. Kenneth, BArch, R.I.B.A., F.S.A.   2 Eldon Road, Reading, Berkshire RG1 4DH
Mann, Mr Arthur   5 Blandford Road, St Albans, Hertfordshire AL1 4JP

LIST OF MEMBERS

Mann, Miss Sylvia   18 Upper Montagu Street, London W1
Marcus, Mr David I.   32 Woodland Gardens, London N10 3UA
Margary, Mr Harry H.   Lympne Castle, Kent
Marks, Mr Stephen N. P., M.A., F.S.A., R.I.B.A., *Vice-President*   Hamilton's, Kilmersdon, near Bath, Somerset
Marle, Mr William J.   White Gates, Wilderness Road, Chislehurst, Kent BR7 5EY
Marshall, Mr W. W., B.A., F.R.G.S., F.G.S.   9 Highshore Road, London SE15 5AA
Mason, Mr A. D.   14 St Martin's Road, London SW9 0SW
Mason, Mr Sydney, F.S.V.A.   100 Park Lane, London W1Y 4AR
Massil, Mr Stephen W.   138 Middle Lane, Crouch End, London N8 7JP
Masters, Miss Betty R., O.B.E., B.A., F.S.A.   9 Robert Avenue, St Albans, Hertfordshire AL1 2QQ
Matthew, Mr Francis, B.A.   Top Flat, 425 Mile End Road, London E3 1PB
Matthews, Mr Marten D.   Flat 4, 59 Mount Ararat Road, Richmond, Surrey TW10 6PL
May, Lt Cdr Lester E., R.N.   24 Reachview Court, London NW1 0TY
McBean, Mr James, A.R.I.C.S.   82 North View, Eastcote, Pinner, Middlesex HA5 1PF
McElligott, Mrs Sheila T., B.A. (Hons)   19 Bracey Street, London N4 3BJ
McKitterick, Mr David J., M.A.   Trinity College, Cambridge CB2 1TQ
McNeill, Mr Peter M.   55 Victoria Parade, Fitzroy, Victoria 3065 Australia
McRory, Mrs Margaret   18009 Highfield Road, Ashton, Maryland 20861, U.S.A.
Medd, Mr Nicholas W.   8 Albury Ride, Cheshunt, Hertfordshire EN8 8XF
Mee, Mr Michael F., A.B., C.F.I.   128 South Nanticoke Avenue, Endicott, New York NY 13760, U.S.A.
Melluish, Mr Derek W.   55 Church Walk, Worthing, West Sussex BN11 2LT
Merrifield, Mr Ralph, Ph.D., B.A., F.S.A., F.M.A., *President*   32 Poplar Walk, Herne Hill, London SE24 0BU
Metcalf, Mr Tertius J. T.   Town Wharf, West Street, Gravesend, Kent
Mickleburgh, Mr Simon P., BSc., MSc.   79–83 North Street, Brighton, East Sussex BN1 1ZA
Middleton, Mr John C.   West Poundgate Manor, Chillies Lane, Crowborough, East Sussex TN6 3TB
Middleton, Mr John R.   2 South Walk, Reigate, Surrey
Miller, Mr David J., M.A.   37 Granville Square, London, WC1X 9PD
Millett, Mr Bernard J.   Green Gables, Cliff Road, Teignmouth, South Devon TQ14 8TW

LIST OF MEMBERS

Millett, Mr Timothy C., B.A.   49 Athledene Road, London SW18 3BN
Mills, Mrs Carole A.   High View, 44B Woodbury Park Road, Tunbridge Wells, Kent TN4 9NG
Milne, Mr A. T.   9 Frank Dixon Close, Dulwich, London SA21 7BD
Moger, Mrs Victoria   76 Manville Road, London SW17 8JL
Moore, Mr J. R.   27 Halliford Road, Sunbury-on-Thames, Middlesex TW16 6DP
Moore, Mr Norman H.   54 Lord Avenue, Clayhall, Ilford, Essex IG5 0HN
Mootham, Sir Orby   25 Claremont Road, Teddington, Middlesex TW11 8DH
Morley, Mr Brian G.   160A Ivydale Road, Nunhead, London SE15 3BT
Morris, Mr Michael R., F.R.I.C.S.   1105 Basil Road, McLean, Virginia VA22101, U.S.A.
Morris, Mr Simon J., M.A., *Publications Secretary*   22 Brooksby Street, Islington, London N1
Moss, Mr Anthony D.   The Bury Farm, Chesham, Buckinghamshire HP5 2JU
Moss, Mrs Suzanne   Southlands Gallery, Pine Cottage, Stoke Wood, Stoke Poges, Buckinghamshire SL2 4AU
Moulder, Mr M. D.   49 Erlstoke Road, Crownhill, Plymouth, Devon PL6 5QN
Mountain, Mrs L. D.   24 Arundel Close, Chivalry Road, London SW11 1HR
Mullinger, Mr J. P. A.   13 Belmont Drive, Maidenhead, Berkshire SL6 6JZ
Mungeam, Miss M. J.   Flat D, 17 St Charles Square, London W10 6EF
Murphy, Mr A. S.   Granary Cottage, Upton, Huntingdon, Cambridgeshire PE17 5YB
Murphy, Mr G. M.   9 Glasnevin Gardens, Tenby, South Pembrokeshire, Dyfed
Musikant, Miss Nicola R., B.A. (Hons)   8 Grange Avenue, Totteridge, London N20 8AD
Myers, Miss Robin   177 Liverpool Road, Islington, London N1 0RF
Nash, Dr Beryl A.   14 O'Donnell Court, Brunswick Centre, London WC1N 1AQ
Neal, Mr John   9 Penshurst Road, Potters Bar, Hertfordshire EN6 5JP
Newman, Dr A. C. C.   51 Basildon Court, 28 Devonshire Street, London W1M 1RH
Newman, Mr John A.   37 Lambardes, New Ash Green, Dartford, Kent DA3 8HX
Newton, Dr John R.   25 Westfield Road, Edgbaston, Birmingham B15 3QF
Nicholson, Mr Robin A.   7 Highbury Place, London N5 1QZ
Nicholson, Miss Sarah L., M.A.   35 Cross Street, London N1 2BA
Nicolson, Mr J. M.   20 Albert Square, London SW8 1BS

LIST OF MEMBERS

Nightingale, Mr Melvin   160 Princes Avenue, London W3 8LU
Noel-Hume, Mrs Audrey, B.A.   PO Box 1711, Williamsburg, Virginia 23187, U.S.A.
Northcott, Mr Barry J., F.C.A.   Servi-Center, Terraza No. 4, 29680 Esteponia, Malaga, Spain
O'Brien, Mr C. I. M.   1B Eastway, Epsom, Surrey KT19 8SE
O'Donnell, Mrs Sally A. L., B.A. (Hons)   53 Iveley Road, Clapham, London SW4 0EN
O'Hara, Mr Michael M., M.A.   20 Mountside, Guildford, Surrey GU2 5JE
O'Neill, Mr J. J.   87 Long Gore, Farncombe, Godalming, Surrey GU7 3TW
Oggins, Mr Robin S.   Department of History, State University of New York, Binghamton, New York NY 13901, U.S.A.
Osborne, Dr Irving   169 Goodman Park, Slough, Middlesex SL2 5NR
Page, Mr James E., F.C.A.   Burgh Beck Meadows, 1 Meadow Way, Briston, Melton Constable, Norfolk NR24 2DJ
Palliser, Mr Charles   Department of English Studies, University of Strathclyde, Livingstone Tower, 26 Richmond Street, Glasgow G1 1XH
Palmer, Mr Geoffrey F.   45A Bedford Court Mansions, Bedford Avenue, London WC1B 3AA
Palmer, Sir John   Hensleigh, Tiverton, Devon
Parker, Miss M. L.   Flat 4, Westward House, Westcliff Parade, Westcliff on Sea, Essex SS0 7QE
Parker, William, T.   30 Albany Park Avenue, Enfield, Middlesex EN3 5NT
Parnell, Mr Geoffrey   HBMCE, Room 410, Fortress House, 23 Savile Row, London W1X 2HE
Pascoe, Mr Graham   Hochstrasse 2, D-8015 Ottenhofen, West Germany
Paterson, Mr A. C.   132 Torriano Avenue, London NW5 2RY
Peach, Mr J. C.   28 Onslow Gardens, Grange Park, London N21 1DX
Peacock, Mr Jonathan H. B.   278 Chester Road, Little Sutton, South Wirral, Merseyside L66 1NL
Pearsall, Mr A. W. H.   71 Parkside, Vanbrugh Park, London SE3 7QF
Peck, Mr John C.   6 Nash House, Park Village East, London NW1 7PY
Peel, Mrs C. M.   Belmore House, Queens Road, Cowes, Isle of Wight PO31 8BQ
Pegg, Mr Denys C. S.   23 Wentworth Avenue, Finchley, London N3 1YA
Pemberton, Mr Jeremy   7 Paultons Square, London SW3 5AP
Pembroke, Mr S. G.   3 Garrick Street, London, WC2E 9AR
Pepper, Dr Simon   21 Warmington Road, London SE24 9LA
Perry, Mr J. G.   183 Prospect Hill Road, Canterbury 3126, Victoria, Australia
Peterson, Mrs Marilyn L. R.   14 Hammond Street, Chestnut Hill, Massachusetts 02167, U.S.A.

LIST OF MEMBERS 343

Peverley, Mr John   21 Western Road, Branksome Park, Poole, Dorset BH13 7BQ
Phillips, Mrs Elizabeth M. R., M.A.   12 The Mall, London SW14 7EN
Phillips, Mr J. F. C., F.S.A., *Member of Council*   92 Rossiter Road, Balham, London SW12 9RX
Philpot, Mr Nicholas A. J.   5 Paper Buildings, Temple, London EC4
Pick, Mr Christopher C., B.A., M.A.   41 Chestnut Road, London SE27 9EZ
Pillinger, Raymond N., B.A. (Hons), F.R.E.S.   58 Gadby Road, Sittingbourne, Kent, ME10 1TJ
Pinching, Mr Albert A.   11 Braemar Avenue, Wood Green, London N22 4BY
Pinhorn, Mr Malcolm, B.A., F.S.G.   Eversleigh, Cranmore, Yarmouth PO41 0XW
Pither, Dr Charles E.   34 Bernard Gardens, Wimbledon, London SW19 7BE
Plant, Cllr John J.   101 Clova Road, London E7 9AG
Port, Professor M. H., M.A., B. Litt. (Oxon)   Department of History, Queen Mary College, Mile End Road, London E1 4NS
Porter, Mr N. G., B.Sc., C.Eng., M.I.C.E.   c/o Glanville & Associates, High Street, Sandridge, St Albans, Hertfordshire
Porter, Dr Roy S.   Wellcome Institute, 183 Euston Road, London NW1 2BP
Porter, Mr Stephen   9 Langham Place, Burlington Lane, London W4 2QL
Post, Ms Margaret   2 Worcester Gardens, Grandison Road, London SW11 6LR
Potter, Commander A. K., R.N.   120 Acacia Avenue, Rockcliffe Park, Ontario K1M 0R1, Canada
Poynter, Mr J. C.   19 The Pallant, Goring by Sea, Worthing, West Sussex BN12 6AZ
Pragnell, Mr H. J.   12 Meadow Road, Canterbury, Kent CN2 8EU
Prebble, Mr T. N.   56 Mahlon Avenue, South Ruislip, Middlesex HA4 6TE
Price, Mr Robin M., M.A., A.L.A.   Flat 2, 5–7 Princedale Road, London W11 4PH
Prockter, Mr Adrian C.   Senior Lecturer, South London College, Knights Hill, London SE27 0TX
Proctor, Mr J. A. J.   25 Rectory Close, Crayford, Dartford, Kent DA1 4RP
Puzey, Mr Anthony S.   9 Fairlawn Gardens, Southend-on-Sea, Essex SS2 6SB
Pye, Mr Ronald G. W., C.Eng, M.I.Mech.E.   12 Dutton Leys, Northleach, Cheltenham, Gloucestershire GL54 3EN
Pyer, Miss Pamela   55 Muncaster Road, London SW11 6NX
Quade, Miss A.   4 Heathway, Woodford Green, Essex

Quarrell, Mr S. E. Soil Consultants Limited, Inkerman Farm, Hazlemere, High Wycombe, Buckinghamshire HP15 7JH
Rains, Mr John H. 14905 Winterwind Drive, Tampa, Florida 33624, U.S.A.
Rates, Mr R. A. Acorns, 7 Mizen Way, Cobham, Surrey KT11 2RG
Rawcliffe, Mr John M. 9 Copley Dene, Bromley, Kent BR1 2PW
Rayment, Mr J. L. 57 Coopers Hill, Ongar, Essex CM5 9LF
Reed, Miss P. H., M.A. 17 Wilmington Square, London WC1X 0ER
Reeder, Dr David A., B.A., B.Sc. (Econ.), M.A., Ph.D. 17 Chantry Hurst, Epsom, Surrey KT18 7BW
Reenen, Ms Marguerite van 5 Queens Mansions, Queens Avenue, London N10 3PD
Reid, Mr David, F.C.A., F.I.Arb., *Honorary member* 40A Ludgate Hill, London EC4M 7DE
Relf, Sqn Ldr Brian R. F. 11 Hanbury Drive, Biggin Hill, Westerham, Kent TN16 3EN
Rendall, Mr John H. 36 Park Road, Twickenham, Middlesex TW1 2PX
Reynolds, Mr John S., B.A. First Floor Flat, 7 Leasowes Road, Leyton, London E10
Richardson, Mr Clifford H. 6 Kinnear Road, Shepherds Bush, London W12 9LE
Richardson, Mr John B. 168 Prescot Road, Aughton, Ormskirk, Lancashire L39 5AG
Richardson, Mr J. C. 32 Ellington Street, London N7
Rickards, Mr Maurice 12 Fitzroy Square, London W1P 5AH
Ridge, Mr T. S. 7 Shepton House, Welwyn Street, London E2 0JN
Ringham, Mr Peter A. 14 Tolsey Mead, Wrotham Road, Borough Green, Sevenoaks, Kent TN15 8EH
Rix, Mrs W. E. 20 New End, London NW3 1JA
Robbins, Mr R. M., C.B.E., M.A., F.S.A. 7 Courthope Villas, Worple Road, Wimbledon, London SW19 4EH
Roberts, Miss Agnes E., M.A. Oak Tree Cottage, Stream Lane, The Moor, Hawkhurst, Kent TN18 4RB
Robertson, Mr James C. Department of History, Washington University, St Louis, Missouri 63130, U.S.A.
Robertson, Miss K. L. 3 Elm Grove Road, Ealing, London W5 3JH
Robinson, Professor David Flat 6, 96–100 New Cavendish Street, London W1M 7FA
Robinson, Mr Leslie J. Camelot, 1 Jeymer Drive, Greenford, Middlesex UB6 8NS
Rose, Mr Martin T., M.A. (Oxon), M.Phil. 25A Willow Road, London NW3 1TL
Rosenfield, Dr Manuel C. Box 395, Mattapoisett, Massachusetts 02739, U.S.A.
Ross, Mrs c/o Shaun Usher, Daily Mail, Northcliffe House, Tudor Street, London EC4Y 0JA

LIST OF MEMBERS 345

Ross, Mr Alastair K.   48 Mount Pleasant Road, Ealing, London W5 1SQ
Rosser, Dr A. G.   School of History, University of Birmingham, Birmingham
Rowbotham, Mr Stanley J.   Flat 6, 27 Brick Lane, London E1 6PU
Rowston, Mr Guy, M.A. (Oxon), Dip.Ed.   6 Kenneth Court, 173 Kennington Road, London SE11 6SS
Roy, Dr Ian, M.A., D.Phil, F.R.Hist.S.   26 The Lane, Blackheath, London SE3 9SL
Roycroft, Mr F. B.   8 Hove Park Villas, Hove, East Sussex BN3 6HG
Ruge-Cope, Mrs Chris   21 Chace Avenue, Potters Bar, Hertfordshire
Russell-Cobb, Mr Trevor   25 Alderney Street, London SW1
Russell-Duff, Mrs J, J.P.   Watch Lane Farm, Moston, Sandbach, Cheshire CW11 9QS
Russett, Mr Alan W. F.   5 Hobury Street, London SW10
Ruston, Mr Alan R.   41 Hamperhill Lane, Oxhey, Watford, Hertfordshire WD1 4NS
Ryall, Mr Ronald B. C.   2 River Court, Taplow, Maidenhead, Berkshire SL6 0AU
Ryan, Miss Caroline, BA, *Member of Council*   26 Advance House, 109 Ladbroke Grove, London W11 1PG
Rye, Mr G. Michael   4 West Lulworth Farm, West Lulworth, Wareham, Dorset BH20 5SP
Samuel, Mr D. E.   57 Southwood Park, Southwood Lawn Road, London N6 5SQ
Sargent, Mr A. J.   33 Coborn Street, Bow, London E3 2AB
Saul, Mr Geoffrey M.   20 West Way, Rickmansworth, Hertfordshire
Saunders, Mrs Ann Loreille, Ph.D., F.S.A., *Honorary Editor*   3 Meadway Gate, London NW11 7LA
Saunders, Mrs Ann S., B.A.   20 Toley Avenue, Wembley, Middlesex HA9 9TD
Saville, Mr J. P.   85 Clifton Hill, London NW8 0JN
Sawyer, Mr A. G.   4 Lebanon Gardens, London SW18 1RG
Schofield, Mr John   Department of Urban Archaeology, Museum of London, London Wall, London EC2Y 5HN
Schurman, Mr Donald M., M.A., Ph.D. (Cantab.)   PO Box 325, Sydenham, Ontario K0H 2T0, Canada
Schwitzer, Dr Joan, Ph.D.   33 Shepherds Hill, Highgate, London N6 5QJ
Scott, Mr Peter F.   35 Rita Road, Vauxhall, London SW8
Scouloudi, Miss Irene, M.Sc., F.S.A., F.R.Hist.S., *Vice-President*   82, 3 Whitehall Court, London SW1R 2EL
Seabury, Mr Roy P.   6 Mimosa Road, Hayes, Middlesex
Searle, Mr Norman P. W.   166 Defoe House, Barbican, London EC2Y 8DN
Selwyn, Mr B.   2nd Floor, 3 Hogarth Road, London SW5 0QH
Sewell, Mr James R., M.A., F.S.A.   120 Addiscombe Road, Croydon, Surrey CR0 5PQ

Shane, Mr D. Grahame 41 Redington Road, London NW3 2UA
Sharp, Mr Dennis C., M.A., A.A.Dip, R.I.B.A. Dennis Sharp Architects, 4 All Saints Street, London N1 9RL
Shaw, Mr Adrian Long Meadow, Vicarage Road, Abbotskerswell, Newton Abbot, Devon TQ12 5PN
Shearer, Mrs B. R. Flat 6, Flaxman House, 1/3 Coleherne Road, London SW10
Sheppard, Dr Francis H. W., M.A., Ph.D. 10 Albion Place, West Street, Henley-on-Thames, Oxfordshire RG9 2DT
Sheridan, Miss Francis W. 58 Norwich Street, Cambridge CB2 1NE
Sherlock, Mr Peter D. 105 Wilberforce Road, Finsbury Park, London N4 2SP
Sherwood, Mrs Doris 16 Maurice Court, Brentford Dock, Middlesex TW8 8QY
Shields, Mrs G. O. 17 Windermere Gardens, Ilford, Essex IG4 5BZ
Shilkret, Professor Robert Department of Psychology and Education, Mount Holyoake College, South Hadley, Massachusetts 01075, U.S.A.
Shooter, Mr Horace V. 68A Marchmont Street, London WC1N 1AB
Sienkiewicz, Mr John, B.Sc., Dip.T.P., M.R.T.P.I. 28 Newbury Gardens, Stoneleigh, Ewell, Surrey KT19 0NU
Silvester-Carr, Miss Denise 1 The Firs, 162 Longlands Road, Sidcup, Kent DA15 7LG
Simms, Mr Dudley J. 10 Woodland Crescent, Bracknell, Berkshire RG12 2LH
Simpson, Ms Karen J. C., B.A. 4A Wembley Park Drive, Wembley, Middlesex HA9 8HA
Skilton, Mr Alan J., A.R.I.B.A. 22A Hoodcote Gardens, Winchmore Hill, London N21 2NE
Slorah, Miss J. E. 84 Shirley Gardens, Ilford, Essex IG11 9XA
Smith, Mr Brian D. 7, 59 Belgrave Road, London SW1V 2BE
Smith, Mr Christopher J. 164 Camberwell Grove, London SE5 8RH
Smith, Mr Eric E. F., F.S.A. 20 Southwood Road, Tankerton, Whitstable, Kent CT5 2PN
Smith, Mr Ronald J., M.B.E., C.Eng.M.I.E.E. 110 Nelson Road, Whitton, Twickenham, Middlesex TW2 7AY
Solman, Mr David I, B.Sc. 68 Greenaway Close, Friern Barnet, London N11 3NT
Somerville, Sir Robert, K.C.V.O., F.S.A. 3 Hunt's Close, Morden Road, London SE3 0AH
Sorrell, Mr Graham H. N. 51 Medway Road, Bow, London E3 5BX
Spence, Mr Craig, B.Sc. Museum of London, London Wall, London EC2Y 5HN
Stacey, Miss Anne 20b Grove Hill, South Woodford, London E18 2JG
Stacey, Miss M. 34 Ambleside Close, Thingwall, Wirral L61 3XQ
Stamp, Mr G. M. 1 St Chad's Street, Kings Cross, London WC1H 8BD

Standing, Mrs Juliet, B.A. 20 Glebe Drive, Brackley, Northamptonshire NN13 5BX
Standing, Mr Roderick T., M.A. Flat 6, 1F Oval Road, London NW1 7EA
Startin, Miss Eileen M. 2 Stanworth Court, Church Road, Heston, Middlesex TW5 0LB
Stein, Miss G. 75 Withy House, Globe Road, London E1 4AL
Stell, Mr C. F., F.S.A. Frognal, 25 Berks Hill, Chorleywood, Hertfordshire WD3 5AG
Stephens, Mr T. R. 1 St Georges Almshouses, Glasshill Street, Southwark, London SE1 0SH
Steppler, Dr Glenn A., B.A., M.A., D.Phil. (Oxon) 88 Sevington Road, Hendon, London NW4 3RS
Stevens, Mr G. A. 84 King's Road, Berkhamsted, Hertfordshire
Stevens, Mr Michael J. 37 Temple Sheen Road, London SW14 7QF
Stevenson, Mr Drew 19 Stratford Villas, London NW1 5SE
Stevenson, Dr J. Westwood, Oak Road, Mottram St Andrew, Macclesfield, Cheshire SK10 4RA
Stewart, Mr Alan B. 256 Stockingstone Road, Luton, Bedfordshire LU2 7DF
Stokes, Mr Malcolm A. 12 Southwood Park, Southwood Lawn Road, Highgate, London N6 5SG
Stokes, Miss M. Veronica 26 Bective Road, Putney, London SW15 2QA
Stone, Mr Michael J., B.A., A.I.F.A. 4 Howard Drive, Letchworth, Hertfordshire SG6 2BW
Stonehouse, Mr Geoffrey F. 7 Parkside Close, East Horsley, Surrey KT24 5BY
Stonier, Dr Peter D. 5 Branstone Road, Kew, Richmond, Surrey TW9 3LB
Summerson, Sir John, C.B.E., F.B.A., A.R.I.B.A., F.S.A. 1 Eton Villas, London NW3 4SK
Swift, Dr A. Katherine The Dower House, Morville Hall, near Bridgnorth, Shropshire WV16 5BN
Sykes, Mr K. G. 36A Cleveland Square, London W2 6DD
Sylvester, Mr Brian D. 3 Russell Court, St Lukes Road, Maidenhead, Berkshire SL6 7AP
Symons, Mrs K. N. Wool House, 2 Kingsbury Square, Wilton, Wiltshire SP2 0BA
Tanner, Mr G. C. A. 107 Camberwell Grove, London SE5 8JH
Taylor, Mr Alan J. Top Flat, 76 Jacksons Lane, Highgate, London N6 5SR
Taylor, Miss Jacqueline M., B.A. 13E Philbeach Gardens, London SW5 9DY
Taylor, Mr John P. 4 Bedale Street, London SE1 9AL
Taylor, Mr Roderick N. 121 Portland Road, Holland Park, London W11

Taylor, Mr Robert T.   99 Usk Road, Aveley, Essex RM15 4NX
Thomas, Mrs Jeanette G., Ph.D.   105 Grasmere Road, Bala-Cynwyd PA-19004, U.S.A.
Thomas, Mr William A. L.   Park House, 23 St Mary's Road, London W5 5RA
Thompson, Ms Anne E., M.A.   Flat B, 4 Howitt Road, Hampstead, London NW3 4LL
Thompson, Mr G. Alan   41 Honley Road, Catford, London SE6 2HY
Thompson, Mrs H. U. M.   34 Well Walk, London NW3 1BX
Thomson, Mr John M.   Box 28–063, Kelburn, Wellington, New Zealand
Thurley, Dr Simon, M.A., Ph.D.   18 Hartington Court, Lansdowne Way, London SW8
Tindall, Ms Gillian   27 Leighton Road, London NW5 2QG
Tite, Mr Graham L., M.A., M.Sc.   Farm Cottage, Sutton, West Sussex RH20 1PN
Towey, Mr P. J.   11 Church Lane, Teddington, Middlesex TW11 8PA
Towner, Mrs Margaret S.   2 Campion Road, Putney, London SW15 6NW
Treadwell, Professor J. M.   Department of English, Trent University, Peterborough, Ontario K9J 7B8, Canada
Tregear, Mr George H. B., M.A. (Cantab.), C.P.A., M.I.T.M.A.   5 Stanley Gardens, London W11
Tribe, Mr Allan, *Honorary member*   1 Maldon Close, Westcote Road, Reading, Berkshire RG3 2DH
Tritton, Mr John A., B.Sc.   4 Boscombe Road, Shepherds Bush, London W12 9HP
Troon, Mr Albert M.   6 Clyro Court, 93 Tollington Park, London N4 3AQ
Trueblood, Mr S. P.   108A Earlham Grove, Forest Gate, London E7 9AS
Tsukada, Naoki   776–17 Komejima, Showamachi, Saitama 344–01, Japan
Tsushima, Mrs Jean   Malmaison, Church Street, Great Bedwyn, Wiltshire SN8 3PE
Tufte, Dr Edward, B.S., M.S., Ph.D.   1161 Sperry Road, Cheshire, Connecticut 06410, U.S.A.
Tuley, Mrs Inge   Wellington House, 63 Guildford Street, Chertsey, Surrey KT16 9AU
Tull, Mr L. C.   6 Starling Close, Buckhurst Hill, Essex IG9 5TN
Turner, Mr J. Horsfall   62 Hyde Vale, Greenwich, London SE10 8HP
Turner, Mrs Philippa D. H., M.A.   40 Stanlake Road, London W12
Turner, Mr R. W.   123 Goldhurst Terrace, London NW6 3EX
Tye, Mr Raymond G.   331 Wimbledon Park Road, London SW19 6NS
Tyrrell, Miss D. M. N.   47 Fairmead Road, London N19
Van Der Merwe, Dr P. T.   c/o National Maritime Museum, Greenwich, London SE10 9HF

Van Sickle, Mr William H. H., B.A., M.Sc. (Pl.) 42 Penrith Road, Basingstoke, Hampshire RG21 1XW
Vance, Miss Peggy-Louise, B.A.Hons (Cantab.) Bestseller Publications Limited, Princess House, 50 Eastcastle Street, London W1N 7AP
Vaughan, Mr Dennis A. 292 Sylvan Road, Upper Norwood, London SE19 2SB
Veale, Miss Elspeth M., Ph.D., *Member of Council* 31 St Mary's Road, Wimbledon, London SW19 7BP
Verdin, Mrs Audrey J., B.Ed 17 Foliat Drive, Wantage, Oxfordshire OX12 7AN
Vine, Mr A. D. de, T.D., B.Sc., C.Eng., M.I.C.E. 177 Cottenham Park Road, Wimbledon, London SW20 0SX
Vine, Mr David W. 19 Gordon Road, London E11 2RA
Waldock, Mrs Patricia, L.R.A.M., A.L.A.M. 206 Twickenham Road, Isleworth, Middlesex TW7 7DR
Walker, Dr Greg Department of English, University of Queensland, St Lucia, Brisbane 4067, Queensland, Australia
Wallace, Mr J. N. 445 West Barnes Lane, New Malden, Surrey
Walliker, Mr Andrew, B.H.(Hons), B.A.(Hons) 26 Avenue Rise, Bushey, Hertfordshire
Wallis, Dr Helen, *Member of Council* Map Library, British Library, Great Russell Street, London WC1B 3DG
Walter, Mr M. A. 29 Broad Lane, Bradmore, Wolverhampton WV3 9BN
Ward, Mr M. J. 80A Seven Sisters Road, London N7 6AE
Ward, Mr R. L. 6 High Kingsdown, Bristol BS2 8EN
Watkins, Mr P. D. 127 Camberwell Road, London SE5
Watson, Mr F. G. M. 51 Mandeville Road, Hertford, SG13 8JJ
Watson, Ms Isobel 29 Stepney Green, London E1 3JX
Watts, Mr Ronald 19 Addington Square, London SE5 7JZ
Watts, Mr R. A. 29 Trent Avenue, Ealing, London W5 4TL
Webb, Mr D. R., B.A., F.L.A., *Member of Council* 21 Meads Lane, Seven Kings, Ilford, Essex IG3 8QS
Weinreb, Mr Ben 16 Millfield Lane, London N6
Weinstein, Mrs Rosemary, BA, *Member of Council* The Museum of London, London Wall, London EC2Y 5HN
Welch, Mr James J. 22 The Park Pale, Tutbury, Burton on Trent, Staffordshire DE13 9LB
Wetters, Mr Basil D. P., C.P.A., E.P.A. 4 Wyvenhoe Drive, Quorn, Loughborough, Leicestershire LE12 8AP
Wheatley, Mr N. R., F.R.I.C.S., F.I.Arb. 177 Hampstead Way, London NW11 7YA
White, Ms Helen, M.A. 53 Ainger Road, London NW3 3AH
Whitehead, Mr A. P. Lauriston Cottage, 6 South Side, Wimbledon Common, London SW19 4TG
Whitehead, Mr J. W. 55 Parliament Hill, London NW3 2TB

Whitelegge, Mr David S. Westport Lodge, Cricket St Thomas, Chard, Somerset TA20 4BY
Whitham, Mr David   Middlesceugh Foot, Ivegill, Carlisle CA4 0WN
Whitting, Mrs Alexena   8 Atwood Road, London W6 0HX
Whytehead, Mr Robert L   27C Colvestone Crescent, London E8 2LG
Wicking, Mr Denis, 78 Thorpedale Road, London N4 3BW
Wiggins, Professor David R. P.   University College, Oxford OX1 4BH
Wilkinson, Mr J. D.   14 North Road Avenue, Hertford SG14 2BT
Williams, Mr John   Whitheath Farm, Corston, Malmesbury, Wiltshire
Williams, Mr J. R.   106 Highgate Road, London NW5 1PB
Willis, Mr A. M. D.   28 Chipstead Street, London SW6 3SF
Willson, Miss E. J., F.L.A.   56 Palewell Park, London SW14 8JH
Wilson, Mr Timothy H.   2 Malvern Road, London E8 3LT
Windas, Mr G. R. H.   58 Scarcroft Road, York YO2 1NF
Winkler, Dr K. T.   Seminar für Mittlere und Neuere Geschichte der Universität, Nikolausberger Weg 9c, 3400 Göttingen, West Germany
Winser, Miss Ann J.   25 Godstone Road, St Margarets, East Twickenham, Middlesex TW1 1JY
Wittich, Mr John C., B.A.   66 St Michael's Street, London W2 1QR
Wolf, Mr Eric W.   6300 Waterway Drive, Falls Church, Virginia 22044, U.S.A.
Wood, Mr Norman J.   20 Jennings Road, Oswestry, Shropshire SY11 1RU
Wooden, Mr Percy J., M.I.F.S.T.   117 Tollers Lane, Old Coulsdon, Surrey CR3 1BG
Woodhead, Mr Peter   65 Aldsworth Avenue, Goring-on-Sea, Worthing, West Sussex BN12 4XG
Woodruff, Mr Charles C.   26 Yeoman's Row, London SW3
Woods, Mr Peter E., A.R.I.C.S., A.R.V.A.   5 Market Place, Kingston upon Thames, Surrey KT1 1JX
Woolfenden, Mr Anthony P.   18 Beresford Road, Cheam, Surrey SM2 6EP
Woollacott, Mr R. J.   185 Gordon Road, Peckham, London SE15 3RT
Woolley, Mr George E., N.D.D.   44 Embercourt Road, Thames Ditton, Surrey KT7 0LQ
Woolley, Mr James D.   Department of English, Lafayette College, Easton PA 18042, U.S.A.
Worms, Mr Laurence J.   J. Ash (Rare Books), 25 Royal Exchange, London EC3
Worrall, Mr Edward S.   21 Merryhills Drive, Enfield, Middlesex EN2 7NS
Worrow, Mr Alan S.   Flat C, 34 Brookfield Road, London E9 5AH
Wright, Mr Martin G.   17 Clarendon Square, Leamington Spa, Warwickshire CV32 5QT
Yule, Mr Brian A.   56A Camden Square, London NW1 9XE

Zavis, Mr William   c/o Foreign Service Lounge, U.S. Information Agency, 301 Fourth Street SW, Washington, D.C. 20547, U.S.A.

Ziegler, Mr John A., Ph.D.   PO Box 1045, Conway, Arkansas 72032, U.S.A.

Zierler, Mr Gerald D. L.   14 Regency Lawn, Croftdown Road, London NW5 1HE

INSTITUTIONAL MEMBERS

Andrews Clark Memorial Library, William, University of California, 2520 Cimarron Street, Los Angeles, California CA 90018, U.S.A.
Andrews Kent & Stone, Messrs, Palladium House, 1/4 Argyll Street, London W1V 2DH
Architectural Association, (The Librarian), 36 Bedford Square, London WC1B 3ES
Ashmolean Library, Oxford OX1 2PH
Athenaeum, The, (The Librarian), Pall Mall, London SW1
Australia, National Library of, Processing Branch NSM 46/70, Canberra, ACT 2600, Australia
Bank of England, Reference Library, Threadneedle Street, London EC2R 8AH
Birmingham, University of, Periodicals Department, The Main Library, PO Box 363, Edgbaston, Birmingham B15 2TT
Birmingham Public Libraries, (The City Librarian), Serials Section, Central Libraries, Birmingham B3 3HQ
Bishopsgate Institute, Reference Library, 230 Bishopsgate, London EC2M 4QH
Boston Athenaeum, Library of the, Acquisitions Department, 10 Beacon Street, Boston, Mass. 02108, U.S.A.
British Library, The, Map Library, Great Russell Street, London WC1B 3DG
Bryn Mawr College Library, The, Bryn Mawr, Pennysylvania 19010, U.S.A.
California, University of, U.C.L.A. Map Library, Los Angeles, California 90024, U.S.A.
Camden, London Borough of, (Local History Librarian), Holborn Library, 32–38 Theobalds Road, London WC1X 8PA
Camden History Society, Honorary Secretary, Swiss Cottage Library, Avenue Road, London NW3 3HA
Canadian Centre for Architecture, Library, 1920 rue Baile, Montreal, Quebec H3H 2S6, Canada
Centre for Urban History, The, Leicester University, Leicester LE1 7RH
Chicago Library, University of, Serials Record Department, 1100 East 57th Street, Chicago, Illinois 60637, U.S.A.
Concordia University Libraries, Acquisitions Department, Sir George Williams Campus, PO Box 2650, Montreal, Quebec H3G 2P7, Canada
Congress, Library of, Continuations Unit, Order Division, Washington D.C. 20540, U.S.A.
Cornell University Libraries, Librarian, Serials Department, Ithaca, New York 14853, U.S.A.

LIST OF MEMBERS 353

Crown Estate Commissioners, The Librarian, Crown Estate Office, 13/15 Carlton House Terrace, London SW1Y 5AH
Croydon Central Library, Katharine Street, Croydon, Surrey CR9 1ET
Cutlers, Worshipful Company of, Cutlers' Hall, Warwick Lane, London EC4
Drapers' Company, (The Clerk), Drapers' Hall, Throgmorton Street, London EC2N 2DQ
Ellis & Moore Consulting Engineers, 9th Floor, Hill House, 17 Highgate Hill, London N19 5NA
Environment, Department of the, Room C 110, P.S.A. Library, Whitgift Centre, Wellesley Road, Croydon CR9 3LY
Folger Shakespeare Memorial Library, 210 East Capitol Street, Washington D.C. 20003, U.S.A.
Goldsmiths' Company, (The Clerk), Goldsmiths' Hall, Foster Lane, London EC2V 6BN
Greater London Records Office, Deputy Librarian, 40 Northampton Square, London EC1R 0HB
Guildhall Library, Aldermanbury, London EC2P 2EJ
Hackney, London Borough of, Hackney Archives Department, Rose Lipman Library, De Beauvoir Road, London N1 5SQ
Hammersmith & Fulham Central Library, Shepherds Bush Road, Hammersmith, London W6 7AT
Hampton Hill Gallery Ltd, 203 High Street, Hampton Hill, Middlesex
Harvard College Library, Serials Division, Cambridge, Mass. 02138, U.S.A.
Hatfield Polytechnic, School of Education, Wall Hall Campus, Aldenhamshire, Watford AL2 8AT
Hounslow, London Borough of, Local Studies Department, Chiswick Public Library, Dukes Avenue, Chiswick, London W4
House of Commons, House of Commons Library, (Reference Division), London SW1
Huntington Library, Henry E, San Marino 2, California 91108, U.S.A.
Illinois at Urbana, University of, Library, 1408 W. Gregory Drive, Urbana, Illinois 61801, U.S.A.
Indiana University Library, Serials Department, Bloomington, Indiana 47401, U.S.A.
Inner Temple Library, Temple, London EC4
Institute of Historical Research, (The Secretary), University of London, Senate House, Malet Street, London WC1E 7HU
Iowa Libraries, University of, Serials Department, Iowa City, Iowa 52242, U.S.A.
Islington, London Borough of, Central Library, 2 Fieldway Crescent, London N5 1PF
Kansas, University of, Department of Special Collections, Kenneth Spencer Research Library, Lawrence, Kansas 66045, U.S.A.
Kensington & Chelsea, Royal Borough of, Central Library, Phillimore Walk, London W8 7RX

King's College London, Department of Geography, Strand, London WC2R 2LS
Kingston Polytechnic Library, Knights Park, Kingston, Surrey KT1 2QJ
Kungliga Biblioteket, Acquisitions Department, Box 50 39 S-102 41, Stockholm 5, Sweden
Lambeth Archives Department, Minet Library, 52 Knatchbull Road, London SE5 9QY
Leiden University Library, Tijdschriften Afd (S339), Post Box 9501, 2300 RA Leiden, Holland
Leverton and Sons Ltd, 212 Eversholt Street, London NW1 1BD
Lewisham, London Borough of, Archives and Local History Department, The Manor House, Old Road, Lee, London SE13 5SY
Liverpool City Library, William Brown Street, Liverpool L3 8EW
London Library, The, 14 St James's Square, London SW1
London Society, The, Room G210, The City University, Northampton Square, London EC1V 0HB
London University Library, (Goldsmiths' Librarian), Senate House, Malet Street, London WC1E 7HU
Marc Fitch Fund, 2a Polstead Road, Oxford, OX2 6TN
McGill University Library, Acquisitions Department, 3459 McTavish Street, Montreal, Quebec H3A 1Y1, Canada
Mercers' Company, (The Clerk), Mercers' Hall, Ironmonger Lane, Cheapside, London EC2
Merton, London Borough of, Wimbledon Library, Wimbledon Hill Road, London SW19 7NB
Michigan State University Library, East Lansing, Michigan 48824, U.S.A.
Michigan University Library, Acquisitions Section, 2 South – Hatcher Library, Ann Arbor, Michigan 48109, U.S.A.
Middlesex Polytechnic, Room 514 McCrae, Queensway, Enfield, Middlesex EN3 4SF
Mills & Whipp, The Finsbury Business Centre, 40 Bowling Green Lane, London EC1R 0NE
Missouri, University of, Ellis Library, Serials Department, Columbia, Missouri 65201-5149, U.S.A.
Museum of London, The, (Librarian), London Wall, London EC2Y 5HN
New York Public Library, f.a.o. Serials, Grand Central Station, Post Office 4154, New York NY10163-4154, U.S.A.
New York at Binghamton, State University of, Library Serials Department, (S403-512), Vestal Parkway East, Binghamton, New York NY 13901, U.S.A.
Newberry Library, 60 West Walton Street, Chicago, Illinois 60610, U.S.A.
Otago Library, University of, PO Box 56, Dunedin, New Zealand
Peabody Department, George, Enoch Pratt Free Library, 17 East Mount Vernon Place, Baltimore, Maryland 21202, U.S.A.

Polytechnic of North London, Department of Teaching Studies, Prince of Wales Road, London NW5

Princeton University Library, Serials Division, Princeton, New Jersey 08544, U.S.A.

Public Record Office, (The Librarian), Ruskin Avenue, Kew, Surrey, TW9 4DU

Queen Mary College, (The Librarian), Mile End Road, London E1 4NS

Queensland, University of, Acquisitions Department, Central Library, St Lucia, Brisbane, Queensland 4067, Australia

RCHM England, Survey of London, Newlands House, 37-40 Berners Street, London W1P 4BP

Reform Club, (The Librarian), Pall Mall, London SW1Y 5EW

Royal Geographical Society, (The Librarian), Kensington Gore, London SW7 2AR

Royal Historical Society, (The Secretary), University College, Gower Street, London WC1E 6BT

Royal Holloway and Bedford New College, Geography Department, Egham Hill, Egham, Surrey TW20 0EX

Royal Institute of British Architects, (Librarian), Periodicals Library, 66 Portland Place, London W1N 4AD

Royal Institute of Chartered Surveyors, (Librarian), 12 Great George Street, Parliament Square, London SW1

Royal Library, Windsor Castle, Berkshire

Society of Antiquaries, (The Librarian), Burlington House, Piccadilly, London W1

Somers Town Urban Studies Centre, Basement, Crowndale Court, Crowndale Road, London NW1

South Australia, State Library of, Attn Acquisitions Section, Box 419 GPO, Adelaide, South Australia 5001, Australia

St Mark & St John, College of, (Librarian), Derriford Road, Plymouth, Devon, PL6 8BH

St Paul's School, (The Librarian), Lonsdale Road, Barnes, London SW13 9JT

Tower Hamlets, London Borough of, Central Library, Bancroft Road, London E1 4DQ

United Oxford and Cambridge, University Club, (Librarian), 71 Pall Mall, London, SW1Y 5HD

University College Library, Gower Street, London, WC1E 6BT

Victoria, State Library of, Serials Section, Technical Services Division, 328 Swanston Street, Melbourne, Victoria 3000, Australia

Victoria and Albert Museum, Keeper of the Library, South Kensington, London SW7 2RI

Waltham Forest, London Borough of, Local Studies Librarian, Vestry House Museum, Vestry Road, Walthamstow, London E17 9NH

Wandsworth, London Borough of, West Hill District Library, West Hill, Wandsworth, London SW18 1RZ

Washington, University of, Libraries, Serials Division, Seattle, Washington 98195, U.S.A.
Watermen and Lightermen, The Company of, (The Clerk), Watermen's Hall, 18 St Mary-at-Hill, London EC3
Western Australia, University of, Reid Library (Periodicals), Nedlands, Western Australia 6009, Australia.
Westfield College Library, Acquisitions Department, Kidderpore Avenue, Hampstead, London NW3 7ST
Westminster City Libraries, Acquisitions Section BSD, Marylebone Library, Marylebone Road, London NW1 5PS
Westminster City Libraries, Central Reference Library, St Martin's Street, London WC2
Wolton Biddell & Co Limited, 52 Borough High Street, Southwark, London SE1 1XN
Yale University Library, Wm Dawson & Sons Ltd, (Yale University – SB), Cannon House, Park Farm Road, Folkestone CT19 5EE
York University Libraries, Central Serial Records, 4700 Keele Street, North York, Ontario M3J 1P3, Canada

INDEXES

compiled by Stephen Powys Marks

a Mere, Thurston 65, 73
A to Z of Regency London, The (Laxton) 214, 215
Account of London by Thomas Pennant 144
Achelay, John 81
advertisements in guide books 144
Alard, Master 6
Alderton, John 22
Alford, Robert 34
Aliard, Robert 8
Allen, John 166
Allen, Thomas 102
Allerton Street, Hoxton 173
almoner of St Paul's 7, 8
almshouses: Finsbury 90; Poplar 168, 170; Southwark 18, 25, 44
altar wars 111
Amsterdam (Netherlands): Zuiderkerk, Westerkerk, Noorderkerk 120
Angel on the Hoop, Southwark 44
Anglo-Saxon London 275
Antelope, The, Barbican 95
Archer, John 127
archers 93
archery practice on Horselydown 22
Architectura Moderna (Salomon de Bray) 120
Ardern, Thomas de, son Thomas 47
Armourers Hall 89
Arnold, Thomas 73
Artillery Yard, Horselydown 22
Arundel House 136
Assumption of the Virgin, fraternity of, Southwark 41
Assumption of the Virgin Mary, guild of 62, 65, 66
Ave Maria Lane 1
Axe, The, Westminster 68

Bailey, Nathan 141
balloons 148
Balls Park 118
Banqueting House, Whitehall 118
Baptist, John 39

Barne, George 92
Barron, Caroline: 'Priscilla Metcalf Ph.D.' 308–311
Baseley, William 35
Battle Abbey 43
Battle Bridge 43
Baulms House 118
Bear Alley, Southwark 39
Bear Tavern, Southwark 39
Becket, Gilbert 9
Bedford, earl of 115
Beech Street, Finsbury 90
Beeston, Cuthbert 24
Bell, The, Westminster 62, 64, 67, 68, 72
Bennet, John 69, 72
Bermondsey Abbey 33
Bermondsey cross 43
Bermondsey House 23
Bermondsey Street 22, 25
Berners, Lord 71
Bew, John 144, 145; widow Jane 145
Billiter Lane, warehouses in 154
Birdwood, Sir George 157
Blackfriars' Playhouse 136
Blackman Street (Blakemannestrete) 17, 33; plan of land at 31–35
Blackwall 154, 170, 172, 264
Blanke, Sir Thomas 15
Blewemeade Alley, Southwark 46
Blue Maid, Southwark 46
bone-working industry, Southwark 47
Bonner, Edmund 86
Boroughside, court at 89
Bostock, – 47; Mrs 35, 47
Boston, William 66, 71
boundaries of manor/parish, Southwark 37, 41–44 *passim*, 46–50 *passim*
Bowle, William 92
bowling alleys, Westminster 66
Boyer, – 45
Brandon, Sir William, Sir Thomas, Charles duke of Suffolk 49
Brayley, E. W. 145
Brewers' Hall 276

brickfields/brickmaking: Finsbury Field 93; Moorfield 83; Southwark 35
Bridge House, Southwark 42
Brightman, Thomas 70
'Broadway Chapel, The, Westminster: a forgotten exemplar', by Peter Guillery 97–133
Broadway Chapel: architect 123–124; burial ground 97; character 114–115; cost 113; description 97–102; font 113; galleries 119; petition to build 110; plan 119–122; site the subject of lawsuit 110; stained-glass window 112; views and plans 97–102
Brook Street, Ratcliff 172
Browne, Adam 112, 113, 116, 118, 123, 124
Buckingham, marquis of 108, 109
building trades in Westminster 63
Bull's Head (bolles hede), Southwark 40
bullring, Southwark 47
Bulrynge Alley, Southwark 47
Burghley, Lord 15
Burgony, Southwark 43
burial ground, Westminster 97, 107
Burntisland, Fife, kirk at 119
Burton, Richard (Nathaniel Crouch) 139
Byston, Adam 22
Byzantine Coins (Whitting) 294
Byzantine history 294

Caddington, prebendary of 6
Cade, Jack 45
'Caesar's Camp', Heathrow 271
Candishe, Christopher 23
Cardew, Lt. Col. G. A. 134
Cardinal's Hat, Southwark 45
Carlin, Martha: 'Four Plans of Southwark in the time of Stow' 15–56
Carter, Edward 116, 124
Carter, John 178
Carter Lane 1, 5
Castle Ashby 116
Cato Street conspiracy 173
Caxton Street 97
'Centre for Metropolitan History', by Heather Creaton 268–270
Chandler, Rebeccah 162
Chapuys, – 66
Chare, William 48

Charing Cross 59
Charité-sur-Loire, La (France) 48
Charles Street, Westminster 162
Charter Act 1833 154
Charterhouse 276
Chiswick Place 40
Christ Church, Lambeth 137
Christ Church, Westminster 97
Christopher, The, Southwark 45
City of London Archaeological Trust 280
city wall 274, 276
Clement, The, Southwark 50
Clitherow, Sir Christopher 153
clockmaker, Westminster 63
Clothale, Robert de 5
clothworkers on Moorfield 81
cobbler's workshop in Westminster 63
Cobham family 40
cockpit, Westminster 66
Cois, Giles 161
Coleshunt, William de 7
Collyn, Alderman 89
Colsoni, François (Francesco Casparo) 140
Colyns, Robert 39
compulsory purchase 66, 162; compensation 67, 71
Cooper's Row, warehouses in 154
Cooper's Company 172
Copley, Sir Roger, wife Elizabeth 22
Copperplate Map 81
copyhold 168
Copyright Act, iniquities of 145
Cornhill, Henry de 5
Corsellis, – 169
Cotgrave, William 12
court Leet 87, 88–89
courts of Finsbury manor 87–88
couturiers in Westminster 63
Cranmer, Thomas 22
Craven, Christopher 23
Craven, earl of 160, 167
Craven House, Leadenhall Street 160
Creaton, Heather: 'Centre for Metropolitan History' 268–270
Cromwell, Thomas 63, 66, 71, 74
Crosby House 153
Cross Keys, Southwark 41
Crosse, John 41
Crouch, Nathaniel 139
Crown Key, Southwark 45

INDEXES

Crucifix Lane 24, 25
Crutched Friars, warehouses in 154
Cumberland Wharf 172
Currer, – 69
Curriers' Company 83
Cutler Street, warehouses in 164
Cutlers Gardens 164

Dalby, William 93
Dale, Alderman 89
Darell, George 108, 109
Davies, Alexander 136
de Keyser, Hendrick 120
de la Hay, William 73
de Latour, Alphonse de Serres 150
de Laune, Thomas 139
Deadman's Place, Southwark 50
Decremps, Henri 150
Denbigh, church at 119
Desmond, Ray 158
Dicers Lane 9
Directory, The, or list of the principal traders in London (Kent) 152
directory 141
Dirty Lane, Southwark 41
ditch, city/Moor ditch 79, 80, 84
Dixon, James 24
Dixon, William 68
Dobner, Alderman 86
'Docklands, The Museum in', by Chris Elmers 264–267
doghouse on Moorfield 81
Donyon 8
Dorset Gardens 137
Drew, John 144
Drury, John 2
Drury's Inn 2
Duck, Richard 62, 69
Duckett, Lionel 91
Duke's Theatre 137
Duke Street, Westminster 162

East India Company 107, 124, 153–176; Committee of Buying and Warehouses 157; Committee of Correspondence 155; committees 154; Court of Directors 155; elevation towards Leadenhall Street 163; inventories 166; maps and plans 158; minute books 155; property documents 157–8; title deeds 166–7; warehouses 154

East India House, The (Foster) 158
East India House 153, 154, 164, 165, 174
Economic and Social Research Council 268
Egan, Pierce 148
Eglesfield, Hugh 22, 24; son Christopher 22
Elizabeth Woodville, Queen 48
Elmers, Chris: 'The Museum in Docklands' 264–267
Endive Brewhouse 70
Endive Lane, Westminster 58, 60, 66, 67, 70; occupants 63
Eton, – 34
Eves, John, wife Edith 40
Excavation of Roman and Medieval London, The (Grimes) 278
Eyre, John 24

Faden, William 214; edition of Horwood's plan 236–259
Falconer, Thomas 80
Falstoff, Sir John 25
Fenchurch Street, house in 167; warehouses in 154, 156, 159, 171
Field Studies Council 280
Fielding, Sir John 145
Finsbury 78; almshouses 90; bailiff 91, 92; Farmer 91, 92; gardens 90, 91; inhabitants 90–91; rent gatherer 92; steward 91, 92; windmill 92, 93; *see* 'Moorfields, Finsbury and the City of London...'
Finsbury Court 89
Finsbury Field(s) 83, 92; brickmaking 93
Finsbury manor 85–93; lease of 85–87; medieval boundary 87; courts 87–88
Finsbury Square 85
fire insurance policies 166
Fishmongers' Company 45
Flege, John, son Henry 11
Folger Library, Washington DC (USA) 179
Folkmoot 10
font by Nicholas Stone 113
footwear modes 63
Fore Street 164, 166
Fores, Samuel 146
Foster, Sir William 158
Foule Lane, Southwark 41

360 INDEXES

Fowler, Christopher 39
fraternities/sisterhood: Assumption of the Virgin 41; St Anne 41; St Anthony 40; St Katherine 40; Virgin & St George 47
Freeman, Susannah 172
French Ordinary Court, warehouses in 154, 156
Froggett, Richard 41

Galloway, Thomas 113
Gardener, Robert 73
Gardeners Lane, Westminster 162, 164
Garlonde, John 68
Gaskyn, Guy 68
Gaynsboroughe, Robert 34
'Gentleman's Magazine, London Illustrations in The, 1746–1863' by Peter Jackson 177–213; see separate index, below
Gentleman's Magazine, maps 178
Gentleman's Magazine, Nichols File of the (Kuist) 179
Gentleman's Magazine Library: Topographical History of London (Gomme) 178
George I, statue of 135
George Inn, Southwark 45
Glean Alley, Southwark 43
Gleanings from Westminster Abbey (Scott) 193, 194
Glene, le, Southwark 43
Glene's rents, Southwark 25
Goat, The, Southwark 50
Godfrey, Walter 140
Golden Lane 90
Gomme, Lawrence 178
Goodyere, Henry 22, 48
Gouge, William 114
Grafton, John de, wife Mary 12
Great Fire 134, 138, 161, 162
Great Liberty manor, Southwark 33, 53
Great Suffolk Street, Southwark 33
Green Dragon, Southwark 40
Greenwich Park Hill 148
Gretehened, John 12
Griffiths, Lewis 64
'Grimes, William Francis, 1905–1988: Obituary', by Ralph Merrifield 271–282
Grimes, W. F.: 'Bibliography' by Ortrun Peyn 283–293; publication of excavations 278–280
Grub Street, Finsbury 90
Grymstead, – 86
'Guide books to London before 1800: a survey', by David Webb 138–152
Guide Books to London before 1900. A History and Bibliography (Webb) 151
guide books to London: Ambulator 144; Angliae Metropolis 139; Antiquities of London and Westminster 141; Brief Description 145; Companion to Every Place of Curiosity and Entertainment 143; Curiosités de Londres 148; Foreigner's Guide 141, 146, 148, 149; Fores' New Guide for Foreigners 146; Guide de Londres 140; Historical Account of the Curiosities 143; Historical Remarques and Observations 139; Kearslys' Stranger's Guide 146; Kurche Beschreibung 140; Leydsman der Vreemdelingen 151; London Adviser and Guide 146, 147; London and its Environs 145; London and Westminster Guide 145; London in Miniature 143; London Pocket Pilot 148; Londres et ses Environs 150; Manuel du Voyageur à Londres 150; New and Universal Guide 146; New Picture of London 145; New View and Observations 139; Observations sur Londre et ses Environs 148; Parisien à Londres 150; Picture of London 148; Pocket Remembrancer 143; Present State of London 139; Remarks on London 141, 142; Sehen-Wurdigkeiten 140; Stranger's Guide 145; Tableau de Londres 148
guides to sordid London life 148
Guildable manor, Southwark 31, 37, 53
guilds: Assumption of the Virgin Mary 62, 65, 66; Name of Jesus 42; Rounceval 65, 74
Gylberd, William 41

Haarlem (Netherlands): Nieuwe Kerk 120, 127
Hackney, manor of 168
Half Moon, Southwark 46
Hammersmith. A History of (Whitting) 294

Hammonds Quay 172
Hamond, Thomas 113, 118
Hampden, John 108
Hampkyn, Richard 67, 69
Hampstead, – 70
Hampstead 136
Hampton Court 58
Harlesden, prebendary of 7
Harris, John 143
Hastings, Lady Wynefride 160
Hawks, Richard 91
Hawksmoor, Nicholas 127
Haydon Square, warehouses in 154, 169
Hayward, Robert 73
Henbury, John 63, 65, 70
Henry, David 143
Henry of Blois, bishop of Winchester 40
Henry VIII, King 22, 45, 49, 57, 58, 66
Herbert, Philip, earl of Pembroke 109
Heritage, Thomas 73
Hermitage, Horselydown 24
Herte/le Herteshed, Southwark 45
Heylyn, Peter 111, 112, 124
Heyse, Henry 67, 69
Highgate 136
Hilliard family, John 90
Hillyard, John 83
Hind, A. M. 134
Hoet, – 169
Hogg Lane, Moorfield 81
'Holborn' gables 107, 118, 125
'Hollar's *Prospect of London and Westminster taken from Lambeth*, Some Notes on', by Peter Jackson 134–37
Holmes College 7
Hornebolte, Luke 73
Horse hede, Southwark 46
Horselydown: archery practice 22; drying and bleaching of laundry 23; pasture of animals 22; plans 17, 18–31; recreation 22; sand and gravel dug 22; used as mustering ground 22
Horselydown Lane 25
Horsemead/Horscimead/Horscidune/Horesdowne 20, 22
Horwood, Richard 216
Horwood, Thomas 216
'Horwood's Plan of London, Richard: a Guide to Editions and Variants, 1792–1819', by Paul Laxton 214–263
Horwood's Plan of London: first edition variants 216–235; main changes in Faden editions 236–259; titles and imprints 259–263
Hyde Abbey 45

ice on Moorfield 79
India Office 153, 154, 155, 157, 158, 162, 165
'India Office Library and Records, Sources for London History at the', by Margaret Makepeace 153–176
Indulgentiae 138
Ingham, Edward 70
Institute of Archaeology 273
Institute of Historical Research 268
Ireland, William 108, 109, 124
Isle of Ducks, Southwark 25
Ivy Lane 9

Jackson, Peter: 'Some Notes on Hollar's *Prospect of London and Westminster taken from Lambeth*' 134–37; 'London Illustrations in *The Gentleman's Magazine*, 1746–1863' 177–213
Jackson, William 33
Jacoobes Garden, Horselydown 24
Jamaica Road, Southwark 18, 24
James I, King 84
Jane Seymour, Queen 49
Jenour, Andrew 160
Jesus College, Cambridge 46
Jewry Street, warehouses in 154, 171
Joceline, Ralph 93
John Company, (Foster) 158
Jonck, Jan 47
Jones, – , owner of Southwark plan 26, 31
Jones, Inigo 102, 111, 115, 116, 118, 122, 123, 124, 125
Juxson, Mrs (Jackson) 33

Katherine de Valois, Queen 48
Keene, Derek 268
Keles, Henry de 11
Kellet, John 70
Kellett, William 50
Kent, Henry 141
Kent Street, Southwark 47
King's/Queen's manor, Southwark 33, 52

King's Bench Prison, Southwark 46
King Street, Westminster: bars 66; property owners 68–70; survey 67; inhabitants 64–5, 73; *see* 'Whitehall Palace and King Street...'
Kinge, John 84
Kings hede, Southwark 45
Kingsland Road, Shoreditch 173
Kitsyn, James 113
Knight, William 24
Knights of St John of Jerusalem 20, 24
Knocker White, The 265
Kuechelbecker, J. B. 140, 141
Kyrton, John and William 46

Lacombe, François 148, 150
Lamb Alley/the Lamb 62, 63, 65, 66, 67, 69, 72
Lambeth marsh 35
Langdon, Ralph de 6
Langton, John of 11
lascars 172–3
latrine on Moorfield 79, 80
Laud, archbishop 108, 110, 111, 112, 114, 124; liturgical innovations of 111–112; Laudian building, window 112; Laudian churches 114
laundry on Horselydown 23
Laxton, Paul: 'Richard Horwood's Plan of London: a Guide to Editions and Variants, 1792–1819' 214–263
laystalls on Moorfield 83, 84
Leadenhall Street 153; warehouses in 154
Leathersellers' Company 48
Leicester Square 135
Lentall, Phillip 60
Lerouge, G. L. 148
Lethieullier, – 169
Levy, Eleanor: 'Moorfields, Finsbury and the City of London in the Sixteenth Century' 78–96
Lilly, Philip 113
Lime Street 153
Lincoln's Inn, chapel 115
Lincoln College, Oxford, chapel 112
Lingfield College, Southwark 40
Literary Society 146
liturgical innovations by archbishop Laud 111–112
lock-up in Southwark 43
Londinopolis (Howell) 139
London before the Fire, by Hollar 134

London Described 135
London Docklands Development Corporation (LDDC) 264, 266
London guidebooks *see* 'Guide books to London...'
'London History, Sources for, at the India Office Library and Records', by Margaret Makepeace 153–176
'London Illustrations in *The Gentleman's Magazine* 1746–1863', by Peter Jackson 177–213
London Museum 273
London Topographical Society publications: Hollar's *Prospect of London and Westminster from Lambeth* 134–37; *Le Guide de Londres* (Colsoni) 140
London Wall 79, 85, 161, 162, 164, 165
Long Lane, Southwark 47
Long Reach 170
Lovel, Fulk 7
Lower Moorfields 85
Ludgate 2
Lumley, Lord 160
Lyde, Sir Lionel 168
Lyons Quay 172

M. V. 140, 141
Mackerell, Thomas 40
MacLeod, Roderick: 'The Topography of St Paul's Precinct, 1200–1500' 1–14
Magdalen College, Oxford 25
Makepeace, Margaret: 'Sources for London History at the India Office Library and Records' 153–176
Malcolm, J. P. 178
Man, William 108, 109, 124
Manfield, John 62
manors in Southwark 31–2, 52–3
maps: in *Gentleman's Magazine* 178; in guide books 139–140, 144, 146
Marbill, Hugh 67, 69
Marshall Yard, Horselydown 22
Marshalsea prison 46
Mary, Queen 49
materials for re-use 71
maternity ward for unwed mothers, Southwark 44
Maynley, Ralph 48
Mazzinghi, John (Giovanni) 146
medicinal herbs 91
Mermaid, The, Southwark 46

INDEXES

Merrifield, Ralph: 'William Francis Grimes, 1905–1988: Obituary' 271–282
Merton priory 50
'Metcalf, Priscilla, Ph.D.', by Caroline Barron 308–311
Mildmay, Sir Walter 93
Millet, Edward 73
Millyng, John 60
Mirabilia 138
Mithras/Mithraeum 274, 275, 277
Mondene 8
Mone (moon) brewhouse, Southwark 46
Monnox, George 42
Montagu, Henry, earl of Manchester 110, 111
Moor/More (Moorfields) 78
Moor lane, sluice at 84
'Moorfields, Finsbury and the City of London in the Sixteenth Century', by Eleanor Levy 78–96
Moorfield(s) 78–85; brickfields 83; clothworkers 81; cow house 83; division 80; doghouse 81; drying 81, 83; gardens 81; ice 79; latrine 79, 80; marshy 79; medieval boundary 87; pasture 81; tentors 81; toll charged 81
Moorgate 80, 83, 84
More, The, Herts 58
More, Thomas 9
Moultons Close, Southwark 35
Muscovy House 160
'Museum in Docklands, The', by Chris Elmers 264–267
Museum of London 264, 265, 266
'Museum on the Move' 266
mustering ground on Horsleydown 22
Mydleton, Ralph 34

Name of Jesus, guild of the 42
Nashe, Lawrence 91
naval stores at Ratcliff 172
Navy Office 161, 167
Needham, James 73
Nevill's Inn 276
New Chapel (Broadway Chapel) 97
New Remarks of London 141
New Street, warehouses in 154
Newell, Robert 111, 112
Newman, Thomas, wife Margaret 24

Nicholls, Sutton 135
Nichols File of the Gentleman's Magazine (Kuist) 179
Noble Street 276
Norden, John 15
Northumberland Alley 166
Norton, Mr (? John Norden) 15
Norwall, Thomas 80
Norwich Place, Charing Cross 49

obituaries: W. F. Grimes 271–282; P. D. Whitting 294–298; Priscilla Metcalf 301–311
Old Change 1
Old Kent Road, Southwark 47
Orseth, Roger de 2
osier crop 60
Ould Thomsons Field, Horsleydown 24
Overton, Henry 135

Page, John 12
Palace Yard, Westminster 116
Palle, Elizabeth 68
Palmer, Herbert 114
Palmer, Will 80
Paravicini, – 169
Pardon Churchawe/Churchyard 9, 10
Paris Garden, Southwark 35; manor 52
Parish Clerks, Company of 95, 141
Park Street, Southwark 41
Parthey, Gustav 134
pasture: on Horsleydown 22; on Moorfield 81
Paternoster Row 1, 9, 12
Paul's Chain 6
Payne, John 73
Pearce, Samuel 166
Pennant, Thomas 144
Pennythorne, Robert 70
pepper warehouse at Blackwall 172
Pepys, Samuel 161
Perseverance IV, The 265
Peter College 7
Peter Guillery: 'The Broadway Chapel, Westminster: a forgotten exemplar' 97–133
Peterborough, Lord 136
Peterborough House 136
Petty Burgen, Southwark 43
Petty Calais/Caleys 71
Petty France 114
Peyn, Ortrun: 'Bibliography' of W. F. Grimes 283–293

Phillipps, Thomas 162
Phillips, Richard 148
Pikot, Nicholas 78
pillory in Southwark 42, 43
plague 85
Pomfrett, John 62, 63, 69
Ponyngs family, Sir Edward 45
Ponyngs Inn, Southwark 45
Pope, Sir Thomas 48
Pope's Head, Southwark 45
Poplar 102, 109, 124; almshouses 168, 170; chapel 168; *Survey of London* 128
Poplar Chapel *see* St Matthias, Poplar
Poplar Fund 170
Port of London Authority 265, 266
posterns 80
Pote, John 141
Poughley, Bucks 71
pound (pynfold), Southwark 34, 49
Princess Alice, The 265
Pye, Sir Robert 108, 109, 111, 113, 114, 115, 123; son Robert 108; son Thomas 109

quit rent 88

Ram, The, Westminster 73
Ram, Thomas atte 79
Ram's Head, Southwark 42
Ratcliff 154, 172; fire at 172
Ratcliff High Street 172
Ratcliff Highway 173
Rawlins, Thomas 67, 69
Raynham Hall 118
recreation on Horselydown 22
Red Lion, Southwark 41
Red Lion, Westminster 63, 65, 70
Rede, John 70
regicides 151
Roach, John 148
Rochester, bishop of 40
Roman and Medieval London Excavation Council 273, 275, 279
Rome, guide to 138
Roper/Rooper Lane/Ropereslane 24, 48
Roper family, Richard 90
Rose, The, Westminster 67, 68, 70
Rosser, Gervase, and Simon Thurley: 'Whitehall Palace and King Street, Westminster' 57–77

Rounceval Guild 65, 74
Rounceval Hospital 65, 66, 72, 74
Row, Samuel 167
Royal East India Volunteers 173–174
Royal Group of Docks 265
Royal Rendezvous public house, Westminster 164
Royal Victoria Dock 264, 266
Russell, Lord, Lord Admiral 40
Russell, William 60, 67, 68
Russell family 64, 65; Richard 64, 65; widow Constance 65; John 68
Rybot, George 276

St Alban Wood Street 275
St Anne, chapel of, Southwark 42
St Anne, Limehouse 127
St Anne, sisterhood of, Southwark 41
St Anne and St Agnes 127
St Anthony, fraternity of, Southwark 40
St Augustine's Gate 4
St Barbara, chapel of, Southwark 42
St Bartholomew's Hospital 85
St Benet, Paul's Wharf 127
St Bride, Fleet Street 275
St Clement, chapel of, Southwark 42
St Faith, parish 10
St George-in-the-East 127
St George's Fields 33
St George the Martyr, Southwark 34, 47
St Gregory 7; cemetery 8; parish 10
St Helen Bishopsgate 116
St James, Garlickhythe 127
St James's Hospital 59
St James's Palace 118
St John's College, Oxford 116, 124
St John of Jerusalem, Knights of 20, 24
St John of Jerusalem's Mill (Saynt Johns Mylle) 22, 23
St John the Evangelist, Great Stanmore 114
St Katharine Cree 114
St Katherine, fraternity of, Southwark 40
St Katherine, hermitage of 72, 73
St Margaret, Southwark 40, 41; churchyard 42
St Margaret's, Westminster 74, 107, 110; font from Broadway Chapel 113; parish boundaries 74

INDEXES

St Martin, Ludgate 127
St Martin-in-the-Fields church 74
St Martin's, Thames Street 137
St Mary-at-Hill 127
St Mary-le-Bow 127, 276
St Mary Magdalen, Bermondsey 48, 125, 126, 127; chapel 39, 40, 41
St Mary Overy, priory 39, 41
St Mary Overy church 39; church door 39; gate to close 39
St Mary Staininglane 7
St Matthias, Poplar (Poplar Chapel) 102–107, 118, 119, 124, 125; copy of Broadway Chapel 107; views and plans 103–106
St Nicholas, Deptford 125, 127
St Olave, Southwark 18, 19, 22, 42
St Olave's Grammer School 19, 20, 22, 24, 31
St Paul, Covent Garden 115–119 passim, 122, 125
St Paul's, Hammersmith 114
St Paul's Cathedral 78, 85, 92, 115, 116, 125, 135, 136; committee for the repair 109; guide 138
St Paul's Cross 10
St Paul's Gate 4
'St Paul's Precinct, The Topography of', by Roderick MacLeod 1–14
St Paul's precinct: bell tower 11, 12; bishop's 'curia' 9; Chancellor's Inn 5; Chapter House 7, 9, 12; Charnel House 11; Cloister 7, 12; College of Petty Canons 9; Dean's House 6, 7; Green Yard 7; John of Gaunt, chantry of 5; Lancaster chantry 5, 6; library 9; New Palace 8; Old Palace (Old Paleys) 5, 6, 9; Precintor's Inn 2, 5, 12; 'Prestehous' 9, 10; 'Purlewe' 8; Treasurer's Inn 5, 6, 12; Tymbar Yard 7; Vicar's Close 8; 'Vicar's stillioury house' 8; walls strengthened 10, 11; Waltham chantry 8
St Paul's School 294, 296
St Saviour, Bermondsey, abbey/priory 48, 276
St Saviour's church, Southwark 40; the chayne gate 40
St Saviour's Dock 23
St Stephen's Court, Westminster 109
St Swithun's, Winchester 40

St Swithun London Stone 275
St Thomas's Hospital, Southwark 41, 44, 45; gate 44
St Thomas's Street, Southwark 44
St Trinity of Pomfrett, College of, Yorks 160
Salcote, William 69
Salisbury House 136
Salter, Nicholas 160
saltpetre warehouse at Ratcliffe 172
Salutation, The, Southwark 50
Sampson, – 48
sand and gravel on Horsleydown 22
Sasanian Coins, Introduction to (Whitting) 294
Savile House 135
Scatcherd, – 145
Schebbelie, Robert Blemmell 97
'Scotland', Westminster 58, 73
Scott, George Gilbert 193
Scraggys, John 48
Seething (Sething) Lane 160, 161, 167; warehouses in 154
Sherington, Walter 9
Skelton, John 57
sluice at Moor Lane 84
Smith, Mary 167
Smith, Robert 94
Smithe, Matthew, wife Edith 40
Smiths Alley, Southwark 43
Smyth, John 160
Smythe, Sir Thomas 153
Snethe, Elias 63, 65
Snow, Jeremiah 172
Social and Economic Study of Medieval London, The 268
Somers Quay 172
South Street, St Paul's 5
'Southwark, Four Plans of, in the time of Stow', by Martha Carlin 15–56
Southwark: courthouse 41; lock-up 43; market place 41; mills 43; pillory 42, 43; plan of central area 35–50
Southwark Cathedral 40
Southwell, Sir Robert 48
Sparowe, John 34
spice warehouse at Blackwall 172
Spode, Josiah 166, 168
Spur, The, Southwark 46
Stafford, Granville Leveson-Gower marquis of 216
Stanton, Charles 125, 127

Stationers' Hall 7
Stationers' Register 15
Stephens, John 69, 72
Stepney, manor of 168
Steynour's Croft, Westminster 60
Stone, Nicholas 109, 113, 120, 123
Stone Stairs, Ratcliff 172
Stonehenge 122
Stortford, Richard de 7
Stow, William 141
Strutton Ground 97
Suffolk, duke of see Brandon
Suffolk Place, Southwark 33, 49
Sugar Loaf Court 167
sugar warehouses in West India Dock 264
Sun Fire Office 166
Sutton, Mr 112
Swakeleys 118
Swallow, Thomas 73
Swan, The, Charing Cross 62
Swan/Swan with Two Necks, Southwark 44

Tabard Inn, Southwark 45
Tabard Street, Southwark 47
Tardy, Abbé 150
taxation, assessment for, in King Street, Westminster 64
Taylor, Charles 216
tennis courts, King Street, Westminster 66
tentors on Moorfield 81
The New State of England (Miège) 141
Thompson, R. H.: 'Publications' of P. D. Whitting 299–307
Thorne, Christopher: 'Philip David Whitting G. M., 1903–1988' 294–298
Thorney, Roger, wife Eleanor 46
Thornton, Richard 162
Thurley, Simon see Rosser, Gervase
Till, Stacey 167
tiltyard, King Street, Westminster 66
tithes paid to rector of St George's, Southwark 34
toll charged on Moorfield 81
Tooley Street, Southwark 26
Tothill (Tuttle) Fields 107, 108; Chapel 97
Tracy, H. 141
Trenet Lane, Southwark 44

Trenthall Hall, Staffs 216
Treswell, Ralph (senior, junior) 17, 18, 73
Tricks of the Town Laid Open 148
Trusler, Dr John 146
Tufton, Mr 108
Tull, Geoffrey 68
Tuscan order 107, 115–120 *passim*, 122, 127

Umfrey, Philip 64
University of London Archaeological Society 276
Upper Moorfields 85

van Campen, Jacob 120
Varlet, The 265
Verr, William 92
Victoria Street 97
Virgin, chapel of the, Southwark 42
Virgin and St George, fraternity of, Southwark 47
Vyncent, Faux 64

warehouses 154, 156, 164, 169, 171
Walbrook 79, 274
Waleys, Henry de 11
Walker, Richard 70
wall, city 274, 276
Wallis, John 144
Walsingham, Sir Francis 167
Watling Street 4
Waverley Place, Southwark 40
Waynflete, William 25
Webb, David: 'Guidebooks to London before 1800: a survey' 138–152
Webb, John 122, 123
Weldon, Edward 23
Welles, Thomas 92
West, Thomas 40, 41
West's Rents 41
West India and Millwall Docks 265
West India Dock sugar warehouses 264
Westminster: assessment for taxation 64; boundaries redrawn 74; bowling alleys 66; building trades 63; clockmaker 63; cobbler's workshop 63; cockpit 66; couturiers 63; increase in population 60; old Palace 57 59; status 62; tennis courts 66; *see* 'Whitehall Palace and King Street...'

INDEXES

Westminster Abbey 58, 59, 72, 107, 109, 110, 111; chantry 60; guide to 138; income 67; waxworks 148
Westminster Bridge 177
Westminster Broadway 97
Westminster, College of 107, 111, 112; committee for 108, 113
Westminster Hall 74, 136
Wheeler, Sir Mortimer 277, 278, 280
Whitaker, – 145
White Hart, Southwark 45
White Horse, Westminster 70, 73
White Horse Inn, Seething Lane 160
White Lion, Charing Cross 62
Whitecross Street, Moorfield 81, 90, 95
'Whitehall Palace and King Street, Westminster', by Gervase Rosser and Simon Thurley 57–77
'Whitting, Philip David, G. M., 1903–1988', by Christopher Thorne 294–298
Whitting, P. D.: 'Publications', by R. H. Thompson 299–307
Whittington, Richard 44

Wilkie, John 145
Williams, Audrey (Grimes) 276
Williams, Dean John 108, 111, 112, 114
Wilton, Wilts 109
Winchester, bishop of: manor 52; park/park gate 33, 49, 50
Winchester House, Southwark 40
windmills in Finsbury 92, 93
Wolff, Geoffrey 42
Wolsey, Cardinal 57, 58, 66
Wolstenholme, John 160
Wool Beam, Calais (France) 92
Wool Staple yard, Westminster 60
Woolwich 170
Worcester House 136
Wren, Sir Christopher 127
Wright, Thomas 168
Wyatt, Henry 68
Wyke, John 5

York, archbishops of 49, 58
York Place/House, Westminster 58, 59, 63, 65, 66, 74

INDEX TO THE CATALOGUE IN 'LONDON ILLUSTRATIONS IN *THE GENTLEMAN'S MAGAZINE*'

Note The names of subjects which are arranged alphabetically in the catalogue are not included in this index, which contains the names of artists and architects and some additional subjects. In the article there are two alphabetical sub-indexes for 'Churches and Chapels' (13 pages) and for 'Maps and Plans' (2 pages).

Allen, T. 186
Argyll, John duke of, monument 192
Audinet, – 189, 191

Basire, J. 180, 187, 189, 192, 196, 203, 212
Bayne, William, monument 192
Bear Garden, Bankside 179
Bedford, F. 184, 186
Bellasyse, John Lord, monument 184

Bigot, C. 202
Billings, B. W. 212
Billings, Robt. Wm. 202, 212
Blackburn, E. L. 195
Blair, William, monument 192
Blore, Edwd. 184
Blore, J. 184, 190
Bolton, Robert 179
Bonnor, T. 208
Briggs, H. P. 184

INDEXES

Brooke, W. H. 203
Buckingham, George Villiers duke of, monument 192
Buckingham, John Sheffield duke of, monument 192
Buckler, G. 190
Buckler, J. C. 187
Buckler, J. Jnr. 180
Burnet, bishop, monument 185

Carlos, E. J. 184, 202
Carter, Benjamin 192
Carter, John 183, 192, 195, 196, 203, 209, 212
Cary, F. 184, 190, 200, 210
Charles I, bust of 184
Chelsea Bun House 181
Chelsea Hospital 181
Cholmeley, Sir Roger, arms 185
Cleghorn, – 189
Cole, B. 203, 204
Colet, Sir Henry, tomb 183, 190
Conde, – 184
Cook, – 188, 196, 199
Cooke, Edward, memorial 182
Cottingham, L. N. 191
Courtenay, W. 195
Crispe, Sir Nicholas 184

Davison, W. 199
Donne, Dr, effigy 189
Dowling, I. 192
Dudley, H. 201

Edward Duke of York and Albany, funeral procession 192
Edward III 186
Edward the Confessor, painting 193
Erskine, Lady, monument 184

Faulkner, T. 190
Fisher, Thomas 180
Flemming, James, monument 191
Francia, L. 199

Globe Theatre, Bankside 179
Gower's monument 190
Greig, J. 199
Grey Friars Monastery 181
Griffith, W. P. 201
Grignion, C 190
Gwilt, G. 212

H, W. T. 188
Harris, J. R. 190
Hart, G. 196
Hawes, William 187
Heath, C. 184, 211
Holden, T. 200
Hollar, W. 186
Hollis, G. 180, 182, 183, 190, 198, 207, 210
Howlett, B. 183
Hulett, – 182, 192, 202

Iliffe, Mary Ann 285
Islip, Abbot, monument 192

Jefferys, T. 197, 211
Jeffreys, W. 183
Jewitt, – 193, 194
Jobbins, J. R. 182

Kearman, Thos 207
King, B. 179
Knight, W. 203
Knyff, L. 179
Knyvet, Thomas Lord, monument 187

Lambe, William 186
Lancaster, John duke of, monument 189
Lane, Rt. 206
Latimer, Lady, effigy 184
LeKeux, J. H. 193, 194
Lewis, S. 210
Lewis, T. Hayter 182
Lodge, J. 199, 207
Longmate, B. 184, 185, 187, 200, 209, 212
Lovell, Gregory, tomb 185

Malcolm, J. P. 183, 189, 195, 196
Manners, Lord Robert, monument 192
Martin, R. 204
Medical Society, Bolt Court 180
Medland, – 209
Metz, – 211
Miller, I. K. 190
Miller, Philip, monument 183
Mills, J. 179
Moore, I. F. 190

Nash, F. 190
Nelson, Lord, monument 189

INDEXES

Newcastle, Cavendish duke of, monument 192
Newcastle, John Holles duke of, monument 192
Nowell, Dean 189

Orme, Lieut 205
Orwade, Nat 189

Paine, J. 206
Parkes, David 188
Phillipa, Queen 186
Pilton's, James, Manufactory 181
Pollard, – 200
Prattent, T. 180, 195, 196, 197, 199, 200, 201, 202, 205, 206, 207, 209, 211, 213

Ranelagh Gardens 181
Ravenhill, – 198
Rawle, S. 210
Raynton family, monument 184
Richmond, Lewis Stuart duke of, monument 192
Roffe, J. 185
Rooker, E. 206
Roper, D. 186
Roubiliac, L. F. 191, 192
Royce, J. 201, 209, 211
Ryley, – 209

Salisbury's, Mr, Botanic Garden 181
Savage, – 191
Scharf, George 189
Schnebbelie, – 182, 184–191 *passim*, 193, 201
Scott, G. G. 193, 194

Sebert's tomb 193
Shaw, Henry 180
Sinnot, A. 185
Sloane, Sir Hans, statue of 179
Smirke, – 191
Smirke, Robert 200
Smith, C. J. 201
Smith, J. 199
Smith, John Thomas 199
Soane, J. 189
Somerset, Lady Ann duchess of, monument 192
Stillingfleet's, Mr, monument 185
Sutton, Thomas 181
Swaine, J. 190, 195

Taylor, W. 208
Thomas, W. 209
Thorpe, Richard de, brass 187

Vinkenboone 210
Vulliamy, – 182

Wagner, Andrew 185
Walker, Ant 191, 192
Wells, Dr W. C., tablet 183
West, R. 188
Whishaw, F. 182
Whitworth, Robt. 181
Wilkinson, W. 195
Wilson, Dr Thomas, monument 190
Windsor, Thomas, tomb 187
World's End Tavern, Chelsea 181
Wyld, J. 205

York, Philippa de Mohun duchess of, monument 192